Feature Papers in *Applied Biosciences* 2023

Feature Papers in *Applied Biosciences* 2023

Editor

Robert Henry

MDPI

Basel • Beijing • Wuhan • Barcelona • Belgrade • Novi Sad • Cluj • Manchester

Editor
Robert Henry
Queensland Alliance for
Agriculture and Food
Innovation
University of Queensland,
Brisbane
Australia

Editorial Office
MDPI AG
Grosspeteranlage 5
4052 Basel, Switzerland

This is a reprint of articles from the Special Issue published online in the open access journal *Applied Biosciences* (ISSN 2813-0464) (available at: https://www.mdpi.com/journal/applbiosci/special_issues/5X6V6313D9).

For citation purposes, cite each article independently as indicated on the article page online and as indicated below:

Lastname, A.A.; Lastname, B.B. Article Title. *Journal Name* **Year**, *Volume Number*, Page Range.

ISBN 978-3-7258-1751-1 (Hbk)
ISBN 978-3-7258-1752-8 (PDF)
doi.org/10.3390/books978-3-7258-1752-8

Contents

applied biosciences

MDPI

Review

Crossiella, a Rare *Actinomycetota* Genus, Abundant in the Environment

Tamara Martin-Pozas [1,†], Jose Luis Gonzalez-Pimentel [2,†], Valme Jurado [3], Leonila Laiz [3], Juan Carlos Cañaveras [4], Angel Fernandez-Cortes [5], Soledad Cuezva [6], Sergio Sanchez-Moral [1] and Cesareo Saiz-Jimenez [3,*]

1 Museo Nacional de Ciencias Naturales, MNCN-CSIC, 28006 Madrid, Spain
2 Centro Andaluz de Biologia del Desarrollo (CABD, UPO-CSIC-JA), Facultad de Ciencias Experimentales, Departamento de Genetica, Universidad Pablo de Olavide, 41013 Sevilla, Spain
3 Instituto de Recursos Naturales y Agrobiologia, IRNAS-CSIC, 41012 Sevilla, Spain
4 Departamento de Ciencias de la Tierra y del Medio Ambiente, Universidad de Alicante, 03080 Alicante, Spain
5 Departamento de Biologia y Geologia, Universidad de Almeria, 04120 Almeria, Spain; acortes@ual.es
6 Departamento de Geologia, Geografia y Medio Ambiente, Universidad de Alcala, 28802 Alcala de Henares, Spain; soledad.cuezva@uah.es
* Correspondence: saiz@irnase.csic.es
† These authors contributed equally to this work.

Abstract: The genus *Crossiella* contains two species, *C. equi*, causing nocardioform placentitis in horses, and *C. cryophila*, an environmental bacterium. Apart from *C. equi*, which is not discussed here, environmental *Crossiella* is rarely reported in the literature; thus, it has not been included among "rare actinobacteria", whose isolation frequency is very low. After *C. cryophila*, only five reports cover the isolation of *Crossiella* strains. However, the frequency of published papers on environmental *Crossiella* has increased significantly in recent years due to the extensive use of next-generation sequencing (NGS) and a huge cascade of data that has improved our understanding of how bacteria occur in the environment. In the last five years, *Crossiella* has been found in different environments (caves, soils, plant rhizospheres, building stones, etc.). The high abundance of *Crossiella* in cave moonmilk indicates that this genus may have an active role in moonmilk formation, as evidenced by the precipitation of calcite, witherite, and struvite in different culture media. This review provides an overview of environmental *Crossiella*, particularly in caves, and discusses its role in biomineralization processes and bioactive compound production.

Keywords: *Crossiella*; biofilms; caves; moonmilk; biomineralization; rhizosphere; soils; bioactive compounds

Citation: Martin-Pozas, T.; Gonzalez-Pimentel, J.L.; Jurado, V.; Laiz, L.; Cañaveras, J.C.; Fernandez-Cortes, A.; Cuezva, S.; Sanchez-Moral, S.; Saiz-Jimenez, C. *Crossiella*, a Rare *Actinomycetota* Genus, Abundant in the Environment. *Appl. Biosci.* **2023**, *2*, 194–210. https://doi.org/10.3390/applbiosci2020014

Academic Editor: Robert Henry

Received: 4 March 2023
Revised: 24 April 2023
Accepted: 27 April 2023
Published: 6 May 2023

1. Introduction

The first strain included in the genus *Crossiella* has a long history of transfers and amendments. Takahashi et al. [1] studied the soil isolate *Nocardiopsis mutabilis*, capable of producing novel antibiotics, and classified it as a new subspecies: *Nocardiopsis mutabilis* subsp. *Cryophilis* based on its growth at low temperatures (8–33 °C). This strain was subsequently transferred by Labeda and Lechevalier [2] to the genus *Saccharothrix* as *Saccharothrix cryophilis* because its morphological and chemotaxonomical properties were more typical of the genus *Saccharothrix* than *Nocardiopsis*. In another study, Labeda [3] erected the genus *Crossiella* to accommodate the species *Saccharothrix cryophilis*, which was misplaced within the genus *Saccharothrix*. The genus only contained the species *Crossiella cryophila*, which was soon accompanied by *Crossiella equi*, responsible for causing abortion cases in horses with equine nocardioform placentitis in Kentucky [4].

In the last 20 years, no other species of *Crossiella* have been described, with scarce reports on the isolation of *Crossiella* strains in the literature. Sánchez-Moral [5] isolated a few strains of *Crossiella* from Altamira Cave, Spain. Adeyemo and Onilude [6] described a strain of *Crossiella* isolated from Nigerian soil with a broad spectrum of antimicrobial activity. Cimermanova et al. [7] isolated nine actinobacteria collected from different heavy metal-contaminated soils and found that one strain could represent a new species within the genus *Crossiella*; however, they did not provide any characterization or description other than its position in the phylogenetic tree and that the strain differed from *Crossiella cryophila* in several biochemical properties. González-Riancho [8] isolated three *Crossiella* strains from white and grey biofilms from Altamira Cave. Finally, Gonzalez-Pimentel et al. [9] studied the genomes of two *Crossiella* strains selected from 13 strains previously isolated from Altamira Cave (unpublished report).

The isolation of *Crossiella* strains on only five occasions in the last 10 years included this genus within the so-called "rare actinobacteria". Oren and Garrity [10] considered *Actinobacteria* a synonym of *Actinomycetota*. They presented the names and formal descriptions of 42 phyla to effect valid publication in their names based on genera as nomenclatural types. However, in this paper, we maintained the original names, as previously published by each author, to avoid mistakes.

"Rare actinobacteria" are non-*Streptomyces* actinobacteria whose isolation frequency is much lower than *Streptomyces* strains, commonly isolated by conventional methods [11,12]. Tiwari and Gupta [13,14] reported 120 new genera of "rare actinobacteria" in the first decade of the 21st century. A total of 40 out of 120 genera were isolated from soils with comparatively lower percentages from other environments: marine and freshwater sediments, marine animals, plants, buildings, etc. A few reports included the rare genera *Actinomadura, Nonomuraea, Micromonospora, Streptosporangium, Nocardiopsis*, and *Pseudonocardia* as most frequent in diverse environments [15–20]. It is noteworthy that an abundance of "rare actinobacteria" is in extreme environments, as exemplified in Atacama [18] and other deserts [21–24]. However, *Crossiella* has not been included among "rare actinobacteria" thus far. We have found that *Crossiella* is an abundant genus in most studied Spanish caves, whether they are gypsum, karstic, or volcanic [9,25–27], and in other terrestrial and aquatic environments.

In this paper, we review the occurrence of environmental *Crossiella*, its relative abundance in Spanish caves and other subterranean environments, as well as its involvement in caves' mineral precipitation. The interest in *Crossiella* lies in its role in biomineralization and potential use in biotechnological processes (stone bioconsolidation, enzyme sources, bioactive compounds, etc.).

2. The Genus *Crossiella* in Caves

Table 1 shows the occurrence of *Crossiella* in different Spanish caves. The high relative abundance of this genus is in moonmilk (Figure 1), either from karstic (Pindal) or volcanic (Fuente de la Canaria and Bucara II) caves, as well as in coloured biofilms (Pindal, Altamira, Castañar, Covadura) is remarkable [27–31]. The relative humidity is near 100% in these caves. In addition, other mineral/biological formations, such as a pink formation in Bucara II, exhibit high relative abundance (38.9%). Similarly, formations such as mucous formations or brown deposits also reach relatively high abundances (6.7–12.8%) [29]. Interestingly, low percentages of *Crossiella* found in the sediments under the moonmilk indicate an aerobic behaviour for this genus [3]. *Crossiella* was also found in phototrophic biofilms from Nerja Cave [30].

Table 1. Occurrence and relative abundance (>1%) of *Crossiella* in Spanish karstic environments, as reported in NGS studies.

Cave	Relative Abundance	Genus	Type of Sample	References
Pindal	16.0–27.1	*Crossiella*	Moonmilk	[26,28,31]
	1.4–1.7	*Crossiella*	Sediment under moonmilk	
	11.3–11.7	*Crossiella*	Top-layer sediments	

Table 1. *Cont.*

Cave	Relative Abundance	Genus	Type of Sample	References
	6.0–9.0	*Crossiella*	Sediments	
	5.3–7.9	*Crossiella*	Yellow biofilm	
	2.0–8.0	*Crossiella*	Grey biofilms	
	7.0–8.0	*Crossiella*	Pink biofilms	
Fuente de la Canaria	12.6–12.8	*Crossiella*	Mucous formations	[29]
	12.3	*Crossiella*	Moonmilk	
	6.7	*Crossiella*	Brown and yellow deposits	
Bucara II	38.9	*Crossiella*	Pink deposit	[29]
	24.9	*Crossiella*	Moonmilk	
Nerja	0.1–1.5	*Crossiella*	Phototrophic biofilms	[30]
Castañar	15.0	*Crossiella*	Grey biofilm	[31]
Altamira	>20.0	*Crossiella*	Grey biofilms	[8]
	27.0	*Crossiella*	White biofilms	
	38.0	*Crossiella*	Yellow biofilms	
Covadura	26.4–54.1	*Crossiella*	White biofilm	Unpublished data
	21.8–51.9	*Crossiella*	Yellow biofilm	
	4.5–19.7	*Crossiella*	Sediments	
Yeso	1.3–13.3	*Crossiella*	Sediments	Unpublished data
Thyssen Museum basement	16.6	*Crossiella*	White biofilm	[32]
	64.2	*Crossiella*	Grey biofilm	
	2.8–7.4	*Crossiella*	Sediment	

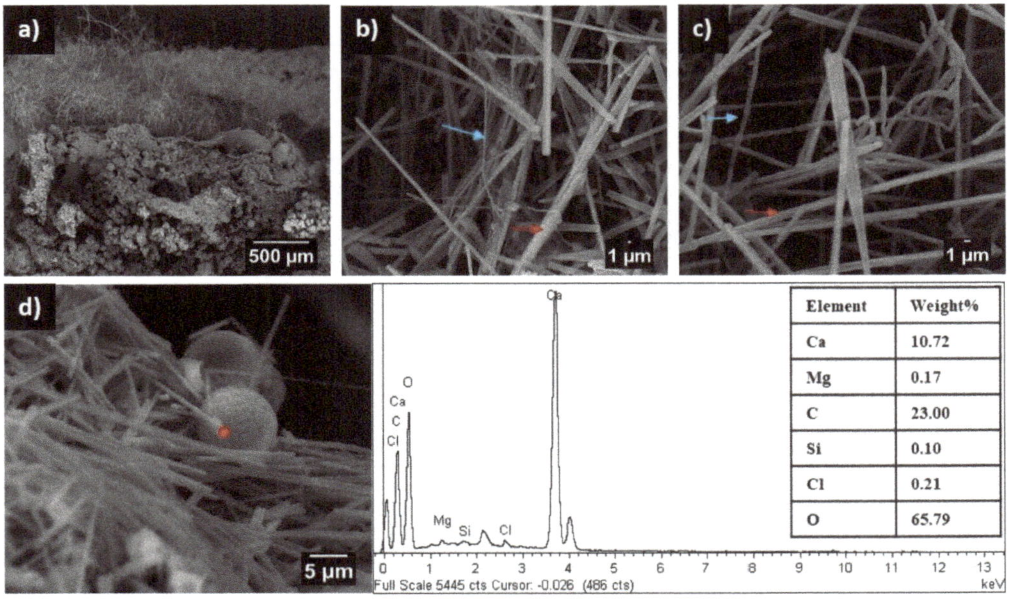

Figure 1. Scanning electron micrographs of moonmilk deposits in Pindal Cave, Spain. (**a**) Longitudinal view of sediment covered by moonmilk. (**b**,**c**) Crystalline calcite fibres (red arrow) and *Actinomycetota* filaments (blue arrow). (**d**) Scanning electron micrographs and EDX spectra of crystalline calcite fibres. Note the swelling of filaments in (**c**), similar to those reported for *Crossiella cryophila* [3].

Crossiella, at a relative abundance of 15.0%, was found in grey biofilms from Castañar Cave [31]. Similar grey biofilms were observed in Altamira Cave [8] and the Thyssen Museum, reaching a relative abundance of 64.2% [32]. Data from a geomicrobiological study of a Roman nymphaeum located in the archaeological basement of the Thyssen Museum in Malaga, Spain, were also included in Table 1 due to its interest.

The environmental conditions of this archaeological basement are special because they mix the characteristics of an environment heavily influenced by the natural underlying karst system with those of an enclosure located in an urban building. Apart from caves, it was remarkable that a subterranean environment, the Roman mortar pavement in the archaeological basement, was colonized by grey biofilms with a high relative abundance of *Crossiella*. This environment is characterized by permanent darkness, the absence of visits, and high relative humidity.

Table 1 shows the occurrence of *Crossiella* in moonmilk, grey, yellow, pink and white biofilms, and sediments from different caves and subterranean environments. *Crossiella* is abundant in different types of rocks, either in volcanic (Fuente de la Canaria, Bucara II), karstic (Pindal, Nerja, Castañar, Altamira) or gypsum (Covadura, Yeso) caves.

In addition to the studies in Table 1, authors have reported the occurrence of *Crossiella* using methodological approaches other than NGS. Stomeo et al. [33] found metabolically active *Crossiella* in white biofilms from Ardales Cave, Malaga, Spain. Portillo and Gonzalez [34] identified *Crossiella* as a major metabolically active bacterium in the black crust of a shelter located in Aragon, Spain, and Sanchez-Moral [5] reported *Crossiella* in Altamira Cave.

Table 2 shows the widespread occurrence of *Crossiella* in caves in the USA, France and China. Less frequent records were found in caves in Italy, Pakistan, Portugal, Serbia, and Thailand, among other countries [35–57].

Table 2. Occurrence and relative abundance (>1%) of *Crossiella* in caves all over the world.

Karstic Caves	**Relative Abundance%**	**Genus**	**Type of Sample (Method)**	**References**
Heshang	n.d.	*Crossiella*	Weathered rocks (NGS)	[35]
Laugerie-Haute	4.0	*Crossiella*	Salt efflorescences (clones)	[36]
Sorcerers	30.0	*Crossiella*	Salt efflorescences (NGS)	[37]
Pillier	n.d.	*Crossiella*	Wall rock (NGS)	[38]
Yixing Shanjuan	3.9	*Crossiella*	Speleothem (NGS)	[39]
Shuanghe	9.5	*Crossiella*	Rock (NGS)	[40]
Manao-Pee	4.1	*Crossiella*	Soil (NGS)	[41]
KN14	27.1–52.3	*Crossiella*	Rock/Clay (NGS)	[42]
RN5	1.0–17.9	*Crossiella*	Rock/Clay/Mud (NGS)	[42]
Maijishan Grottoes	n.d.	*Crossiella*	Walls paintings (NGS)	[43]
Heshang	n.d.	*Crossiella*	Weathered rocks (NGS)	[44]
Kashmir and Tiser	11.9–36.6	*Crossiella*	Soil (NGS)	[45]
Zhijin	4.1	*Crossiella*	Wall rock (NGS)	[46]
Rouffignac	~70.0	*Crossiella*	Wall rock (NGS)	[47]
Stiffe	9.9	*Crossiella*	Biofilms (NGS)	[48]
Heshang	n.d.	*Crossiella*	Weathered rocks (NGS)	[49]
Cave Church	0.1–4.9	*Crossiella*	Fresco (NGS)	[50]
Volcanic Caves				
Azorean caves	18.6	*Crossiella*	Biofilms (clones)	[51]

Table 2. *Cont.*

Karstic Caves	Relative Abundance%	Genus	Type of Sample (Method)	References
Hawaiian caves	n.d.	*Crossiella*	Biofilms (NGS)	[52]
Californian caves	n.d.	*Crossiella*	Biofilms (NGS)	[53]
Idahoan caves	n.d.	*Crossiella*	Biofilms (NGS)	[54]
Sicilian caves	62.5–77.6	*Crossiella*	Biofilms (NGS)	[55]
Other Cave Types				
Carlsbad Cavern	n.d.	*Crossiella*	Rocks (clones)	[56]
Imawarì Yeuta	n.d.	*Crossiella*	Patina/Speleothems (NGS)	[57]

n.d. Not determined.

Apart from the high abundance in Spanish caves, the high relative abundance of Crossiella in Italian caves is also remarkable. In this regard, Nicolosi et al. [55] recorded high relative abundances in four Etna volcano caves. One of them ranged from 62.5 to 77.6%. Other notable abundances were found in the salt efflorescences of a French shelter [37] and in caves in the USA [42], France [47], Pakistan [45], and the Azores, Portugal [51].

Crossiella has been identified as one of the dominant bacterial phylotypes, with an increasing prevalence when global humidity conditions rise, in a research covering 1050 cave microbiomes worldwide (manuscript in preparation).

3. *Crossiella* in the Environment

Supplementary Table S1 lists papers in the literature that use the keyword "*Crossiella*", including *Crossiella* misspelling [58–146]. The occurrence of the genus *Crossiella* in different environments is significant. Papers reporting *Crossiella equi* and its involvement in animal diseases [4] were excluded.

The papers listed in Supplementary Table S1 rely on molecular methods, except for five articles describing the properties of isolated strains [5–9]. The genus *Crossiella* shows a ubiquitous and extensive geographical distribution on all continents, including Antarctica, but not in Australia/Oceania, likely due to a lack of relevant studies.

Fewer reports locate *Crossiella* in mines and reclaimed mine soils [7,120–124]. However, the number of studies on its presence in soils and the rhizospheres of diverse plants is considerable. Several *Crossiella* findings in stones and building stones were also confirmed [125–138]. Finally, a few records in sea sediments and freshwaters were significant [139–146] because they included aquatic environments among *Crossiella* habitats.

From Supplementary Table S1, we can conclude that *Crossiella*, in addition to caves, is relatively abundant in diverse environments, namely soils, plant rhizospheres, mines, building stones, and other occasional habitats, but is rarely isolated.

Considering the abundance of reports on *Crossiella* in soils [58–96] and plant rhizospheres [97–119], the presence of this genus in caves and other subterranean environments could be attributed to its transport to the subsurface via percolation waters. In this regard, *Crossiella* in percentages <1% have been found in drip waters from Pindal Cave [28]. It may be possible that once transported to the caves, the environmental conditions favour and increase the colonization and growth of *Crossiella* on different mineral substrata.

4. *Crossiella* Isolates

Only five reports shed light on *Crossiella* isolates. A screening of Nigerian soils resulted in the isolation of *Crossiella* sp. strain EK18. The 16S rRNA nucleotide sequence showed 98% similarity to *C. equi*. This strain grew well in different culture media and exhibited broad-spectrum antimicrobial activity [6]. The authors studied the effects of pH, temperature, carbon and nitrogen sources, sodium chloride concentration, and incubation time on antimicrobial activity. In addition, they reported a list of 12 so-called antimicrobial metabolites, including alkanes, alkenes, commonly synthesized by bacteria, and phthalates, which are contaminants from materials and impurities from products used in culture media. Therefore, no conclusive data on the real bioactive compounds produced by the *Crossiella* strain can be derived from this study.

Cimermanova et al. [7] isolated *Crossiella* sp., strain S2, from mining wastes, with a 16rRNA gene sequence similarity of 99.1% to *C. cryophila*. The authors suggested that it may represent a novel, never described species, based on its location in the phylogenetic tree. The strain also exhibited high heavy metal resistance.

González-Riancho [8] found relative abundances of *Crossiella* >20% in white, yellow, and grey biofilms from Altamira Cave. She isolated two strains from white and one strain from grey biofilms with similarities of 99.0–100.0 to *C. cryophila* using the medium Actinomycete Isolation Agar (AIA).

Gonzalez-Pimentel et al. [9] studied two of the thirteen *Crossiella* strains previously isolated from grey biofilms colonizing Altamira Cave, Spain (unpublished report). In vitro and in silico analyses showed the inhibition of pathogenic bacteria and fungi. The exclusive combination of gene clusters involved in the synthesis of lanthipeptides, lasso peptides, nonribosomal peptides and polyketides indicates that these two strains represent a source of new bioactive compounds. The taxonomical distance of both strains from their closest relative, *C. cryophila*, suggests that they represent a new species of *Crossiella*, which will be described in future works.

So far, the low number of isolated *Crossiella* strains indicates that most of the culture media used are inadequate to reproduce their growth in the laboratory. The environmental conditions of their ecological niche should also be considered when designing specific culture media, which are superior to conventional ones.

5. Biomineralization in Caves Induced by *Crossiella*

Biomineralization or crystal formation is a general phenomenon caused by soil bacteria, as reported by Boquet et al. [147]. These authors isolated 210 bacteria that could form calcite crystals in a medium with calcium acetate and stated that their occurrence depended on the composition of the medium used.

The role of bacteria in speleogenesis has been discussed for decades. Barton and Northup [148] stated that in the 1960s, a few authors proposed that microbes played a role in forming cave deposits. Banks et al. [149] confirmed the link between calcium metabolism in bacteria and calcification using cave isolates. They suggested that the toxicity of Ca^{2+} ions to bacteria promoted the need to remove Ca^{2+} ions from the cell via calcification as a detoxification mechanism.

Further evidence of biomineralization has been reported in recent decades that sheds light on microbially induced mineral precipitation [150–153]. This precipitation has been attributed to several causes: the modulation of environmental pH, nucleation sites on cell surfaces, or enzymatically driven processes involving carbonic anhydrase, urease, etc. [154].

Grey biofilms from Altamira Cave were studied, and scanning electron microscopy (SEM) revealed an abundance of bioinduced calcite crystals in addition to moonmilk [152]. The biofilms mainly comprised *Actinomycetota* filaments promoting carbon dioxide uptake and formation of calcite deposits. A model for bioinduced calcite formation, supported by scanning and transmission electron microscopy data, was proposed by Cuezva et al. [152].

Apart from the precipitation of calcite by *Crossiella*, another experiment (Figure 2) with two strains of *Crossiella* isolated from Altamira Cave [9] revealed that both strains induced the formation of different crystals when incubated in a culture medium with barium acetate, yeast extract, and agar (Ba1). Two crystal types were identified on the plates: witherite (barium carbonate) and struvite (magnesium ammonium phosphate), with distinct abundances that were higher for witherite and scarcer for struvite. Witherite precipitation is due to an abundance of barium in the medium. Occasional struvite crystals can be derived from the amino acids and minor amounts of phosphorus and magnesium in the yeast extract [155].

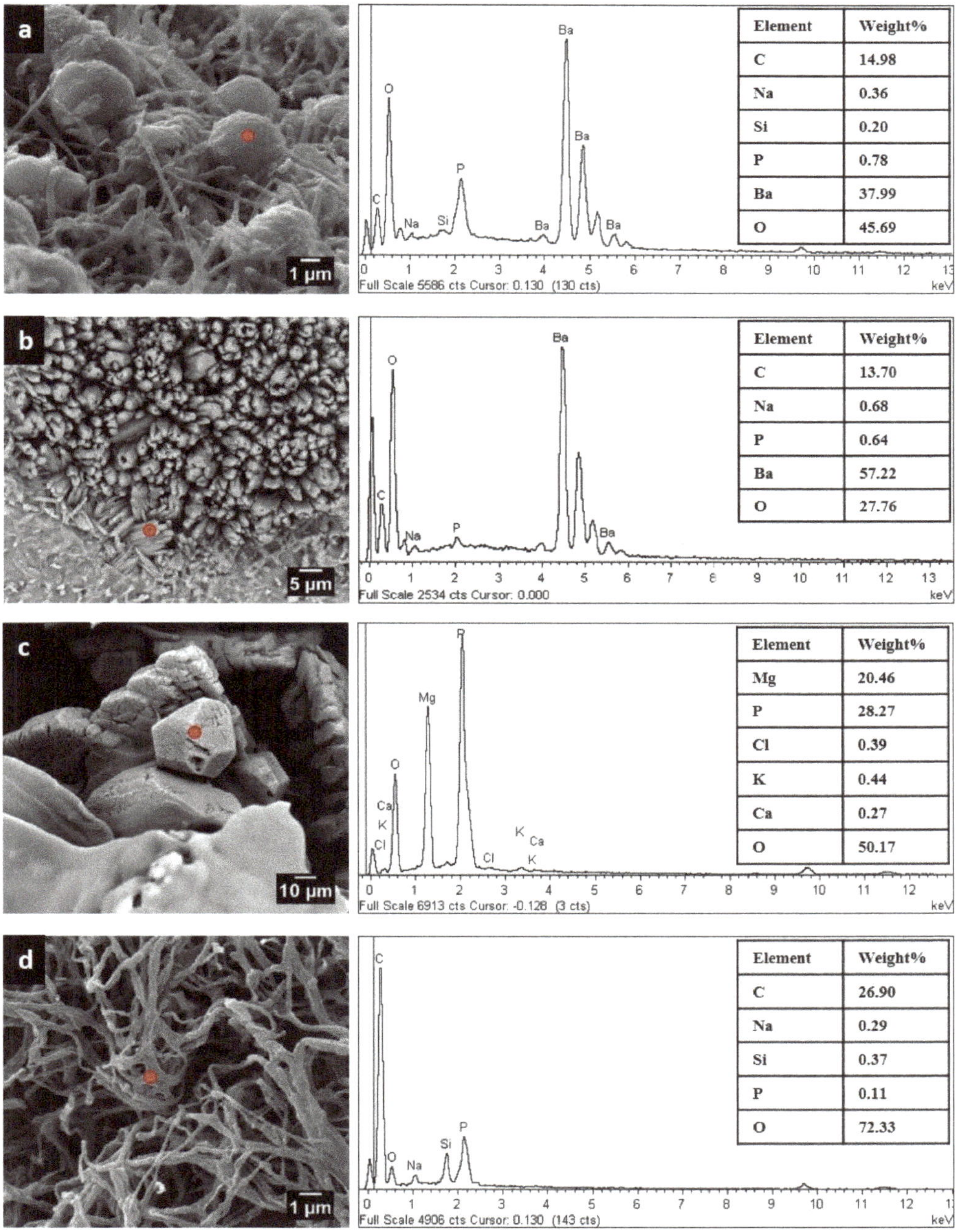

Figure 2. Scanning electron micrographs and EDX spectra of barium carbonate and phosphate crystals from two strains of *Crossiella* sp. (Cross-1 and Cross-2) [7]. (**a**) Witherite (barium carbonate) crystals and *Crossiella* filaments in culture medium Ba1 (Cross-1). (**b**) Witherite crystals in culture medium Ba1 (Cross-2). (**c**) Struvite (magnesium ammonium phosphate) crystals in culture medium Ba1 (Cross-1). (**d**) *Crossiella* filaments without crystal formation in tryptic soy agar medium (Cross-2).

Baryte (barium sulphate) precipitation by bacteria was previously reported by Joubert et al. [156]. Sanchez-Moral et al. [157] found that baryte was associated with filamentous bacteria in altered volcanic rocks. *Agromyces* spp., *Bacillus* spp., *Lysobacter* spp., *Ralstonia* sp., *Stenotrophomonas maltophilia*, and *Streptomyces* sp. were isolated from the volcanic rocks and precipitated witherite or calcite/vaterite in media with barium or calcium acetate, respectively. The occurrence of baryte, but not witherite, in volcanic rocks was due to the presence of sulphate ions that easily transform witherite into barite. This transformation was not produced on the *Crossiella* plates without sulphate ions.

Struvite precipitation is not as widespread as calcite or witherite in bacteria. Sanchez-Moral et al. [158] tested eight bacteria isolated from the Roman catacombs of St. Callixtus and Domitilla and reported that *Agromyces ramosus* precipitated calcite, magnesium calcite, witherite, and struvite, depending on the media composition. Other bacteria precipitating struvite were *Bacillus* sp. and *Ralstonia metallidurans*.

Rivadeneyra et al. [159] found that only 20.8% of the tested bacterial isolates precipitated struvite and that calcium acetate appeared to inhibit struvite precipitation in culture media, whereas ammonium ions triggered it [160]. Manzoor et al. [161] stated that urease-producing bacteria play a key role in struvite precipitation, controlling nucleation, and modulating crystalline phases and crystal shapes. Urease is present in the strain type of *Crossiella*, *C. cryophila* [3], and urease genes have been identified in the genomes of the two *Crossiella* strains from Altamira Cave [9].

Sánchez-Román et al. [162] reported that carbon and phosphorus cycles are interrelated during biomineralization. They also demonstrated the co-precipitation of carbonate and struvite, which we also found in *Crossiella* strains.

The data reported show that biomineralization by *Crossiella* is an active process in the presence of different ions, confirming its role in moonmilk formation. *Crossiella* strains' ability to induce carbonate precipitation, which is used to consolidate cultural heritage stones and buildings, should be explored.

6. Moonmilk Formation

Moonmilk formation has been discussed in the literature for a long time [150,163–168]. The question: Is moonmilk an abiotic process driven physicochemically, or is it biotic, mediated by microorganisms? A biological origin currently prevails; even a combination of physicochemical and biogenic processes is being considered [166]. Cañaveras et al. [150,165] indicated that bacteria influenced the physicochemistry of calcite precipitation. They observed that cave moonmilk comprises a network of calcite crystals and active filamentous bacteria. They also found that hydromagnesite and needle-fibre aragonite deposits were associated with bacteria in Altamira Cave, predominantly *Streptomyces*, for which they demonstrated their ability to precipitate calcite/vaterite in the laboratory. The association between bacteria and mineral crystals was reported in other papers and described using SEM [152,153,165,168–170]. In addition, different bacterial genera such as *Agromyces*, *Amycolatopsis*, *Brachybacterium*, *Nocardioides*, *Nocardiopsis*, *Paenibacillus*, and *Rothia* precipitated vaterite/calcite and Mg-calcite [171].

Maciejewska et al. [153] found that all the *Streptomyces* strains tested could promote calcification and biomineralization. The metabolic activities involved in the precipitation were amino acids ammonification and ureolysis, which increased environmental pH. Sanchez-Moral et al. [169] stated that microbial activity induces carbonate precipitation in the early stages of deposition. However, as carbonate accumulates, a progressive decline in microbial activity occurs, as deduced from the RNA/DNA ratio, which is used as a marker of metabolic activity. The decreased metabolic activity is due to the progressive accumulation of carbonate and bacterial entrapment in mineral deposits.

The high relative abundance of *Crossiella* in moonmilk from different caves indicates that this genus is active in moonmilk formation (Table 1). Enzymatic processes induce this mineralization, and several enzymes have been linked to *Crossiella* activity in moonmilk. Martin-Pozas et al. [26] suggested that moonmilk formation is related to syntrophic relationships between *Crossiella* and nitrifying bacteria, and Cuezva et al. [27] associated *Crossiella* with the ability to capture CO_2 from the atmosphere and precipitate calcium carbonate as a by-product of carbonic anhydrase action, as observed in cave moonmilk.

7. Is Moonmilk a Source of Bioactive Compounds?

Caves and moonmilk are colonized by complex bacterial communities. Maciejewska et al. [153] reported that *Proteobacteria* was the dominant phylum of moonmilk from a Belgian cave, followed by *Actinobacteria, Acidobacteria, Chloroflexi, Nitrospirae, Gemmatimonadetes*, and *Planctomycetes*. These seven phyla accounted for 85.8–90.2% of the total community. Martin-Pozas et al. [26] investigated the moonmilk composition from Pindal Cave in Spain. They found that *Proteobacteria* and *Actinobacteria* dominated the community with over 30% of relative abundance for each phyla, followed by *Acidobacteria, Chloroflexi, Planctomycetes, Gemmatimonadetes*, and *Nitrospirae*. These phyla accounted for 93.1–93.9% of the total community. The similarities between the phyla compositions of moonmilk from two different caves are remarkable. Moonmilk from a geographically distant cave [172] and another subterranean environment [173] also showed relatively similar phyla compositions.

Moonmilk has been a promising reservoir for novel bacteria producing bioactive compounds, and a few novel species have been isolated, namely *Streptomyces lunaelactis* [174], *Pseudomonas karstica*, and *Pseudomonas spelaei* [175]. Several studies have stressed the great diversity of unknown bacteria inhabiting moonmilk and the isolates' production of bioactive compounds [176–178].

The high abundance of *Actinomycetota* (=*Actinobacteria*) and *Pseudomonadota* (=*Proteobacteria*) in moonmilk has prompted researchers to test a series of strategies to isolate hard-to-culture "rare actinobacteria" and discover novel bioactive compounds [13,14]. Adam et al. [176] obtained 40 isolates represented by *Agromyces, Amycolatopsis, Kocuria, Micrococcus, Micromonospora, Nocardia, Streptomyces*, and *Rhodococcus* species. *Streptomyces* isolates displayed strong inhibitory activities against Gram-positive and Gram-negative bacteria and fungi [177–179]. Genome mining of *Streptomyces lunaelactis* revealed 42 biosynthetic gene clusters [180] and the production of the antibiotics bagremycins and lunaemycins [180,181]. The genome of *Crossiella*, abundant in moonmilk [26], showed the presence of a combination of gene clusters involved in synthesising different bioactive compounds [9]. The data suggest the possibility of finding other moonmilk bacteria involved in synthesising bioactive compounds.

8. Conclusions

The genus *Crossiella* is widely distributed in all environments, reaching a relative abundance of up to 78% in a Sicilian cave. Its occurrence in soils, plant rhizospheres and caves is especially important. The last case is probably due to its transport to the subsurface by percolating waters. Despite this abundance, the strains isolated were scarce.

The data suggest that more environmental *Crossiella* species are waiting to be described, apart from *Crossiella cryophila* and *Crossiella equi*. The increasing number of metagenomic sequence data from all environments offers clear opportunities to guide the isolation and cultivation of *Crossiella*. Therefore, further efforts are required to design suitable isolation culture media. They should consider the environmental conditions of the niches where *Crossiella* thrives, namely alkaline pH and high mineral concentrations.

Crossiella has an important role in carbon sequestration in subterranean environments. Metagenomic studies and isolating more *Crossiella* strains and/or species are the only way to advance knowledge of *Crossiella* functions in different ecosystems. Furthermore, its role in biomineralization and moonmilk formation is also apparent.

Finally, *Crossiella* appears to be a promising source of active compounds, and the isolated strains deserve more attention regarding their potential use in biotechnological processes.

Supplementary Materials: The following supporting information can be downloaded at: https://www.mdpi.com/article/10.3390/applbiosci2020014/s1, Table S1: Occurrence of the genus *Crossiella* in different environments.

Author Contributions: Conceptualization, C.S.-J. and S.S.-M.; investigation, T.M.-P., J.L.G.-P., V.J., L.L., J.C.C., A.F.-C. and S.C.; writing—original draft preparation, C.S.-J., S.S.-M. and T.M.-P.; writing—review and editing, C.S.-J. and S.S.-M. All authors have read and agreed to the published version of the manuscript.

Funding: This research was funded by the projects PID2020-114978GB-I00 and PID2019-110603RB-I00. The Malaga City Council financed data from the archaeological basement of the Thyssen Museum of Malaga through a conservation contract for this Roman site.

Institutional Review Board Statement: Not applicable.

Informed Consent Statement: Not applicable.

Data Availability Statement: All data reported in this review can be found in the relevant papers cited. Scanning electron micrographs, EDX spectra, and unpublished data are available on request from Sergio Sanchez-Moral.

Acknowledgments: This is a contribution from CSIC Interdisciplinary Thematic Platform Open Heritage: Research and Society (PTI-PAIS).

Conflicts of Interest: The authors declare no conflict of interest.

References

1. Takahashi, A.; Hotta, K.; Saito, N.; Morioka, M.; Okami, Y.; Umezawa, H. Production of novel antibiotic, dopsisamine, by a new subspecies of *Nocardiopsis mutabilis* with multiple antibiotic resistance. *J. Antibiot.* **1986**, *39*, 175–183. [CrossRef] [PubMed]
2. Labeda, D.P.; Lechevalier, M.P. Amendment of the genus *Saccharothrix* Labeda et al. 1984 and descriptions of *Saccharothrix espanaensis* sp. nov., *Saccharothrix cryophilis* sp. nov., and *Saccharothrix mutabilis* comb. nov. *Int. J. Syst. Bacteriol.* **1989**, *39*, 420–423. [CrossRef]
3. Labeda, D.P. *Crossiella* gen. nov., a new genus related to *Streptoalloteichus*. *Int. J. Syst. Evol. Microbiol.* **2001**, *51*, 575–579. [CrossRef]
4. Donahue, J.M.; Williams, N.M.; Sells, S.F.; Labeda, D.P. *Crossiella equi* sp. nov., isolated from equine placentas. *Int. J. Syst. Evol. Microbiol.* **2002**, *52*, 2169–2173.
5. Sánchez-Moral, S. *Estudio Integral del Estado de Conservación de la Cueva de Altamira y su Arte Paleolítico (2007–2009). Perspectivas Futuras de Conservación*; Monografías N° 24; Museo Nacional y Centro de Investigación de Altamira: Santillana del Mar, Spain, 2014.
6. Adeyemo O.M.; Onilude, A.A. Antimicrobial potential of a rare actinomycete isolated from soil: *Crossiella* sp.-EK18. *J. Adv. Microbiol.* **2018**, *11*, 1–15. [CrossRef]
7. Cimermanova, M.; Pristas, P.; Piknova, M. Biodiversity of actinomycetes from heavy metal contaminated technosols. *Microorganisms* **2021**, *9*, 1635. [CrossRef]
8. González-Riancho Fernández, C. Análisis Descriptivo y Funcional de Las Colonias Microbianas Visibles Que Crecen en la Cueva de Altamira, Enfocado al Diseño de Medidas de Control. Ph.D. Thesis, Universidad de Cantabria, Santander, Spain, 2021.
9. Gonzalez-Pimentel, J.L.; Dominguez-Moñino, I.; Jurado, V.; Laiz, L.; Caldeira, A.T.; Saiz-Jimenez, C. The rare actinobacterium *Crossiella* sp. is a potential source of new bioactive compounds with activity against bacteria and fungi. *Microorganisms* **2022**, *10*, 1575. [CrossRef]
10. Oren, A.; Garrity, G.M. Valid publication of the names of forty-two phyla of prokaryotes. *Int. J. Syst. Evol. Microbiol.* **2021**, *71*, 005056. [CrossRef] [PubMed]
11. Berdy, J. Bioactive microbial metabolites. *J. Antibiot.* **2005**, *58*, 1–26. [CrossRef]
12. Subramani, R.; Aalbersberg, W. Culturable rare Actinomycetes: Diversity, isolation and marine natural product discovery. *Appl. Microbiol. Biotechnol.* **2013**, *97*, 9291–9321. [CrossRef]
13. Tiwari, K.; Gupta, R.K. Rare actinomycetes: A potential storehouse for novel antibiotics. *Crit. Rev. Biotechnol.* **2012**, *32*, 108–132. [CrossRef]
14. Tiwari, K.; Gupta, R.K. Diversity and isolation of rare actinomycetes: An overview. *Crit. Rev. Microbiol.* **2013**, *39*, 256–294. [CrossRef] [PubMed]
15. Seong, C.N.; Choi, J.H.; Baik, K.-S. An improved selective isolation of rare actinomycetes from forest soil. *J. Microbiol.* **2001**, *39*, 17–23.
16. Bredholdt, H.; Galatenko, O.A.; Engelhardt, K.; Fjærvik, E.; Terekhova, L.P.; Zotchev, S.B. Rare actinomycete bacteria from the shallow water sediments of the Trondheim Fjord, Norway: Isolation, diversity and biological activity. *Environ. Microbiol.* **2007**, *9*, 2756–2764. [CrossRef]

17. Fang, B.-Z.; Salam, N.; Han, M.-X.; Jiao, J.-Y.; Cheng, J.; Wei, D.-Q.; Xiao, M.; Li, W.-J. Insights on the effects of heat pretreatment, pH, and calcium salts on isolation of rare actinobacteria from karstic caves. *Front. Microbiol.* **2017**, *8*, 1535. [CrossRef] [PubMed]
18. Goodfellow, M.; Nouioui, I.; Sanderson, R.; Xie, F.; Bull, A.T. Rare taxa and dark microbial matter: Novel bioactive actinobacteria abound in Atacama Desert soils. *Anton. Leeuw.* **2018**, *111*, 1315–1332. [CrossRef]
19. Benhadj, M.; Gacemi-Kirane, D.; Menasria, T.; Guebla, K.; Ahmane, Z. Screening of rare actinomycetes isolated from natural wetland ecosystem (Fetzara Lake, northeastern Algeria) for hydrolytic enzymes and antimicrobial activities. *J. King Saud Univ. Sci.* **2019**, *31*, 706–712. [CrossRef]
20. Zamora-Quintero, A.Y.; Torres-Beltrán, M.; Guillén Matus, D.G.; Oroz-Parra, I.; Millán-Aguiñaga, N. Rare actinobacteria isolated from the hypersaline Ojo de Liebre Lagoon as a source of novel bioactive compounds with biotechnological potential. *Microbiology* **2022**, *168*, 001144. [CrossRef]
21. Tiwari, K.; Upadhyay, D.J.; Mösker, E.; Süssmuth, R.; Gupta, R.K. Culturable bioactive actinomycetes from the Great Indian Thar Desert. *Ann. Microbiol.* **2015**, *65*, 1901–1914. [CrossRef]
22. Mohammadipanah, F.; Wink, J. Actinobacteria from arid and desert habitats: Diversity and biological activity. *Front. Microbiol.* **2016**, *6*, 1541. [CrossRef]
23. Gacem, M.A.; Ould-El-Hadj-Kheli, A.; Abd-Elsalam, K.A.; Wink, J. Actinobacteria in the Algerian Sahara: Diversity, adaptation mechanism and special unexploited biotopes for the isolation of novel rare taxa. *Biologia* **2001**, *76*, 3787–3799. [CrossRef]
24. Hui, M.L.-Y.; Tan, L.T.-H.; Letchumanan, V.; He, Y.-W.; Fang, C.-M.; Chan, K.-G.; Law, J.W.-F.; Lee, L.-H. The extremophilic Actinobacteria: From microbes to medicine. *Antibiotics* **2021**, *10*, 682. [CrossRef] [PubMed]
25. Gonzalez-Pimentel, J.L.; Martin-Pozas, T.; Jurado, V.; Miller, A.Z.; Caldeira, A.T.; Fernandez-Lorenzo, O.; Sanchez-Moral, S.; Saiz-Jimenez, C. Prokaryotic communities from a lava tube cave in La Palma Island (Spain) are involved in the biogeochemical cycle of major elements. *PeerJ* **2021**, *9*, e11386. [CrossRef] [PubMed]
26. Martin-Pozas, T.; Cuezva, S.; Fernandez-Cortes, A.; Cañaveras, J.C.; Benavente, D.; Jurado, V.; Saiz-Jimenez, C.; Janssens, I.; Seijas, N.; Sanchez-Moral, S. Role of subterranean microbiota in the carbon cycle and greenhouse gas dynamics. *Sci. Total Environ.* **2022**, *831*, 154921. [CrossRef]
27. Cuezva, S.; Martin-Pozas, T.; Fernandez-Cortes, A.; Cañaveras, J.C.; Janssens, I.; Sanchez-Moral, S. On the role of cave-soil in the carbon cycle. A first approach. *EGU Gen. Assem.* **2020**. Available online: https://presentations.copernicus.org/EGU2020/EGU2020-21793_presentation.pdf. (accessed on 22 April 2022).
28. Martin-Pozas, T.; Fernandez-Cortes, A.; Cuezva, S.; Cañaveras, J.C.; Benavente, D.; Duarte, E.; Saiz-Jimenez, C.; Sanchez-Moral, S. New insights into the structure, microbial diversity and ecology of yellow biofilms in a Paleolithic rock art cave (Pindal Cave, Asturias, Spain). *Sci. Total Environ.* **2023**, *882*. *in press*.
29. González-Pimentel, J.L. Microorganismos de las Cuevas Volcánicas de La Palma (Islas Canarias). Diversidad y Potencial Uso Biotecnológico. Ph.D. Thesis, Universidad Pablo Olavide, Sevilla, Spain, 2019.
30. Jurado, V.; del Rosal, Y.; Gonzalez-Pimentel, J.L.; Hermosin, B.; Saiz-Jimenez, C. Biological control of phototrophic biofilms in a show cave: The case of Nerja Cave. *Appl. Sci.* **2020**, *10*, 3448. [CrossRef]
31. Martin-Pozas, T. Papel de los Microorganismos en Procesos de Captación y Emisión de Gases de Efecto Invernadero en Ambientes Subterráneos. Ph.D. Thesis, Universidad Complutense de Madrid, Madrid, Spain, 2023.
32. Sánchez-Moral, S.; Martín-Pozas, T.; Seijas Morales, N.; Fernández-Cortés, A.; Benavente García, D.; Cañaveras Jiménez, J.C.; Cuezva, S. *Instalación de Sensores en Sótano Arqueológico del Museo Carmen Thyssen de Málaga Para la Toma de Datos, el Análisis y Adopción de Medidas Correctoras del Deterioro Del Recinto*; Unpublished Report; MNCN: Madrid, Spain, 2021.
33. Stomeo, F.; Portillo, M.C.; Gonzalez, J.M.; Laiz, L.; Saiz-Jimenez, C. *Pseudonocardia* in white colonizations in two caves with Paleolithic paintings. *Int. Biodeter. Biodegr.* **2008**, *62*, 483–486. [CrossRef]
34. Portillo, M.C.; Gonzalez, J.M. Microbial community diversity and the complexity of preserving cultural heritage. In *Biocolonization of Stone: Control and Preventive Methods*; Charola, A.E., McNamara, C., Koestler, R.J., Eds.; Smithsonian Institution, Scholarly Press: Washington, DC, USA, 2011; pp. 19–28.
35. Yun, Y.; Wang, H.; Man, B.; Xiang, X.; Zhou, J.; Qiu, X.; Duan, Y.; Engel, A.S. The relationship between ph and bacterial communities in a single karst ecosystem and its implication for soil acidification. *Front. Microbiol.* **2016**, *7*, 1955. [CrossRef] [PubMed]
36. Lepinay, C.; Mihajlovski, A.; Seyer, D.; Touron, S.; Bousta, F.; Di Martino, P. Biofilm communities survey at the areas of salt crystallization on the walls of a decorated shelter listed at UNESCO World cultural Heritage. *Int. Biodeter. Biodegr.* **2017**, *122*, 116–127. [CrossRef]
37. Lepinay, C.; Mihajlovski, A.; Touron, S.; Seyer, D.; Bousta, F.; Di Martino, P. Bacterial diversity associated with saline efflorescences damaging the walls of a French decorated prehistoric cave registered as a World Cultural Heritage Site. *Int. Biodeter. Biodegr.* **2018**, *130*, 55–64. [CrossRef]
38. Alonso, L.; Pommier, T.; Kaufmann, B.; Dubost, A.; Chapulliot, D.; Doré, J.; Douady, C.J.; Moënne-Loccoz, Y. Anthropization level of Lascaux Cave microbiome shown by regional-scale comparisons of pristine and anthropized caves. *Mol. Ecol.* **2019**, *28*, 3383–3394. [CrossRef] [PubMed]
39. Li, M.; Fang, C.; Kawasaki, S.; Huang, M.; Achal, V. Bio-consolidation of cracks in masonry cement mortars by *Acinetobacter* sp. SC4 isolated from a karst cave. *Int. Biodeter. Biodegr.* **2019**, *141*, 94–100. [CrossRef]

40. Long, Y.; Jiang, J.; Hu, X.; Zhou, J.; Hu, J.; Zhou, S. Actinobacterial community in Shuanghe Cave using culture-dependent and -independent approaches. *World J. Microbiol. Biotechnol.* **2019**, *35*, 153. [CrossRef]
41. Wiseschart, A.; Mhuantong, W.; Tangphatsornruang, S.; Chantasingh, D.; Pootanakit, K. Shotgun metagenomic sequencing from Manao-Pee cave, Thailand, reveals insight into the microbial community structure and its metabolic potential. *BMC Microbiol.* **2019**, *19*, 144. [CrossRef] [PubMed]
42. Frazier, V.E. Carbon Metabolism in Cave Subaerial Biofilms. Master's Thesis, University of Tennessee, Knoxville, TN, USA, 2020.
43. He, D.; Wu, F.; Ma, W.; Zhang, Y.; Gu, J.-D.; Duan, Y.; Xu, R.; Feng, H.; Wang, W.; Li, S.-W. Insights into the bacterial and fungal communities and microbiome that causes a microbe outbreak on ancient wall paintings in the Maijishan Grottoes. *Int. Biodeter. Biodegr.* **2021**, *163*, 105250. [CrossRef]
44. Ma, L.; Huang, X.; Wang, H.; Yun, Y.; Cheng, X.; Liu, D.; Lu, X.; Qiu, X. Microbial interactions drive distinct taxonomic and potential metabolic responses to habitats in karst cave ecosystem. *Microbiol. Spect.* **2021**, *9*, e01152-21. [CrossRef]
45. Zada, S.; Xie, J.; Yang, M.; Yang, X.; Sajjad, W.; Rafiq, M.; Hasan, F.; Hu, Z.; Wang, H. Composition and functional profiles of microbial communities in two geochemically and mineralogically different caves. *Appl. Microbiol. Biotechnol.* **2021**, *105*, 8921–8936. [CrossRef]
46. Ai, J.; Guo, J.; Li, Y.; Zhong, X.; Lv, Y.; Li, J.; Yang, A. The diversity of microbes and prediction of their functions in karst caves under the influence of human tourism activities—A case study of Zhijin Cave in Southwest China. *Environ. Sci. Pollut. Res.* **2022**, *29*, 25858–25868. [CrossRef]
47. Buresova-Faitova, A.; Kopecky, J.; Sagova-Mareckova, M.; Alonso, L.; Vautrin, F.; Moënne-Loccoz, Y.; Rodriguez-Nava, V. Comparison of Actinobacteria communities from human-impacted and pristine karst caves. *MicrobiologyOpen* **2022**, *11*, e1276. [CrossRef]
48. Djebaili, R.; Mignini, A.; Vaccarelli, I.; Pellegrini, M.; Spera, D.M.; Del Gallo, M.; D'Alessandro, A.M. Polyhydroxybutyrate-producing cyanobacteria from lampenflora: The case study of the "Stiffe" caves in Italy. *Front. Microbiol.* **2022**, *13*, 933398. [CrossRef]
49. Cheng, X.; Xiang, X.; Yun, Y.; Wang, W.; Wang, H.; Bodelier, P.L.E. Archaea and their interactions with bacteria in a karst ecosystem. *Front. Microbiol.* **2023**, *14*, 1068595. [CrossRef] [PubMed]
50. Dimkic, I.; Copic, M.; Petrovic, M.; Stupar, M.; Savkovic, Ž.; Kneževic, A.; Simic, G.S.; Grbic, M.L.; Unkovic, N. Bacteriobiota of the cave church of Sts. Peter and Paul in Serbia—Culturable and non-culturable communities' assessment in the bioconservation potential of a peculiar fresco painting. *Int. J. Mol. Sci.* **2023**, *24*, 1016. [CrossRef]
51. Riquelme, C.; Rigal, F.; Hathaway, J.J.M.; Northup, D.E.; Spilde, M.N.; Borges, P.A.V.; Gabriel, R.; Amorin, I.R.; Dapkevicius, M.L.N.E. Cave microbial community composition in oceanic islands: Disentangling the effect of different colored mats in diversity patterns of Azorean lava caves. *FEMS Microbiol. Ecol.* **2015**, *91*, fiv141. [CrossRef] [PubMed]
52. Spilde, M.N.; Northup, D.E.; Caimi, N.A.; Boston, P.J.; Stone, F.D.; Smith, S. Microbial mat communities in Hawaiian lava caves. *Int. Symp. Vulcanospeleol.* **2016**. Available online: https://www.cavepics.com/IVS17/SPILDE.pdf. (accessed on 29 October 2022).
53. Lavoie, K.H.; Winter, A.S.; Read, K.J.H.; Hughes, E.M.; Spilde, M.N.; Northup, D.E. Comparison of bacterial communities from lava cave microbial mats to overlying surface soils from Lava Beds National Monument, USA. *PLoS ONE* **2017**, *12*, e0169339. [CrossRef] [PubMed]
54. Weng, M.M.; Zaikova, E.; Millan, M.; Williams, A.J.; McAdam, A.C.; Knudson, C.A.; Fuqua, S.R.; Wagner, N.Y.; Craft, K.; Nawotniak, S.K.; et al. Life underground: Investigating microbial communities and their biomarkers in Mars-analog lava tubes at Craters of the Moon National Monument and Preserve. *J. Geophys. Res. Planets* **2022**, *127*, e2022JE007268. [CrossRef]
55. Nicolosi, G.; Gonzalez-Pimentel, J.L.; Piano, E.; Isaia, M.; Miller, A.Z. First insights into the bacterial diversity of Mount Etna volcanic caves. *Microb. Ecol.* **2023**, *85*. *in press*. [CrossRef] [PubMed]
56. Barton, H.A.; Taylor, N.M.; Kreate, M.P.; Springer, A.C.; Oehrle, S.A.; Bertog, J.L. The impact of host rock geochemistry on bacterial community structure in oligotrophic cave environments. *Int. J. Speleol.* **2007**, *36*, 93–104. [CrossRef]
57. Ghezzi, D.; Sauro, F.; Columbu, A.; Carbone, C.; Hong, P.-Y.; Vergara, F.; De Waele, J.; Cappelletti, M. Transition from unclassified *Ktedonobacterales* to *Actinobacteria* during amorphous silica precipitation in a quartzite cave environment. *Sci. Rep.* **2021**, *11*, 3921. [CrossRef]
58. Weber, C.F. Reduced vertical stratification of soil bacterial community structure and composition is associated with *Bromus tectorum* invasion of sagebrush steppe. *J. Arid Environ.* **2015**, *115*, 90–99. [CrossRef]
59. Osman, J.R.; Fernandes, G.; Regeard, C.; Jaubert, C.; DuBow, M.S. Examination of the bacterial biodiversity of coastal eroded surface soils from the Padza de Dapani (Mayotte Island). *Geomicrobiol. J.* **2018**, *35*, 355–365. [CrossRef]
60. Lambrechts, S.; Willems, A.; Tahon, G. Uncovering the uncultivated majority in Antarctic soils: Toward a synergistic approach. *Front. Microbiol.* **2019**, *10*, 242. [CrossRef]
61. Li, J.; Wu, Z.; Yuan, J. Impact of agro-farming activities on microbial diversity of acidic red soils in a *Camellia Oleifera* Forest. *Rev. Bras. Cien. Solo* **2019**, *43*, e0190044. [CrossRef]
62. Zhenqing, Z.; Binglin, Z.; Wei, Z.; Guangxiu, L.; Tuo, C.; Yang, L.; Jingwei, C.; Mao, T. Distribution characteristics and anti-radiation activity of culturable bacteria in black gobi ecosystem of the Hexi Corridor. *J. Desert Res.* **2020**, *40*, 52–62.
63. Bossolani, J.; Crusciol, C.A.C.; Leite, M.F.A.; Merloti, L.F.; Moretti, L.G.; Pascoaloto, I.M.; Kuramae, E.E. Modulation of the soil microbiome by long-term Ca-based soil amendments boosts soil organic carbon and physicochemical quality in a tropical no-till crop rotation system. *Soil Biol. Biochem.* **2021**, *156*, 108188. [CrossRef]

64. Chen, B.; Jiao, S.; Luo, S.; Ma, B.; Qi, W.; Cao, C.; Zhao, Z.; Du, G.; Ma, X. High soil pH enhances the network interactions among bacterial and archaeal microbiota in alpine grasslands of the Tibetan Plateau. *Environ. Microbiol.* **2021**, *23*, 464–477. [CrossRef]
65. Liu, X.; Liu, Y.; Zhang, L.; Yin, R.; Wu, G.-L. Bacterial contributions of bio-crusts and litter crusts to nutrient cycling in the Mu Us Sandy Land. *Catena* **2021**, *199*, 105090. [CrossRef]
66. Liu, Z.; Yang, Y.; Ji, S.; Dong, D.; Li, Y.; Wang, M.; Han, L.; Chen, X. Effects of elevation and distance from highway on the abundance and community structure of bacteria in soil along Qinghai-Tibet highway. *Int. J. Environ. Res. Public Health* **2021**, *18*, 13137. [CrossRef]
67. Schulze-Makuch, D.; Lipus, D.; Arens, F.L.; Baqué, M.; Bornemann, T.L.V.; de Vera, J.-P.; Flury, M.; Frösler, J.; Heinz, J.; Hwang, Y.; et al. Microbial hotspots in lithic microhabitats inferred from DNA fractionation and metagenomics in the Atacama Desert. *Microorganisms* **2021**, *9*, 1038. [CrossRef]
68. Xie, J.; Wu, Z.; Zhang, X.; Peng, T.; Yang, C.; Zhang, J.; Liang, J. Diversity and structural characteristics of soil microbial communities in different habitats of wild *Lilium regale* Wilson in Wenchuan area. *Bioengineered* **2021**, *12*, 10457–10469. [CrossRef] [PubMed]
69. Benaud, N.; Chelliah, D.S.; Wong, S.Y.; Ferrari, B.C. Soil substrate culturing approaches recover diverse members of *Actinomycetota* from desert soils of Herring Island, East Antarctica. *Extremophiles* **2022**, *26*, 24. [CrossRef] [PubMed]
70. Guerra, V.A.; Beule, L.; Mackowiak, C.L.; Dubeux, J.C.B., Jr.; Blount, A.R.S.; Wang, X.-B.; Rowland, D.L.; Liao, H.-L. Soil bacterial community response to rhizoma peanut incorporation into Florida pastures. *J. Environ. Qual.* **2022**, *51*, 55–65. [CrossRef] [PubMed]
71. Ke, M.; Xu, N.; Zhang, Z.; Qiu, D.; Kang, J.; Lu, T.; Wang, T.; Peijnenburg, W.J.G.M.; Sun, L.; Hu, B.; et al. Development of a machine-learning model to identify the impacts of pesticides characteristics on soil microbial communities from high-throughput sequencing data. *Environ. Microbiol.* **2022**, *24*, 5561–5573. [CrossRef]
72. Sun, H.; Zhang, H. Alien species invasion of deep-sea bacteria into terrestrial soil. *J. Clean. Product.* **2022**, *371*, 133662. [CrossRef]
73. Sun, H.; Peng, Q.; Guo, J.; Zhang, H.; Bai, J.; Mao, H. Effects of short-term soil exposure of different doses of ZnO nanoparticles on the soil environment and the growth and nitrogen fixation of alfalfa. *Environ. Pollut.* **2022**, *309*, 119817. [CrossRef]
74. Topalovic, O.; Santos, S.S.; Heuer, H.; Nesme, J.; Kanfra, X.; Hallmann, J.; Sørensen, S.J.; Vestergård, M. Deciphering bacteria associated with a pre-parasitic stage of the root-knot nematode *Meloidogyne hapla* in nemato-suppressive and nemato-conducive soils. *Appl. Soil Ecol.* **2022**, *172*, 104344. [CrossRef]
75. Wang, L.; Peng, C.; Gong, B.; Yang, Z.; Song, J.; Li, L.; Xu, L.; Yue, T.; Wang, X.; Yang, M.; et al. Actinobacteria community and their antibacterial and cytotoxic activity on the Weizhou and Xieyang volcanic islands in the Beibu Gulf of China. *Front. Microbiol.* **2022**, *13*, 911408. [CrossRef]
76. Feng, Z.; Sun, H.; Qin, Y.; Zhou, Y.; Zhu, H.; Yao, Q. A synthetic community of siderophore-producing bacteria increases soil selenium bioavailability and plant uptake through regulation of the soil microbiome. *Sci. Total Environ.* **2023**, *871*, 162076. [CrossRef]
77. Conte, A. Phylogenetic Diversity and Metabolic Potential of Prokaryotic Communities in Permafrost and Brine Pockets of Perennially Frozen Antarctic Lakes (Northern Victoria Land). Ph.D. Thesis, Università Degli Studi di Messina, Messina, Italy, 2017.
78. Perez-Mon, C.; Stierli, B.; Plötze, M.; Frey, B. Fast and persistent responses of alpine permafrost microbial communities to in situ warming. *Sci. Total Environ.* **2022**, *807*, 150720. [CrossRef]
79. Narendrula, R. Biochemical and Molecular Characterization of Microbial Communities from a Metal Contaminated and Reclaimed Region. Ph.D. Thesis, Laurentian University, Sudbury, ON, Canada, 2017.
80. Lin, J.; He, F.; Su, B.; Sun, M.; Owens, G.; Chen, Z. The stabilizing mechanism of cadmium in contaminated soil using green synthesized iron oxide nanoparticles under long-term incubation. *J. Hazard. Mater.* **2019**, *379*, 120832. [CrossRef]
81. Dong, S.; Liu, S.; Cui, S.; Zhou, X.; Gao, Q. Responses of soil properties and bacterial community to the application of sulfur fertilizers in black and sandy soils. *Pol. J. Environ. Stud.* **2022**, *31*, 35–47. [CrossRef] [PubMed]
82. Pu, C.; Liu, H.; Ding, G.; Sun, Y.; Yu, X.; Chen, J.; Ren, J.; Gong, X. Impact of direct application of biogas slurry and residue in fields: In situ analysis of antibiotic resistance genes from pig manure to fields. *J. Hazard. Mater.* **2018**, *344*, 441–449. [CrossRef] [PubMed]
83. Deng, J.; Zhang, Y.; Yin, Y.; Zhu, X.; Zhu, W.; Zhou, Y. Comparison of soil bacterial community and functional characteristics following afforestation in the semi-arid areas. *PeerJ* **2019**, *7*, e7141. [CrossRef]
84. Deng, J.; Zhou, Y.; Zhu, W.; Yin, Y. Effects of afforestation with *Pinus sylvestris* var. *mongolica plantations combined with enclosure management on soil microbial community. PeerJ* **2020**, *8*, e8857. [PubMed]
85. Liu, K.; Ding, X.; Wang, J. Soil metabolome correlates with bacterial diversity and co-occurrence patterns in root-associated soils on the Tibetan Plateau. *Sci. Total Environ.* **2020**, *735*, 139572. [CrossRef]
86. Jiang, H.; Chen, Y.; Hu, Y.; Wang, Z.; Lu, X. Soil bacterial communities and diversity in alpine grasslands on the Tibetan Plateau based on 16S rRNA gene sequencing. *Front. Ecol. Evol.* **2021**, *9*, 630722. [CrossRef]
87. Solon, A.J.; Mastrangelo, C.; Vimercati, L.; Sommers, P.; Darcy, J.L.; Gendron, E.M.S.; Porazinska, D.L.; Schmidt, S.K. Gullies and moraines are islands of biodiversity in an arid, mountain landscape, Asgard Range, Antarctica. *Front. Microbiol.* **2021**, *12*, 654135. [CrossRef]

88. Rodríguez-Berbel, N.; Ortega, R.; Lucas-Borja, M.E.; Solé-Benet, A.; Miralles, I. Long-term effects of two organic amendments on bacterial communities of calcareous mediterranean soils degraded by mining. *J. Environ. Manag.* **2020**, *271*, 110920. [CrossRef]
89. Chuvochina, M.S.; Alekhina, I.A.; Normand, P.; Petit, J.-R.; Bulat, S.A. Three events of Saharan dust deposition on the Mont Blanc glacier associated with different snow-colonizing bacterial phylotypes. *Microbiology* **2011**, *80*, 125–131. [CrossRef]
90. Hui, N.; Sun, N.; Du, H.; Umair, M.; Kang, H.; Liu, X.; Romantschuk, M.; Liu, C. Karst rocky desertification does not erode ectomycorrhizal fungal species richness but alters microbial community structure. *Plant Soil* **2019**, *445*, 383–396. [CrossRef]
91. Reverdy, A.; Hathaway, D.; Jha, J.; Michaels, G.; Sullivan, J.; Diaz Mac-Adoo, D.; Riquelme, C.; Chai, Y.; Godoy, V.G. Insights into the diversity and survival strategies of soil bacterial isolates from the Atacama Desert. *bioRxiv* **2020**. [CrossRef]
92. Biderre-Petit, C.; Hochart, C.; Gardon, H.; Dugat-Bony, E.; Terrat, S.; Jouan-Dufournel, I.; Paris, R. Analysis of bacterial and archaeal communities associated with Fogo volcanic soils of different ages. *FEMS Microbiol. Ecol.* **2020**, *96*, fiaa104. [CrossRef] [PubMed]
93. Zhang, X.; Du, W.; Xu, Y.; Wang, Y.L. Soil bacterial diversity and function in semi-arid forest parks in Baotou City. *Biodivers. Sci.* **2022**, *30*, 22245. [CrossRef]
94. Li, Y.; Gao, W.; Wang, C.; Gao, M. Distinct distribution patterns and functional potentials of rare and abundant microorganisms between plastisphere and soils. *Sci. Total Environ.* **2023**, *873*, 162413. [CrossRef]
95. Zhang, S.; Pei, L.; Zhao, Y.; Shan, J.; Zheng, X.; Xu, G.; Sun, Y.; Wang, F. Effects of microplastics and nitrogen deposition on soil multifunctionality, particularly C and N cycling. *J. Hazard. Mater.* **2023**, *451*, 131152. [CrossRef]
96. Lin, Y.; Yang, L.; Chen, Z.; Gao, Y.; Kong, J.; He, Q.; Su, Y.; Li, J.; Qiu, Q. Seasonal variations of soil bacterial and fungal communities in a subtropical *Eucalyptus* plantation and their responses to throughfall reduction. *Front. Microbiol.* **2023**, *14*, 1113616. [CrossRef]
97. De Tender, C.; Haegeman, A.; Vandecasteele, B.; Clement, L.; Cremelie, P.; Dawyndt, P.; Maes, M.; Debode, J. Dynamics in the strawberry rhizosphere microbiome in response to biochar and *Botrytis cinerea* leaf infection. *Front. Microbiol.* **2016**, *7*, 2062. [CrossRef]
98. Echeverría Molinar, A. Efecto de Factores Abióticos y Bióticos Sobre la Estructura de la Comunidad Microbiana del Suelo en un Ambiente Oligotrófico. Master's Thesis, Instituto Potosino de Investigación Científica y Tecnológica, San Luis Potosí, México, 2017.
99. Visioli, G.; Sanangelantoni, A.M.; Vamerali, T.; Dal Cortivo, C.; Blandino, M. 16S rDNA profiling to reveal the influence of seed-applied biostimulants on the rhizosphere of young maize plants. *Molecules* **2018**, *23*, 1461. [CrossRef]
100. Gao, X.; Wu, Z.; Liu, R.; Wu, J.; Zeng, Q.; Qi, Y. Rhizosphere bacterial community characteristics over different years of sugarcane ratooning in consecutive monoculture. *BioMed Res. Int.* **2019**, *2019*, 4943150. [CrossRef]
101. Chen, Y.; Tian, W.; Shao, Y.; Li, Y.-J.; Lin, L.-A.; Zhang, Y.-J.; Han, H.; Chen, Z.-J. *Miscanthus* cultivation shapes rhizosphere microbial community structure and function as assessed by Illumina MiSeq sequencing combined with PICRUSt and FUNGUIld analyses. *Arch. Microbiol.* **2020**, *202*, 1157–1171. [CrossRef] [PubMed]
102. López-Lozano, N.E.; Echeverría Molinar, A.; Ortiz Durán, E.A.; Hernández Rosales, M.; Souza, V. Bacterial diversity and interaction networks of *Agave lechuguilla* rhizosphere differ significantly from bulk soil in the oligotrophic basin of Cuatro Cienegas. *Front. Plant Sci.* **2020**, *11*, 1028. [CrossRef] [PubMed]
103. Monteiro, L.C.P. Comunidades Microbianas Rizosféricas de Plantas em Coexistência Sob Diferentes Condições Edáficas. Ph.D. Thesis, Universidade Federal de Viçosa, Viçosa, Brasil, 2020.
104. Bettermann, A.; Zethof, J.H.T.; Babin, D.; Cammeraat, L.H.; Solé-Benet, A.; Lázaro, R.; Luna, L.; Nesme, J.; Sorensen, S.J.; Kalbitz, K.; et al. Importance of microbial communities at the root-soil interface for extracellular polymeric substances and soil aggregation in semiarid grasslands. *Soil Biol. Biochem.* **2021**, *159*, 108301. [CrossRef]
105. Deng, Q.; Zhang, T.; Xie, D.; Yang, Y. Rhizosphere microbial communities are significantly affected by optimized phosphorus management in a slope farming system. *Front. Microbiol.* **2021**, *12*, 739844. [CrossRef]
106. Engelbrecht, G.; Claassens, S.; Mienie, C.M.S.; Fourie, H. Screening of rhizosphere bacteria and nematode populations associated with soybean roots in the Mpumalanga Highveld of South Africa. *Microorganisms* **2021**, *9*, 1813. [CrossRef]
107. Li, C.; Chen, G.; Zhang, J.; Zhu, P.; Bai, X.; Hou, Y.; Zhang, X. The comprehensive changes in soil properties are continuous cropping obstacles associated with American ginseng (*Panax quinquefolius*) cultivation. *Sci. Rep.* **2021**, *11*, 5068. [CrossRef]
108. Ye, F.; Wang, X.; Wang, Y.; Wu, S.; Wu, J.; Hong, Y. Different pioneer plant species have similar rhizosphere microbial communities. *Plant Soil* **2021**, *464*, 165–181. [CrossRef]
109. Zuo, J.; Zu, M.; Liu, L.; Song, X.; Yuan, Y. Composition and diversity of bacterial communities in the rhizosphere of the Chinese medicinal herb *Dendrobium*. *BMC Plant Biol.* **2021**, *21*, 127. [CrossRef]
110. Gu, Y.-Y.; Zhang, H.-Y.; Liang, X.-Y.; Fu, R.; Li, M.; Chen, C.-J. Impact of biochar and bioorganic fertilizer on rhizosphere bacteria in saline–alkali soil. *Microorganisms* **2022**, *10*, 2310. [CrossRef]
111. He, C.; Wang, R.; Ding, W.; Li, Y. Effects of cultivation soils and ages on microbiome in the rhizosphere soil of *Panax ginseng*. *Appl. Soil Ecol.* **2022**, *174*, 104397. [CrossRef]
112. Jara-Servin, A.; Silva, A.; Barajas, H.; Cruz-Ortega, R.; Tinoco-Ojanguren, C.; Alcaraz, L.D. The Allelopathic Buffelgrass and Its Rhizosphere Microbiome. Available online: https://papers.ssrn.com/sol3/papers.cfm?abstract_id=4263505 (accessed on 28 February 2023).
113. Kushwaha, P.; Neilson, J.W.; Maier, R.M.; Babst-Kostecka, A. Soil microbial community and abiotic soil properties influence Zn and Cd hyperaccumulation differently in *Arabidopsis halleri*. *Sci. Total Environ.* **2022**, *803*, 150006. [CrossRef] [PubMed]

114. Lin, Y.; Zhang, Y.; Liang, X.; Duan, R.; Yang, L.; Du, Y.; Wu, L.; Huang, J.; Xiang, G.; Bai, J.; et al. Assessment of rhizosphere bacterial diversity and composition in a metal hyperaccumulator (*Boehmeria nivea*) and a nonaccumulator (*Artemisia annua*) in an antimony mine. *J. Appl. Microbiol.* **2022**, *132*, 3432–3443. [CrossRef] [PubMed]
115. Lizano-Bastardín, A.L.; Villadas, P.J.; Pulido-Suárez, L.; Fernández-López, M.; León-Barrios, M. The Rhizosphere Microbiome Associated with the Legume *Spartocytisus supranubius* in the High Mountain Ecosystem of Teide N.P. First Spanish-Portuguese Congress on Beneficial Plant-Microbe Interactions, Poster S1-P-03. Available online: https://events.iniav.pt/bemiplant/images/book-abstracts_15-10-2022.pdf (accessed on 9 February 2023).
116. Ren, H.; Islam, M.S.; Wang, H.; Guo, H.; Wang, Z.; Qi, X.; Zhang, S.; Guo, J.; Wang, Q.; Li, B. Effect of humic acid on soil physical and chemical properties, microbial community structure, and metabolites of decline diseased bayberry. *Int. J. Mol. Sci.* **2022**, *23*, 14707. [CrossRef] [PubMed]
117. Taniguchi, T.; Isobe, K.; Imada, S.; Eltayeb, M.; Akaji, Y.; Nakayama, M.; Allen, M.F.; Aronson, E. Root endophytic bacterial and fungal communities in a natural hot desert are differentially regulated in dry and wet seasons by stochastic processes and functional traits. Available online: https://ssrn.com/abstract=4363629 (accessed on 25 February 2023).
118. Zhou, Y.; Pang, Z.; Jia, H.; Yuan, Z.; Ming, R. Responses of roots and rhizosphere of female papaya to the exogenous application of GA_3. *BMC Plant Biol.* **2023**, *23*, 35. [CrossRef] [PubMed]
119. Obieze, C.C.; George, P.B.L.; Boyle, B.; Khasa, D.P. Black pepper rhizomicrobiome: Spectrum of plant health indicators, critical environmental factors and community compartmentation in Vietnam. *Appl. Soil Ecol.* **2023**, *187*, 104857. [CrossRef]
120. Deng, J.; Yin, Y.; Zhu, W.; Zhou, Y. Response of soil environment factors and microbial communities to phytoremediation with *Robinia pseudoacacia* in an open-cut magnesite mine. *Land Degrad. Dev.* **2020**, *31*, 2340–2355. [CrossRef]
121. Ezeokoli, O.T.; Bezuidenhout, C.C.; Maboeta, M.S.; Khasa, D.P.; Adeleke, R.A. Structural and functional differentiation of bacterial communities in post-coal mining reclamation soils of South Africa: Bioindicators of soil ecosystem restoration. *Sci. Rep.* **2020**, *10*, 1759. [CrossRef]
122. Thompson, E.; Erickson, M.; Malik, N.; Mettler, R.; Reman, B.; Ren, Y.; Bergmann, D. Culture-independent characterization of "cave silver" biofilms from the 1470 m level of the Sanford Underground Research Facility, Lead, SD. *Proc. South Dak. Acad. Sci.* **2020**, *99*, 29–55.
123. Arif, S.; Nacke, H.; Schliekmann, E.; Reimer, A.; Arp, G.; Hoppert, M. Composition and niche-specific characteristics of microbial consortia colonizing Marsberg copper mine in the Rhenish Massif. *Biogeosciences* **2021**, *19*, 4883–4902. [CrossRef]
124. Shi, A.; Hu, Y.; Zhang, X.; Zhou, D.; Xu, J.; Rensing, C.; Zhang, L.; Xing, S.; Ni, W.; Yang, W. Biochar loaded with bacteria enhanced Cd/Zn phytoextraction by facilitating plant growth and shaping rhizospheric microbial community. *Environ. Pollut.* **2023**, *327*, 121559. [CrossRef]
125. Diaz-Herraiz, M.; Jurado, V.; Cuezva, S.; Laiz, L.; Pallecchi, P.; Tiano, P.; Sanchez-Moral, S.; Saiz-Jimenez, C. Deterioration of an Etruscan tomb by bacteria from the order *Rhizobiales*. *Sci. Rep.* **2014**, *4*, 3610. [CrossRef] [PubMed]
126. Zacharenka, F. Study of the Bacterial Diversity on Ancient Mural Paintings from Kalliroi's Fountain and Pana's Sanctuary. Master's Thesis, University of Thessaly, Volos, Greece, 2014.
127. Duan, Y.; Wu, F.; Wang, W.; He, D.; Gu, J.-D.; Feng, H.; Chen, T.; Liu, G.; An, L. The microbial community characteristics of ancient painted sculptures in Maijishan Grottoes, China. *PLoS ONE* **2017**, *12*, e0179718.
128. Li, Q.; Zhang, B.; Wang, L.; Ge, Q. Distribution and diversity of bacteria and fungi colonizing ancient Buddhist statues analyzed by high-throughput sequencing. *Int. Biodeter. Biodegr.* **2017**, *117*, 245–254. [CrossRef]
129. Li, Q.; Zhang, B.; Yang, X.; Ge, Q. Deterioration-associated microbiome of stone monuments: Structure, variation, and assembly. *Appl. Environ. Microbiol.* **2018**, *84*, e02680-17. [CrossRef] [PubMed]
130. Zhang, X.; Ge, Q.; Zhu, Z.; Deng, Y.; Gu, J.-D. Microbiological community of the Royal Palace in Angkor Thom and Beng Mealea of Cambodia by Illumina sequencing based on 16S rRNA gene. *Int. Biodeter. Biodegr.* **2018**, *134*, 127–135. [CrossRef]
131. Dyda, M.; Pyzik, A.; Wilkojc, E.; Kwiatkowska-Kopka, B.; Sklodowska, A. Bacterial and fungal diversity inside the medieval building constructed with sandstone plates and lime mortar as an example of the microbial colonization of a nutrient-limited extreme environment (Wawel Royal Castle, Krakow, Poland). *Microorganisms* **2019**, *7*, 416. [CrossRef]
132. Jroundi, F.; Elert, K.; Ruiz-Agudo, E.; Gonzalez-Muñoz, M.T.; Rodriguez-Navarro, C. Bacterial diversity evolution in Maya plaster and stone following a bio-conservation treatment. *Front. Microbiol.* **2020**, *11*, 599144. [CrossRef]
133. Louati, M.; Ennis, N.J.; Ghodhbane-Gtari, F.; Hezbri, K.; Sevigny, J.L.; Fahnestock, M.F., Cherif-Silini, H.; Bryce, J.G.; Tisa, L.S.; Gtari, M. Elucidating the ecological networks in stone-dwelling microbiomes. *Environ. Microbiol.* **2020**, *22*, 1467–1480. [CrossRef]
134. Schröer, L.; De Kock, T.; Cnudde, V.; Boon, N. Differential colonization of microbial communities inhabiting Lede stone in the urban and rural environment. *Sci. Total Environ.* **2020**, *733*, 139339. [CrossRef]
135. Yang, S.; Wu, L.; Wu, B.; Zhang, Y.; Wang, H.; Tan, X. Diversity and structure of soil microbiota of the Jinsha earthen relic. *PLoS ONE* **2020**, *15*, e0236165. [CrossRef]
136. Coelho, C.; Mesquita, N.; Costa, I.; Soares, F.; Trovão, J.; Freitas, H.; Portugal, A.; Tiago, I. Bacterial and archaeal structural diversity in several biodeterioration patterns on the limestone walls of the Old Cathedral of Coimbra. *Microorganisms* **2021**, *9*, 709. [CrossRef] [PubMed]
137. Ipekci, E. Evaluation of Stone Deterioration Problems of Anavarza Archaeological Site for the Purpose of Conservation. Ph.D. Thesis, İzmir Institute of Technology, Urla, Turkey, 2021.

138. Sansupa, C.; Suphaphimol, N.; Nonthijun, P.; Ronsuek, T.; Yimklan, S.; Semakul, N.; Khrueraya, T.; Suwannarach, N.; Purahong, W.; Disayathanoowat, T. The microbiome of a 13th century Lan Na mural painting: Diversity, taxonomic distribution and their biodeterioration potentials. *Microorganisms* **2023**, *11*. *in press*.
139. Chen, P.; Zhang, L.; Guo, X.; Dai, X.; Liu, L.; Xi, L.; Wang, J.; Song, L.; Wang, Y.; Zhu, Y.; et al. Diversity, biogeography, and biodegradation potential of actinobacteria in the deep-sea sediments along the Southwest Indian Ridge. *Front. Microbiol.* **2016**, *7*, 1340. [CrossRef] [PubMed]
140. Pietrzak, K.; Otlewska, A.; Dybka, K.; Danielewicz, D.; Pangallo, D.; Demnerová, K.; Ďurovič, M.; Kraková, L.; Scholtz, V.; Bučková, M.; et al. A modern approach to biodeterioration assessment and disinfection of historical book. In *A Modern Approach to Biodeterioration Assessment and the Disinfection of Historical Book Collections*; Gutarowska, B., Ed.; Institute of Fermentation Technology and Microbiology, Łódź University of Technology: Łódź, Poland, 2016; pp. 81–125.
141. Gozari, M.; Bahador, N.; Jassbi, A.R.; Mortazavi, M.S.; Hamzehei, S.; Eftekhar, E. Isolation, distribution and evaluation of cytotoxic and antioxidant activity of cultivable actinobacteria from the Oman Sea sediments. *Acta Oceanol. Sin.* **2019**, *38*, 84–90. [CrossRef]
142. Van Assche, A. Characterization of the Bacterial Community Composition in Drinking Water Production and Distribution Systems, Emphasizing *Acinetobacter* species. Ph.D. Thesis, University of Leuven, Leuven, Belgium, 2019.
143. Argudo Fernández, M. Microbial Communities Responses in Fluvial Biofilms under Metal Stressed Scenarios. Ph.D. Thesis, Universitat de Girona, Girona, Spain, 2020.
144. Cao, W.; Xiong, Y.; Zhao, D.; Tan, H.; Qu, J. Bryophytes and the symbiotic microorganisms, the pioneers of vegetation restoration in karst rocky desertification areas in southwestern China. *Appl. Microbiol. Biotechnol.* **2020**, *104*, 873–891. [CrossRef] [PubMed]
145. Caffrey, P.; Hogan, M.; Song, Y. New glycosylated polyene macrolides: Refining the ore from genome mining. *Antibiotics* **2022**, *11*, 334. [CrossRef]
146. Yang, Y.; Qiu, J.; Wang, X. Exploring the dynamic of bacterial communities in Manila clam (*Ruditapes philippinarum*) during refrigerated storage. *Front. Microbiol.* **2022**, *13*, 882629. [CrossRef]
147. Boquet, E.; Boronat, A.; Ramos-Cormenzana, A. Production of calcite (calcium carbonate) crystals by soil bacteria is a general phenomenon. *Nature* **1973**, *246*, 527–528. [CrossRef]
148. Barton, H.A.; Northup, D.E. Geomicrobiology in cave environments: Past, current and future perspectives. *J. Cave Karst Stud.* **2007**, *69*, 163–178.
149. Banks, E.D.; Taylor, N.M.; Gulley, J.; Lubbers, B.R.; Giarrizzo, J.G.; Bullen, H.A.; Hoehler, T.M.; Barton, H.A. Bacterial calcium carbonate precipitation in cave environments: A function of calcium homeostasis. *Geomicrobiol. J.* **2010**, *27*, 444–454. [CrossRef]
150. Cañaveras, J.C.; Hoyos, M.; Sanchez-Moral, S.; Sanz-Rubio, E.; Bedoya, J.; Soler, V.; Groth, I.; Schumann, P.; Laiz, L.; Gonzalez, I.; et al. Microbial communities associated to hydromagnesite and needle fiber aragonite deposits in a karstic cave (Altamira, Northern Spain). *Geomicrobiol. J.* **1999**, *16*, 9–25.
151. Sanchez-Moral, S.; Cañaveras, J.C.; Laiz, L.; Saiz-Jimenez, C.; Bedoya, J.; Luque, L. Biomediated precipitation of calcium carbonate metastable phases in hypogean environments: A short review. *Geomicrobiol. J.* **2003**, *20*, 491–500. [CrossRef]
152. Cuezva, S.; Fernandez-Cortes, A.; Porca, E.; Pasic, L.; Jurado, V.; Hernandez-Marine, M.; Serrano-Ortiz, P.; Cañaveras, J.C.; Sanchez-Moral, S.; Saiz-Jimenez, C. The biogeochemical role of Actinobacteria in Altamira Cave, Spain. *FEMS Microbiol. Ecol.* **2012**, *81*, 281–290. [CrossRef] [PubMed]
153. Maciejewska, M.; Adam, D.; Naômé, A.; Martinet, L.; Tenconi, E.; Calusinska, M.; Delfosse, P.; Hanikenne, M.; Baurain, D.; Compère, P.; et al. Assessment of the potential role of *Streptomyces* in cave moonmilk formation. *Front. Microbiol.* **2017**, *8*, 1181. [CrossRef] [PubMed]
154. Hoffmann, T.D.; Reeksting, B.J.; Gebhard, S. Bacteria-induced mineral precipitation: A mechanistic review. *Microbiology* **2021**, *167*, 001049. [CrossRef]
155. Tomé, D. Yeast extracts: Nutritional and flavoring food ingredients. *ACS Food Sci. Technol.* **2021**, *1*, 487–494. [CrossRef]
156. Joubert, J.J.; van Rensburg, E.J.; Pitout, M.J. A plate method for demonstrating the breakdown of heparin and chrondoitin sulphate by bacteria. *J. Microbiol. Methods* **1984**, *2*, 197–202. [CrossRef]
157. Sanchez-Moral, S.; Luque, L.; Cañaveras, J.C.; Laiz, L.; Jurado, V.; Saiz-Jimenez, C. Bioinduced barium precipitation in St. Callixtus and Domitilla catacombs. *Ann. Microbiol.* **2004**, *54*, 1–12.
158. Sanchez-Moral, S.; Bedoya, J.; Luque, L.; Cañaveras, J.C.; Jurado, V.; Laiz, L.; Saiz-Jimenez, C. Biomineralization of different crystalline phases by bacteria isolated from catacombs. In *Molecular Biology and Cultural Heritage*; Saiz-Jimenez, C., Ed.; Balkema: Lisse, The Netherlands, 2003; pp. 179–185.
159. Rivadeneyra, M.A.; Ramos-Cormenzana, A.; García-Cervigón, A. Bacterial formation of struvite. *Geomicrobiol. J.* **1983**, *3*, 151–163. [CrossRef]
160. Rivadeneyra, M.A.; Pérez-García, I.; Ramos-Cormenzana, A. Influence of ammonium ion on bacterial struvite production. *Geomicrobiol. J.* **1992**, *10*, 125–137. [CrossRef]
161. Manzoor, M.A.P.; Singh, B.; Agrawal, A.K.; Arun, A.B.; Mujeeburahiman, M.; Rekha, P.-D. Morphological and micro-tomographic study on evolution of struvite in synthetic urine infected with bacteria and investigation of its pathological biomineralization. *PLoS ONE* **2018**, *13*, e0202306. [CrossRef] [PubMed]
162. Sánchez-Román, M.; Rivadeneyra, M.A.; Vasconcelos, C.; McKenzie, J.A. Biomineralization of carbonate and phosphate by moderately halophilic bacteria. *FEMS Microbiol. Ecol.* **2007**, *61*, 273–284. [CrossRef]

163. Cañaveras, J.C.; Cuezva, S.; Sanchez-Moral, S.; Lario, J.; Laiz, L.; Gonzalez, J.M.; Saiz-Jimenez, C. On the origin of fiber calcite crystals in moonmilk deposits. *Naturwissenschaften* **2006**, *93*, 27–32. [CrossRef] [PubMed]
164. Onac, B.P.; Ghergari, L. Moonmilk mineralogy in some Romanian and Norwegian Caves. *Cave Sci.* **1993**, *20*, 106–120.
165. Cañaveras, J.C.; Sanchez–Moral, S.; Soler, V.; Saiz-Jimenez, C. Microorganisms and microbially induced fabrics in cave walls. *Geomicrobiol. J.* **2001**, *18*, 223–240.
166. Baskar, S.; Baskar, R.; Routh, J. Biogenic evidences of moonmilk deposition in the Mawmluh Cave, Meghalaya, India. *Geomicrobiol. J.* **2011**, *28*, 252–265.
167. Jones, B.; Peng, X. Abiogenic growth of needle-fiber calcite in spring towers at Shiqiang, Yunnan province, China. *J. Sediment. Res.* **2014**, *84*, 1021–1040. [CrossRef]
168. Cuezva, S.; Sanchez-Moral, S.; Saiz-Jimenez, C.; Cañaveras, J.C. Microbial communities and associated mineral fabrics in Altamira Cave, Spain. *Int. J. Speleol.* **2009**, *38*, 83–92.
169. Sanchez-Moral, S.; Portillo, M.C.; Janices, I.; Cuezva, S.; Fernández-Cortés, A.; Cañaveras, J.C.; Gonzalez, J.M. The role of microorganisms in the formation of calcitic moonmilk deposits and speleothems in Altamira Cave. *Geomorphology* **2012**, *139–140*, 285–292. [CrossRef]
170. Maciejewska, M.; Calusinska, M.; Cornet, L.; Adam, D.; Pessi, I.S.; Malchair, S.; Delfosse, P.; Baurain, D.; Barton, H.; Carnol, M.; et al. High-throughput sequencing analysis of the actinobacterial spatial diversity in moonmilk deposits. *Antibiotics* **2018**, *7*, 27. [CrossRef]
171. Groth, I.; Schumann, P.; Laiz, L.; Sanchez–Moral, S.; Cañaveras, J.C.; Saiz-Jimenez, C. Geomicrobiological study of the Grotta dei Cervi, Porto Badisco, Italy. *Geomicrobiol. J.* **2001**, *18*, 241–258. [CrossRef]
172. Park, S.; Cho, Y.-J.; Jung, D.; Jo, K.; Lee, E.-J.; Lee, J.-S. Microbial diversity in moonmilk of Baeg-nyong Cave, Korean CZO. *Front. Microbiol.* **2020**, *11*, 613. [CrossRef] [PubMed]
173. Cirigliano, A.; Tomassetti, M.C.; Di Pietro, M.; Mura, F.; Maneschi, M.L.; Gentili, M.D.; Cardazzo, B.; Arrighi, C.; Mazzoni, C.; Negri, R.; et al. Calcite moonmilk of microbial origin in the Etruscan Tomba degli Scudi in Tarquinia, Italy. *Sci. Rep.* **2018**, *8*, 15839. [CrossRef] [PubMed]
174. Maciejewska, M.; Pessi, I.S.; Arguelles-Arias, A.; Noirfalise, P.; Luis, G.; Ongena, M.; Barton, H.; Carnol, M.; Rigali, S. *Streptomyces lunaelactis* sp. nov., a novel ferroverdin A-producing *Streptomyces* species isolated from a moonmilk speleothem. *Anton. Leeuw.* **2015**, *107*, 519–531. [CrossRef] [PubMed]
175. Svec, P.; Kosina, M.; Zeman, M.; Holochová, P.; Králová, S.; Nemcová, E.; Micenková, L.; Urvashi, M.; Gupta, V.; Sood, U.; et al. *Pseudomonas karstica* sp. nov. and *Pseudomonas spelaei* sp. nov., isolated from calcite moonmilk deposits from caves. *Int. J. Syst. Evol. Microbiol.* **2020**, *70*, 5131–5140. [CrossRef]
176. Adam, D.; Maciejewska, M.; Naômé, A.; Martinet, L.; Coppieters, W.; Karim, L.; Baurain, D.; Rigali, S. Isolation, Characterization, and antibacterial activity of hard-to-culture Actinobacteria from cave moonmilk deposits. *Antibiotics* **2018**, *7*, 28. [CrossRef]
177. Jaroszewicz, W.; Bielanska, P.; Lubomska, D.; Kosznik-Kwasnicka, K.; Golec, P.; Grabowski, Ł.; Wieczerzak, E.; Dróżdz, W.; Gaffke, L.; Pierzynowska, K.; et al. Antibacterial, antifungal and anticancer activities of compounds produced by newly isolated *Streptomyces* strains from the Szczelina Chochołowska Cave (Tatra Mountains, Poland). *Antibiotics* **2021**, *10*, 1212. [CrossRef]
178. Jaroszewicz, W.; Bielanska, P.; Lubomska, D.; Kosznik-Kwasnicka, K.; Golec, P.; Grabowski, Ł.; Wieczerzak, E.; Dróżdz, W.; Gaffke, L.; Pierzynowska, K.; et al. Antimicrobial activities of compounds produced by newly isolated *Streptomyces* strains from Mountain Caves. *Med. Sci. Forum* **2022**, *12*, 7.
179. Maciejewska, M.; Adam, D.; Martinet, L.; Naômé, A.; Całusińska, M.; Delfosse, P.; Carnol, M.; Barton, H.A.; Hayette, M.-P.; Smargiasso, N.; et al. A Phenotypic and genotypic analysis of the antimicrobial potential of cultivable *Streptomyces* isolated from cave moonmilk deposits. *Front. Microbiol.* **2016**, *7*, 1455. [CrossRef]
180. Martinet, L.; Naômé, A.; Rezende, L.C.D.; Tellatin, D.; Pignon, B.; Docquier, J.-D.; Sannio, F.; Baiwir, D.; Mazzucchelli, G.; Frédérich, M.; et al. Lunaemycins, new cyclic hexapeptide antibiotics from the cave moonmilk-dweller *Streptomyces lunaelactis* MM109T. *Int. J. Mol. Sci.* **2023**, *24*, 1114. [CrossRef]
181. Martinet, L.; Naômé, A.; Baiwir, D.; De Pauw, E.; Mazzucchelli, G.; Rigali, S. On the risks of phylogeny-based strain prioritization for drug discovery: *Streptomyces lunaelactis* as a case study. *Biomolecules* **2020**, *10*, 1027. [CrossRef] [PubMed]

applied biosciences

MDPI

Article

Caffeic Acid and Biopesticides Interactions for the Control of Storage Beetles

Chrysanthi Zarmakoupi [1], Konstantinos Mpistiolis [1], George Pantazis [1], Panagiota Psatha [1], Despoina Dimitriadi [2], Foteini Kitsiou [1], Panagiotis Eliopoulos [3,*], George Patakioutas [1,*] and Spiridon Mantzoukas [1,*]

[1] Department of Agriculture, University of Ioannina, 45100 Ioannina, Greece
[2] Karvelas AVEE, 80 km N.R. Athens-Lamia, 32200 Thiva, Greece
[3] Laboratory of Plant Health Management, Department of Agrotechnology, University of Thessaly, Geopolis, 45100 Larissa, Greece
* Correspondence: eliopoulos@uth.gr (P.E.); gpatakiu@uoi.gr (G.P.); sdmantzoukas1979@gmail.com (S.M.)

Abstract: Infestations of stored-product pests cause significant losses of agricultural produce every year. Despite various environmental and health risks, chemical insecticides are now a ready-to-use solution for pest control. Against this background and in the context of Integrated Pest Management research, the present study focuses on the potential insecticidal effect of caffeic acid at five different concentrations (250, 500, 750, 1500 and 3000 ppm), and their combination with *Cydia pomonella* Granulovirus (CpGV), *Bacillus thuringiensis* subsp. tenebrionis and *Beauveria bassiana* strain GHA on three major insect stored-product beetle species, *Tribolium confusum* (Coleoptera: Tenebrionidae), *Cryptolestes ferrugineus* (Coleoptera: Laemophloeidae) and *Trogoderma granarium* Everts (Coleoptera: Dermestidae). Treatment efficacy was expressed as mortality in relation to exposure time and adult species number. Compared to the control, the results showed a clear dose-dependent pesticidal activity, expressed as significant adult mortality at a high-dose application, although some of the combinations of caffeic acid concentrations with the other substances acted positively (synergistically and additively) and some negatively. Based on our results, bioinsecticides can be combined with plant compounds such as caffeic acid and be integrated with other modern IPM tools in storage facilities.

Keywords: caffeic acid; biopesticides; *Cydia pomonella* Granulovirus; *Bacillus thuringiensis* subsp. tenebrionis; *Beauveria bassiana*; interactions; stored pests

Citation: Zarmakoupi, C.; Mpistiolis, K.; Pantazis, G.; Psatha, P.; Dimitriadi, D.; Kitsiou, F.; Eliopoulos, P.; Patakioutas, G.; Mantzoukas, S. Caffeic Acid and Biopesticides Interactions for the Control of Storage Beetles. *Appl. Biosci.* **2023**, *2*, 211–221. https://doi.org/10.3390/applbiosci2020015

Academic Editor: Robert Henry

Received: 10 January 2023
Revised: 27 March 2023
Accepted: 4 May 2023
Published: 8 May 2023

1. Introduction

Storage pests can cause significant economic losses by contaminating stored products, resulting in both quantitative and qualitative deterioration. The deterioration of stored commodities is caused not only by the consumption of the product, but also by the contamination of dead skin, excreta and dead insects, that can be dangerous for human health because they cause allergic reactions [1,2]. Moreover, the presence of insect populations in stored products can considerably increase relative humidity, which promotes secondary fungal infestations [3]. Most agricultural products can be affected by such infestations, resulting in annual losses of 9–20% [4].

Practices such as sanitation, aeration cooling, drying and controlled atmospheres are implemented, but are not sufficient to effectively control insect infestations in storage facilities [3]. Until now, fumigation with synthetic insecticides such as phosphine was primarily applied in storage facilities for disinfestation, but the increasing hazards to human health and the environment restricted their use [5,6]. Needless to say, the overreliance on these substances all these years has led to resistance development, [7] and the neglect of research into alternative control methods [6].

Due to the above facts, new investigations have recently emerged aimed at finding more ecological methods for the management of storage pests, by utilizing natural plant compounds or more specific products of plants' secondary metabolism such as essential oils. Apart from the fact that they do not pollute the environment, they are very effective against insects due to their volatility [8]. Substances derived from metabolic reactions of plants can be bioactive towards insects, as they are part of their natural defense mechanisms and include compounds such as terpenes, flavonoids, alkaloids, polyphenols, quinones, and others [9]. Plant extracts and essential oils can exert a wide range of actions against insects, such as toxicity, repellency, inhibition of respiration, oviposition, growth or feeding and a reduction in adult emergence and abnormalities in larvicidal transitions [10–12].

Phenolic acids such as salicylic, coumaric, caffeic and chlorogenic acids are ubiquitously present in plants and mostly participate in plant defense mechanisms [13]. Some of these substances have already been investigated to utilize the natural immunity of plants in the concept of biological control in agriculture. Caffeic acid (CA) is an early intermediate of phenylpropanoid metabolism, and a precursor for structural polyphenols and many biologically active secondary compounds that are important in the plant defense mechanisms [14,15]. This specific phenolic compound has been attributed to antifungal, antibacterial and insecticidal properties [15].

Another promising aspect of insect biological control is the use of entomopathogens. This approach has been thoroughly investigated lately as they offer a great alternative in the context of integrated pest management (IPM). Viruses, bacteria and fungi have been described as effective against various insect species [16–18]. These insect pathogens are not hazardous as they already exist in nature and so have a very low environmental impact and low mammalian toxicity [19,20]. There have been some studies that investigated the synergistic effect of insect pathogens with biopesticides, and the results have varied between a lesser, zero or enhanced efficacy against arthropods.

In this context, the present study aimed to investigate the efficacy of CA, in combination with commercially available biopesticides (fungal, viral and bacterial) on three major insect stored-product beetle species. All tested species are globally distributed stored-product pests and cause serious quantitative and qualitative losses in a vast range of commodities. Our results are discussed in the context of enhancing the use of insect pathogens as a key component of integrated pest management against stored-product pests.

2. Materials and Methods

2.1. Insect Rearing

Three important stored-product beetle species were selected for experimentation. The insect species tested were *T. confusum*, *C. ferrugineus* and *T. granarium*. Insects were reared in incubators (PHC Europe/Sanyo/Panasonic Biomedical MLR-352-PE) at 27.5 °C and 75% relative humidity (r.h.). *T. granarium* was kept on whole wheat, *C. ferrugineus* on rolled oats with 5% brewer's yeast, and *T. confusum* on whole wheat flour with 10% brewer's yeast. Adults of uniform age (<2 weeks old) and mixed sex were used for experimentation.

2.2. Caffeic Acid Solution and Biopesticides

The solution was obtained for Karvelas AVEE with lot number 15038821. The composition of the tested solution was natural caffeic acid at 1120 mg/kg, conductivity 97.9 mS/cm, pH 4.62 and density 1.215 g/cm^3.

Biopesticides tested during the present study were commercial formulations obtained from the market. Specifically, we used Madex® (*Cydia pomonella* granulovirus (CpGV) (Hellafarm, Athens, Greece), Novodor® FC (*Bacillus thuringiensis* subsp. *Tenebrionis* 3%) (BIOFA Germany, Bad Boll, Germany) and Botanigard® 10.7SC (*Beauveria bassiana* strain GHA 10.735%) (K&N Efthymiadis Single Member S.A., Thessaloniki, Greece).

2.3. Experimental Procedure

500 g of wheat (var. Mexa) were divided into separate lots and filled into 0.45 L cylinder jars. Since it is difficult for these species to reproduce on intact grains, the wheat used had 5% broken kernels. The wheat was stored for 28 days under ambient conditions to adjust the moisture content (m.c.) to 12%.

Experimentation included five concentrations of CA solution (Karvellas AVEE, Thiva, Greece) (250 ppm, 500 ppm, 750, ppm, 1500 ppm and 3000 ppm) and one (3000 ppm) for commercial biopesticides. The solvent used to prepare all solutions was distilled water. Twenty 10 g wheat samples were taken from the jars and placed in 9 cm Petri dishes. Following this, ten adult beetles of each species, of uniform age (<2 weeks old) and mixed sex, were transferred to each Petri dish. The inner "neck" of the Petri dish was covered with fluon to prevent insect escape (Northern Products, Woonsocket, RI, USA). A Potter spray tower (Burkard Manufacturing Co., Ltd., Rickmansworth, Hertfordshire, UK) was used to apply the solutions to the products at a rate of 1 kgf cm^2. For separate doses testing, the experimental adults were sprayed once with 2 mL of the CA or biopesticide. Conversely, for the combined treatments, spraying was performed twice, once with 2 mL of the CA solution and once with 2 mL of the biopesticide solution, each 2 s apart. The Petri dishes were then transferred to Toshiba incubators (PHC Europe/Sanyo/Panasonic Biomedical MLR-352-PE) and set at 27.5 °C and 75% relative humidity. The beetles were observed daily, and mortality was recorded 7, 14, 21, and 28 days after treatment.

The entire procedure was repeated twenty times by preparing new batches of treated and untreated grains at each replicate (separate treatments: 9 × 3 × 20 = 540 Petri dishes for each dose × insect species × replicate, combined treatments: 15 × 3 × 20 = 900 Petri dishes for each dose × insect species × replicate).

2.4. Mathematical Estimation and Statistical Analysis

The interaction between the CA and the biopesticides was estimated using the formula of Robertson and Preisler:

$$P_E = P_0 + (1 - P_0) \times (P_1) + (1 - P_0) \times (1 - P_1) \times (P_2),$$

where: P_E is the expected mortality induced by the combined treatment; P_0 is the mortality of the control; P_1 is the mortality caused by the CA; P_2 is the mortality caused by the biopesticide.

Distribution was determined by the chi-square formula: $x^2 = (L_0 - L_E)^2/L_E + (D_0 - D_E)^2/D_E$ where L_0 is the number of living adults, D_0 is the number of dead larvae, L_E is the expected number of live larvae, and D_E is the expected number of dead larvae. The formula was used to test the hypothesis independent–simultaneous relationship (1 df, $p = 0.05$). If $x^2 < 3.84$, the ratio is defined as additive (A); if $x^2 > 3.84$ and the observed mortality is higher than expected, the relationship is defined as synergistic (S). On the contrary, if $x^2 > 3.84$ and the observed mortality is less than expected, the relationship is defined as competitive (C).

The general linear model of SPSS (version 23.0, IBM Corp., Armonk, NY, USA) was then used to evaluate the data using a three-way ANOVA (IBM 2014). The Bonferroni test was used to compare means in cases where there were substantial F values.

3. Results

The results of the laboratory bioassays on adults of *T. granarium*, *C. ferrugineus*, and *T. confusum* showed that separate treatments with CA and all pathogens caused varying degrees of time-, treatment- and dose-dependent mortality. Adult mortality of *T. granarium* was 57–73%, of *C. ferrugineus* was 43–67%, and of *T. confusum* was 27–67% twenty-eight days after treatment with CA solution at the highest dose (3000 ppm). After twenty-eight days, the application of *B. thuringiensis* caused 67% mortality in *T. granarium* adults, 73% in *C. ferrugineus*, and 69% in *T. confusum*. After twenty-eight days of CpGV treatment, the observed mortality of adults of *T. granarium*, *C. ferrugineus*, and *T. confusum* was 70%, 43%, and 47%, respectively. The mortality of *T. confusum*, *C. ferrugineus*, and *T. granarium* after twenty-eight days of treatment with *B. bassiana* was 93%, 77%, and 93%, respectively. In all of the tested insects, the control mortality was less than 3%.

According to results of the combined bioassays, all combinations tested induced various levels of time- and dose-dependent mortality (Table 1). The results of the combined treatments showed a distinct interaction between treatments, as follows: for *T. granarium* adults, the interaction between the pathogens was additive in nine combinations the first seven days, synergistic in two and antagonistic in five. The following fourteen days, the interactions proved to be additive in seven combinations, synergistic in one and antagonistic in six. After twenty-one days, the interaction was additive in eight combinations and competitive in seven (Table 1). Finally, twenty-eight days later, the interaction was characterized as additive in seven combinations and competitive in eight (Table 1). Adult *T. granarium* mortality was between 37 and 100% (F: 19.764; df: 654.2360; p: <0.001) (overall 15 treatments).

Interactions between treatments on *T. confusum* for seven days were additive in ten combinations, synergistic in four combinations and competitive in one combination. For fourteen days, the interactions between treatments were all additive. At twenty-one days, the interaction between treatments was additive in fourteen combinations and synergistic in one combination (Table 2). As for the twenty-eighth day, the interaction between treatments was additive in fourteen combinations and synergistic in one combination (Table 2). Adult *T. confusum* mortality ranged from 27 to 100% (F: 20.764; df: 654.2360; p: <0.001) (overall 15 treatments).

Table 1. Percentage of observed and expected mortality of *T. granarium* adults at seven, fourteen, twenty-one and twenty-eight days of the experiment, treated with treatments in several combinations, and their interactions (n = 100).

Treatment		Mortality Observed% 1	Mortality Expected% 2	x^2	Interaction 3	Mortality Observed%	Mortality Expected%	x^2	Interaction	Mortality Observed%	Mortality Expected%	x^2	Interaction	Mortality Observed%	Mortality Expected%	x^2	Interaction
Entomopathogen (3000 ppm)	caffeic acid (ppm)	7 days				14 days				21 days				28 days			
Bacillus thuringiensis 4	250	37	55	7.0100	C	37	75	18.7841	C	43	84	46.0425	C	47	89	30.6352	C
	500	37	63	7.0169	C	37	75	18.5344	C	47	86	40.2042	C	47	89	30.0133	C
	750	37	61	7.0094	C	40	75	17.4215	C	47	84	34.6456	C	53	89	29.2205	C
	1500	40	61	7.0292	C	47	74	15.5904	C	47	85	29.3120	C	57	89	23.8527	C
	3000	43	63	7.0450	C	57	79	4.3251	C	60	85	11.9527	C	67	93	18.5516	C
Cydia pomonella Granulovirus (CpGV) 5	250	40	42	0.0435	A	57	67	0.7815	A	62	78	3.2709	A	69	90	7.6806	C
	500	57	53	0.3601	A	62	67	0.0967	A	70	80	1.9105	A	77	90	2.7879	A
	750	60	50	1.6065	A	65	67	0.1029	A	70	78	0.8179	A	77	90	2.7879	A
	1500	67	50	9.3082	S	77	72	0.5404	A	71	79	0.8291	A	81	90	3.2932	A
	3000	67	53	2.7871	A	79	65	3.1823	A	83	79	0.6247	A	89	94	0.5655	A
Beauveria bassiana strain GHA 6	250	47	53	0.5122	A	57	87	15.8274	C	69	97	66.4130	C	84	98	16.5679	C
	500	50	62	1.4289	A	83	87	0.0438	A	87	97	9.3253	C	90	98	4.1772	C
	750	52	59	0.3538	A	97	87	3.5782	A	93	97	1.0476	A	93	98	1.0648	A
	1500	57	59	0.0292	A	100	89	4.4242	S	100	97	1.0261	A	100	98	1.0154	A
	3000	60	62	0.0162	A	100	86	5.6963	S	100	97	1.0261	A	100	98	1.0070	A

1: Percentage of dead adults recorded during experiments. 2: Mortality calculated according to Robertson and Preisler. 3: A = Additive, C = Competitive, S = Synergistic. 4: Novodor® FC (BIOFA Germany). 5: Madex® (Hellafarm, Athens. Greece). 6: Botanigard® 10.7SC (K&N Efthymiadis Single Member S.A., Thessaloniki, Greece).

Table 2. Percentage of observed and expected mortality of *T. confusum* adults at seven, fourteen, twenty-one and twenty-eight days of the experiment, treated with treatments in several combinations, and their interactions (A = Additive, C = Competitive, S = Synergistic) (n = 100). Expected mortality calculated according to Robertson and Preisler [20].

Treatment		Mortality Observed% 1	Mortality Expected% 2	x^2	Interaction 3	Mortality Observed%	Mortality Expected%	x^2	Interaction	Mortality Observed%	Mortality Expected%	x^2	Interaction	Mortality Observed%	Mortality Expected%	x^2	Interaction
Entomopathogen (3000 ppm)	caffeic acid (ppm)	7 days				14 days				21 days				28 days			
Bacillus thuringiensis 4	250	20	19	0.0558	A	37	50	1.4805	A	43	61	3.7692	A	63	78	3.2719	A
	500	30	22	1.3274	A	37	51	2.2317	A	50	64	2.4033	A	80	83	0.0296	A
	750	35	25	1.6483	A	50	56	0.2562	A	63	68	0.2309	A	90	85	0.8895	A
	1500	47	19	15.6913	S	50	58	0.6202	A	77	70	1.1977	A	93	88	0.9887	A
	3000	47	25	7.7754	S	57	60	0.1210	A	83	70	3.1762	A	97	90	1.6296	A
Cydia pomonella Granulovirus (CpGV) 5	250	23	28	0.3229	A	40	48	0.6507	A	50	53	0.0373	A	63	62	0.0688	A
	500	33	30	0.1053	A	57	50	0.8984	A	60	56	0.3524	A	80	71	1.3673	A
	750	33	34	0.0037	A	57	55	0.1206	A	60	61	0.0050	A	87	74	2.7530	A
	1500	47	28	5.5592	S	60	56	0.3524	A	67	63	0.4454	A	90	79	2.7072	A
	3000	47	34	2.4437	S	63	59	0.3567	A	77	63	3.1074	A	97	83	5.0396	S

Table 2. *Cont.*

Treatment		Mortality Observed% [1]	Expected% [2]	x^2	Interaction [3]	Mortality Observed%	Expected%	x^2	Interaction	Mortality Observed%	Expected%	x^2	Interaction	Mortality Observed%	Expected%	x^2	Interaction
Beauveria bassiana strain GHA [6]	250	10	31	5.9015	C	43	51	0.6802	A	77	85	0.9549	A	87	95	2.0092	A
	500	23	33	1.1206	A	50	53	0.0492	A	87	86	0.2435	A	90	96	0.6173	A
	750	23	36	1.9048	A	67	57	1.2404	A	87	88	0.0046	A	90	97	0.6355	A
	1500	37	31	0.5979	A	67	59	1.0598	A	97	88	2.6542	A	93	97	1.2666	A
	3000	40	36	0.3271	A	70	62	1.1024	A	100	88	4.4576	S	100	98	1.0154	A

[1]: Percentage of dead adults recorded during experiments. [2]: Mortality calculated according to Robertson and Preisler. [3]: A = Additive, C = Competitive, S = Synergistic. [4]: Novodor® FC (BIOFA Germany). [5]: Madex® (Hellafarm, Athens, Greece). [6]: Botanigard® 10.7SC (K&N Efthymiadis Single Member S.A., Thessaloniki, Greece). The interaction between treatments for *C. ferrugineus* was additive in ten combinations and competitive in five combinations over the first seven days. After fourteen and twenty-one days, the interactions between the treatments were all additive. At last, for twenty-eight days, the interaction between the treatments was additive in fourteen combinations and synergistic in one combination (Table 3). Adult *C. ferrugineus* mortality was 10–100% (F: 15.164; df: 654.2360; *p*: <0.001) (overall 15 treatments).

Table 3. Percentage of observed and expected mortality of *C. ferrugineus* adults at seven, fourteen, twenty-one and twenty eight days of the experiment, treated with treatments in several combinations, and their interactions (A = Additive, C = Competitive, S = Synergistic) (n = 100). Expected mortality calculated according to Robertson and Preisler [20].

Treatment		Mortality Observed% [1]	Expected% [2]	x^2	Interaction [3]	Mortality Observed%	Expected%	x^2	Interaction	Mortality Observed%	Expected%	x^2	Interaction	Mortality Observed%	Expected%	x^2	Interaction
Entomopathogen (3000 ppm)	caffeic acid (ppm)	7 days				14 days				21 days				28 days			
Bacillus thuringiensis [4]	250	10	46	14.4489	C	57	60	0.1155	A	70	79	0.8927	A	90	86	0.8651	A
	500	20	48	9.0018	C	67	64	0.1496	A	73	79	0.3189	A	93	86	1.8692	A
	750	27	51	6.5817	C	70	64	0.5271	A	81	81	0.0151	A	93	86	1.8692	A
	1500	33	55	5.3190	C	70	68	0.1276	A	83	84	0.0029	A	97	88	2.6660	A
	3000	33	55	5.3190	C	77	72	0.6169	A	91	84	1.2601	A	100	92	3.2246	A
Cydia pomonella Granulovirus (CpGV) [5]	250	27	20	0.9188	A	53	60	0.6191	A	67	64	0.1496	A	73	70	0.4055	A
	500	27	23	0.3924	A	60	64	0.1692	A	73	64	1.1362	A	80	70	2.0486	A
	750	30	27	0.1455	A	60	64	0.1692	A	77	68	1.3612	A	87	70	4.9951	S
	1500	37	33	0.3555	A	63	68	0.2114	A	83	72	2.2449	A	87	75	2.6684	A
	3000	40	33	0.8740	A	67	72	0.1750	A	83	72	2.2449	A	90	83	1.7042	A
Beauveria bassiana strain GHA [6]	250	20	20	0.0000	A	53	67	2.5699	A	73	75	0.0142	A	83	88	0.3815	A
	500	23	23	0.0337	A	56	70	2.8425	A	73	75	0.0142	A	93	88	0.9946	A
	750	30	27	0.1455	A	67	70	0.1390	A	80	77	0.1733	A	93	88	0.9946	A
	1500	37	33	0.3555	A	73	73	0.0902	A	87	80	0.9188	A	97	90	2.5684	A
	3000	37	33	0.3555	A	77	77	0.1522	A	87	80	0.9100	A	97	93	1.5232	A

[1]: Percentage of dead adults recorded during experiments. [2]: Mortality calculated according to Robertson and Preisler. [3]: A = Additive, C = Competitive, S = Synergistic. [4]: Novodor® FC (BIOFA Germany). [5]: Madex® (Hellafarm, Athens. Greece). [6]: Botanigard® 10.7SC (K&N Efthymiadis Single Member S.A., Thessaloniki, Greece).

Overall, all the main effects of examined factors (insect species, exposure time, treatment) and their interactions proved to be significant as was demonstrated by a 3-way analysis of variance (Table 4).

Table 4. An analysis of variance (3-way ANOVA) for the main effects and interactions for the mortality of *T. granarium*, *T. confusum* and *C. ferrugineus* adults exposed to separate and combined treatments with CA and biopesticides.

	Separate Treatments			Combined Treatments		
Source	df	F	Sig.	df	F	Sig.
Exposure time	3	11.838	<0.001	3	8.142	<0.001
Insect species	2	10.099	<0.001	2	6.499	<0.001
Treatment	3	16.476	<0.001	4	3.702	<0.001
Exposure time * Insect species	6	11.109	<0.001	6	7.288	<0.001
Exposure time * Treatment	9	11.540	<0.001	12	11.534	<0.001
Insect Species * Treatment	6	13.829	<0.001	8	5.420	<0.001
Exposure time * Insect species * Treatment	16	14.950	<0.001	24	9.946	<0.001
Error	210			380		
Total	280			400		
Corrected total	279			399		

4. Discussion

As chemical insecticides are being more and more neglected, many studies now focus on alternatives, investigating compounds derived from nature. Plant chemicals can act as insecticides by preventing insects from feeding or by demonstrating repellent and growth inhibition effects [21,22]. The insecticidal potential of phenolic plant compounds such as CA has been well documented [23–28]. In our bioassays, adult beetles treated only with CA showed noteworthy mortality (up to 70%). The lethal effect of CA on insects has been also verified for the tobacco cutworm, *Spodoptera litura* (Fabricius) [29] and the cotton bollworm, *Helicoverpa armigera* (Hübner) (Lepidoptera: Noctuidae) [30]. Apart from mortality effects, various studies have demonstrated that CA and other plant phenolic compounds may have negative effects on insect feeding, larval growth rate and reproduction [31–35]. Pacifico et al. [35] investigated the effect of CA on the larvae of *Phthorimaea operculella* and recorded sublethal effects and anti-nutrient action as it inhibited larval growth.

A possible explanation for these results may lie in the interaction of the phenolic compounds with digestive proteins of the insects leading to a decrease in nutritional quality. The way phenolic compounds affect the interaction of plants with bacteria and fungi has already been investigated even though little is known about the toxicity of phenolics against insects [36].

As expected, separate treatments with biopesticides caused high mortality in all tested species. There are several main factors that can influence the efficacy of biopesticides, such as the type of biopathogen, the dose applied, temperature, relative humidity and the type of product [20,37–43]. Moreover, the insecticidal efficacy of biopesticides can be highly influenced by a host's physiology, morphology and behavior, the population density, age, nutrition, and genetic information [39].

Our original hypothesis was that the interaction between CA and biopesticides either leads to additional efficacy or plays only a supporting role. Based on our results, the interaction was additive in *T. confusum* in most combinations. On the other hand, it was negative in four treatments in some combinations for *T. granarium* and *C. ferrugineus* adults, especially in the first 7 days of the experiment when the bacterial insecticide was applied. A negative interaction refers to the competitive relationship between CA and the pathogen. The nature of this competition is not precisely known. Entomopathogenic microorganisms have also shown increased efficacy when applied in combination treatments not only with other entomopathogens but also with synthetic insecticides [44]. Regarding their coexistence with plant extracts, entomopathogenic microorganisms have shown both an inhibitory effect [45] and a positive interaction as Neem seed cake improved the pathogenicity of the fungus *Metarhizium anisopliae* against the Black Vine Weevil [46]. The entomopathogenic fungus *M. anisopliae* has been successfully combined with plant extracts for the control of ticks [47], whereas other plant extracts showed compatible capacity with entomopathogenic bacteria against aphids [48]. To the best of our knowledge, there are no data available concerning the interaction of CA or other plant phenolic metabolites with entomopathogens.

In general, combinations of feeding stimulants and deterrents affect the feeding response of phytophagous insects [49,50]. It has been suggested that the Colorado potato beetle selects its hosts among solanaceous plants based on the presence of deterrents such as alkaloid glycosides rather than on the presence of feeding stimulants [51,52]. Various types of sesquiterpene lactones are present in Asteraceae and deter numerous phytophagous insects from feeding on the plants [53]. Caffeic acid derivatives play an important role in plant defense [54]. Chlorogenic acid has been reported to inhibit larval development of some Lepidoptera, such as *H. armigera*, the corn earworm *Heliothis zea* (Boddie), and the fall armyworm *Spodoptera frugiperda* (J.E. Smith) (Lepidoptera: Noctuidae) [55–58] and deters feeding in leaf beetles *Lochmaea caprea* (L.) [59], and *Agelastica alni* (L.) (Coleoptera: Chrysomelidae) [60,61].

To conclude, the interactions between tested insecticidal agents could be positive or negative, acting synergistically (increasing host mortality compared to single pathogen infections) [20,62,63] or antagonistically (reducing the observed host mortality compared to single pathogen infections) [64]. Needless to say, pest mortality can be affected by genotype, dose and sequence of infection [65,66].

5. Conclusions

Based on our results, the combined application of plant extracts and entomopathogenic microorganisms may become an effective strategy for eco-friendly pest management in storage facilities. However, special attention should be paid to the selection of the combined agents as the additive or synergistic effect is not always valid. Our study has shown the significant insecticidal action of CA alone or in combination with biopesticides. Further research is needed to clarify the effects of various factors, such as pest species, storage environment, application dose, time interval, stored product type, etc., and to enhance the use of plant compounds in stored-product IPM.

Author Contributions: Conceptualization, S.M. and D.D.; methodology, S.M.; software, S.M.; validation, S.M., G.P. (Georgios Parakioutas) and P.E.; formal analysis, S.M.; investigation, C.Z., K.M., G.P. (Georgios Pantazis), P.P. and F.K.; resources, S.M.; data curation, S.M.; writing—original draft preparation, S.M., G.P. (Georgios Parakioutas) and P.E.; writing—review and editing, S.M., G.P. (Georgios Parakioutas), P.E. and F.K.; visualization, S.M.; supervision, S.M.; project administration, S.M. All authors have read and agreed to the published version of the manuscript.

Funding: This research received no external funding.

Institutional Review Board Statement: Not applicable.

Informed Consent Statement: Not applicable.

Data Availability Statement: Not applicable.

Conflicts of Interest: The authors declare no conflict of interest.

References

1. Kumar, S.; Mohapatra, D.; Kotwaliwale, N.; Singh, K. Vacuum hermetic fumigation: A review. *J. Stored Prod. Res.* **2017**, *71*, 47–56. [CrossRef]
2. Neethirajan, S.; Karunakaran, C.; Jayas, D.; White, N. Detection techniques for stored-product insects in grain. *Food Control* **2007**, *18*, 157–162. [CrossRef]
3. Nayak, M.K.; Daglish, G.J. Importance of stored product insects. In *Recent Advances in Stored Product Protection*; Athanassiou, C., Arthur, F., Eds.; Springer: Berlin/Heidelberg, Germany, 2018; pp. 1–17.
4. Phillips, T.W.; Throne, J.E. Biorational approaches to managing stored-product insects. *Annu. Rev. Entomol.* **2010**, *55*, 375–397. [CrossRef]
5. Navarro, S. New global challenges to the use of gaseous treatments in stored products. In Proceedings of the 9th International Working Conference on Stored Product Protection, Fundo, Brazil, 15–18 October 2006; Brazilian Post-Harvest Association–ABRAPOS: Passo Fundo, RS, Brazil, 2006; pp. 495–509.
6. Morrison, W.R., III; Scully, E.D.; Campbell, J.F. Towards developing areawide semiochemical-mediated, behaviorally-based integrated pest management programs for stored product insects. *Pest Manag. Sci.* **2021**, *77*, 2667–2682. [CrossRef]
7. Nayak, M.K.; Falk, M.G.; Emery, R.N.; Collins, P.J.; Holloway, J.C. An analysis of trends, frequencies and factors influencing the development of resistance to phosphine in the red flour beetle *Tribolium castaneum* (Herbst) in Australia. *J. Stored Prod. Res.* **2017**, *72*, 35–48. [CrossRef]
8. Chaudhari, A.K.; Singh, V.K.; Kedia, A.; Das, S.; Dubey, N.K. Essential oils and their bioactive compounds as eco-friendly novel green pesticides for management of storage insect pests: Prospects and retrospects. *Environ. Sci. Pollut. Res.* **2021**, *28*, 18918–18940. [CrossRef] [PubMed]
9. Souto, A.L.; Sylvestre, M.; Tölke, E.D.; Tavares, J.F.; Barbosa-Filho, J.M.; Cebrián-Torrejón, G. Plant-derived pesticides as an alternative to pest management and sustainable agricultural production: Prospects, applications and challenges. *Molecules* **2021**, *26*, 4835. [CrossRef]
10. Sarwar, M.; Salman, M. Toxicity of oils formulation as a new useful tool in crop protection for insect pests control. *Int. J. Chem. Biomol. Sci.* **2015**, *1*, 297–302.
11. El-Sheikh, T.M.; Al-Fifi, Z.I.; Alabboud, M.A. Larvicidal and repellent effect of some *Tribulus terrestris* L., (Zygophyllaceae) extracts against the dengue fever mosquito, *Aedes aegypti* (Diptera: Culicidae). *J. Saudi Chem. Soc.* **2016**, *20*, 13–19. [CrossRef]
12. Ali, M.A.; Doaa, S.M.; El-Sayed, H.S.; Asmaa, M.E. Antifeedant activity and some biochemical effects of garlic and lemon essential oils on *Spodoptera littoralis* (Boisduval) (Lepidoptera: Noctuidae). *J. Entomol. Zool.* **2017**, *5*, 1476–1482.
13. Laura, A.; Moreno-Escamilla, J.O.; Rodrigo-García, J.; Alvarez-Parrilla, E. Phenolic compounds. In *Postharvest Physiology and Biochemistry of Fruits and Vegetables*; Yahia, E., Carrillo-Lopez, A., Eds.; Woodhead Publishing: Cambridge, UK, 2019; pp. 253–271.
14. Summers, C.B.; Felton, G.W. Prooxidant effects of phenolic acids on the generalist herbivore *Helicoverpa zea* (Lepidoptera: Noctuidae): Potential mode of action for phenolic compounds in plant anti-herbivore chemistry. *Insect Biochem. Mol. Biol.* **1994**, *24*, 943–953. [CrossRef]
15. Harrison, H.F.; Peterson, J.K.; Snook, M.E.; Bohac, J.R.; Jackson, D.M. Quantity and potential biological activity of caffeic acid in sweet potato [*Ipomoea batatas* (L.) Lam.] storage root periderm. *J. Agric. Food Chem.* **2003**, *51*, 2943–2948. [CrossRef] [PubMed]
16. Mantzoukas, S.; Kitsiou, F.; Natsiopoulos, D.; Eliopoulos, P.A. Entomopathogenic Fungi: Interactions and Applications. *Encyclopedia* **2002**, *2* 646–656. [CrossRef]
17. Valicente, F.H. Entomopathogenic viruses. In *Natural Enemies of Insect Pests in Neotropical Agroecosystems*; Souza, B., Vázquez, L.L., Marucci, R.C., Eds.; Springer: Berlin, Germany, 2019; pp. 137–150.
18. Glare, T.R.; Jurat-Fuentes, J.L.; O'Callaghan, M. Basic and applied research: Entomopathogenic bacteria. In *Microbial Control of Insect and Mite Pests*; Lacey, L., Ed.; Academic Press: London, UK, 2017; pp. 47–67.
19. Del Rincón-Castro, M.C.; Ibarra, J.E. Entomopathogenic viruses. In *Biological Cotrol of Insect Pests*; Rosas-Garcia, N.M., Ed.; Studium Press: New Delhi, India, 2011; pp. 29–64.
20. Mantzoukas, S.; Milonas, P.; Kontodimas, D.; Angelopoulos, K. Interaction between the entomopathogenic bacterium *Bacillus thuringiensis* subsp. kurstaki and two entomopathogenic fungi in bio-control of *Sesamia nonagrioides* (Lefebvre) (Lepidoptera: Noctuidae). *Ann. Microbiol.* **2013**, *63*, 1083–1091. [CrossRef]
21. Usha Rani, P.; Devanand, P. Biological potency of certain plant extracts in management of two lepidopteran pests of *Ricinus communis* L. *J. Biopestic.* **2008**, *1*, 170–176.
22. Usha Rani, P.; Rajasekharreddy, P. Toxic and antifeedant activities of *Sterculia foetida* (L.) seed crude extract against *Spodoptera litura* (F.) and *Achaea janata* (L.). *J. Biopestic.* **2009**, *2*, 161–164.
23. Elu, A.; Ezhang, Q.; Ezhang, J.; Eyang, B.; Ewu, K.; Exie, W.; Eluan, Y.-X.; Eling, E. Insect prophenoloxidase: The view beyond immunity. *Front. Physiol.* **2014**, *5*, 252.
24. Salminen, J.-P.; Karonen, M. Chemical ecology of tannins and other phenolics: We need a change in approach. *Funct. Ecol.* **2011**, *25*, 325–338. [CrossRef]
25. Kubo, I. New concept to search for alternate insect control agents from plants. In *Naturally Occurring Bioactive Compounds 3*; Rai, M., Carpinella, M., Eds.; Elsevier: Amsterdam, The Netherland, 2006; pp. 61–80.
26. Łukasik, I.; Goławska, S.; Wojcicka, A.; Goławski, A. Effect of host plants on antioxidant system of pea aphid *Acyrthosiphon pisum*. *Bull Insect.* **2011**, *64*, 153–158.

27. Agrawal, A.A.; Kearney, E.E.; Hastings, A.P.; Ramsey, T.E. Attenuation of the Jasmonate Burst, Plant Defensive Traits, and Resistance to Specialist Monarch Caterpillars on Shaded Common Milkweed (*Asclepias syriaca*). *J. Chem. Ecol.* **2012**, *38*, 893–901. [CrossRef]
28. War, A.R.; Paulraj, M.G.; War, M.Y.; Ignacimuthu, S. Differential defensive response of groundnut germplasms to *Helicoverpa armigera* (Hubner) (Lepidoptera: Noctuidae). *J. Plant Interact.* **2012**, *7*, 45–55. [CrossRef]
29. Punia, A.; Singh, V.; Thakur, A.; Chauhan, N.S. Impact of caffeic acid on growth, development and biochemical physiology of insect pest, *Spodoptera litura* (Fabricius). *Heliyon* **2023**, *9*, e14593. [CrossRef] [PubMed]
30. Joshi, R.S.; Wagh, T.P.; Sharma, N.; Mulani, F.A.; Sonavane, U.; Thulasiram, H.V.; Joshi, R.; Gupta, V.S.; Giri, A.P. Way toward "dietary pesticides": Molecular investigation of insecticidal action of caffeic acid against *Helicoverpa armigera*. *J. Agric. Food Chem.* **2014**, *62*, 10847–10854. [CrossRef] [PubMed]
31. Punia, A.; Chauhan, N.; Singh, R.; Kaur, S.; Sohal, S. Growth disruptive effects of ferulic acid against *Spodoptera litura* (Fabricius) and its parasitoid *Bracon hebetor* (Say). *Allelopath. J.* **2022**, *55*, 79–92. [CrossRef]
32. Nakhaie, B.M.; Mikani, A.; Moharramipour, S. Effect of caffeic acid on feeding, α-amylase and protease activities and allatostatin—A content of Egyptian cotton leafworm, *Spodoptera littoralis* (Lepidoptera: Noctuidae). *J. Pest. Sci.* **2018**, *43*, 73–78. [CrossRef]
33. Mattar, V.T.; Borioni, J.L.; Hollmann, A.; Rodriguez, S.A. Insecticidal activity of the essential oil of *Schinus areira* against *Rhipibruchus picturatus* (F.) (Coleoptera: Bruchinae), and its inhibitory effects on acetylcholinesterase. *Pest. Biochem. Physiol.* **2022**, *185*, 105134. [CrossRef]
34. Divekar, P.A.; Narayana, S.; Divekar, B.A.; Kumar, R.; Gadratagi, B.G.; Ray, A.; Singh, A.K.; Rani, V.; Singh, V.; Singh, A.K.; et al. Plant secondary metabolites as defense tools against herbivores for sustainable crop protection. *Int. J. Mol. Sci.* **2022**, *23*, 2690. [CrossRef]
35. Pacifico, D.; Musmeci, S.; del Pulgar, J.S.; Onofri, C.; Parisi, B.; Sasso, R.; Mandolino, G.; Lombardi-Boccia, G. Caffeic acid and α-chaconine influence the resistance of potato tuber to *Phthorimaea operculella* (Lepidoptera: Gelechiidae). *Am. J. Potato Res.* **2019**, *96*, 403–413. [CrossRef]
36. Szatmári, Á.Á.; Zvara, Á.M.; Móricz, E.; Besenyei, E.; Szabó; Ott, P.G.; Puskas, L.G.; Bozso, Z. Pattern triggered immunity (PTI) in tobacco: Isolation of activated genes suggests role of the Phenylpropanoid pathway in inhibition of bacterial pathogens. *PLoS ONE* **2014**, *9*, e102869. [CrossRef]
37. Moino, A., Jr.; Alves, S.B.; Pereira, R.M. Efficacy of *Beauveria bassiana* (Balsamo) Vuillemin isolates for control of stored-grain pests. *J. Appl. Entomol.* **1998**, *122*, 301–305. [CrossRef]
38. Mantzoukas, S.; Eliopoulos, P.A. Endophytic entomopathogenic fungi: A valuable biological control tool against plant pests. *Appl. Sci.* **2020**, *10*, 360. [CrossRef]
39. Fargues, J.; Goettel, M.S.; Smits, N.; Ouedraogo, A.; Vidal, C.; Lacey, L.A.; Rougier, M. Variability in susceptibility to simulated sunlight of conidia among isolates of entomopathogenic Hyphomycetes. *Mycopathologia* **1996**, *135*, 171–181. [CrossRef]
40. Hallsworth, J.E.; Magan, N. Water and temperature relations of growth of the entomogenous fungi *Beauveria bassiana, Metarhizium anisopliae*, and *Paecilomyces farinosus*. *J. Invertebr. Pathol.* **1999**, *74*, 261–266. [CrossRef] [PubMed]
41. Luz, C.; Fargues, J. Temperature and moisture requirements for conidial germination of an isolate of *Beauveria bassiana*, pathogenic to *Rhodnius prolixus*. *Mycopathologia* **1997**, *138*, 117–125. [CrossRef]
42. Luz, C.; Fargues, J. Factors Affecting Conidial Production of *Beauveria bassiana* from Fungus-Killed Cadavers of *Rhodnius prolixus*. *J. Invertebr. Pathol.* **1998**, *72*, 97–103. [CrossRef]
43. Padin, S.; Bello, G.D.; Fabrizio, M. Grain Loss Caused by *Tribolium castaneum, Sitophilus oryzae* and *Acanthoscelides obtectus* in Stored Durum Wheat and Beans Treated with *Beauveria bassiana*. *J. Stored Prod. Res.* **2002**, *38*, 69–74. [CrossRef]
44. Furlong, M.J.; Groden, E. Evaluation of synergistic interactions between the Colorado potato beetle (Coleoptera: Chrysomelidae) pathogen *Beauveria bassiana* and the insecticides, imidacloprid, and cyromazine. *J. Econ. Entomol.* **2001**, *94*, 344–356. [CrossRef] [PubMed]
45. Mann, A.J.; Davis, T.S. Plant secondary metabolites and low temperature are the major limiting factors for *Beauveria bassiana* (Bals.-Criv.) Vuill. (Ascomycota: Hypocreales) growth and virulence in a bark beetle system. *Biol. Control* **2020**, *141*, 104130. [CrossRef]
46. Shah, F.A.; Gaffney, M.; Ansari, M.A.; Prasad, M.; Butt, T.M. Neem seed cake enhances the efficacy of the insect pathogenic fungus *Metarhizium anisopliae* for the control of black vine weevil, *Otiorhynuchs sulcatus* (Coleoptera: Curculionidae). *Biol. Control* **2008**, *44*, 111–115. [CrossRef]
47. Nana, P.; Maniania, N.K.; Maranga, R.O.; Boga, H.I.; Kutima, H.L.; Eloff, J.N. Compatibility between *Calpurnia aurea* leaf extract, attraction aggregation, and attachment pheromone and entomopathogenic fungus *Metarhizium anisopliae* on viability, growth, and virulence of the pathogen. *J. Pest Sci.* **2012**, *85*, 109–115. [CrossRef]
48. Noureldeen, A.; Kumar, U.; Asad, M.; Darwish, H.; Alharthi, S.; Fawzy, M.A.; Al-Barty, A.M.; Alotaibi, S.S.; Fallatah, A.; Alghamdi, A.; et al. Aphicidal activity of five plant extracts applied singly or in combination with entomopathogenic bacteria, *Xenorhabdus budapestensis* against rose aphid, *Macrosiphum rosae* (Hemiptera: Aphididae). *J. King Saud Univ. Sci.* **2022**, *34*, 102306. [CrossRef]
49. Dethier, V.G. Mechanism of host-plant recognition. *Entomol. Exp. Appl.* **1982**, *31*, 49–56. [CrossRef]

50. Schoonhoven, L.M.; van Loon, J.J.A. An inventory of taste in caterpillars: Each species its own key. *Acta Zool. Acad. Sci. Hung.* **2002**, *48*, 215–263.
51. Hsiao, T.H.; Fraenkel, G. Isolation of phagostimulative substances from the host plant of the colorado potato beetle. *Ann. Entomol. Soc. Am.* **1968**, *61*, 476–484. [CrossRef]
52. Hsiao, T.H.; Fraenkel, G. The role of secondary plant substances in the food specificity of the Colorado potato beetle. *Ann. Entomol. Soc. Am.* **1968**, *61*, 485–493. [CrossRef]
53. Bernays, E.A.; Chapman, R.F. Chemicals in plants. In *Host-Plant Selection by Phytophagous Insects*; Bernays, E.A., Chapman, R.F., Eds.; Chapman & Hall: New York, NY, USA, 1994; pp. 14–60.
54. Dixon, R.A.; Achnine, L.; Kota, P.; Liu, C.J.; Reddy, M.S.S.; Wang, L. The phenylpropanoid pathway and plant defence–A genomics perspective. *Mol. Plant Pathol.* **2002**, *3*, 371–390. [CrossRef]
55. Elliger, C.A.; Wong, Y.; Chan, B.G.; Waiss, A.C., Jr. Growth inhibitors in tomato (Lycopersicon) to tomato fruitworm (*Heliothis zea*). *J. Chem. Ecol.* **1981**, *7*, 753–758. [CrossRef]
56. Isman, M.B.; Duffey, S.S. Toxicity of tomato phenolic compounds to the fruitworm, *Heliothis zea*. *Entomol. Exp. Appl.* **1982**, *31*, 370–376. [CrossRef]
57. Kimmins, F.M.; Padgham, D.E.; Stevenson, P.C. Growth inhibition of the cotton bollworm (*Helicoverpa armigera*) larvae by caffeoylquinic acids from the wild groundnut. *Arachis Paraguariensis. Insect Sci. Appl.* **1995**, *16*, 363–368. [CrossRef]
58. Wiseman, B.R.; Gueldner, R.C.; Lynch, R.E.; Severson, R.F. Biochemical activity of centipedegrass against fall armyworm larvae. *J. Chem. Ecol.* **1990**, *16*, 2677–2690. [CrossRef] [PubMed]
59. Matsuda, K.; Sembo, S. Chlorogenic acid as a feeding deterrent for the Salicaceae-feeding leaf beetle, *Lochmaeae capreae* cribrata (Coleoptera: Chrysomelidae) and other species of leaf beetles. *Appl. Entomol. Zool.* **1986**, *21*, 411–416. [CrossRef]
60. Ikonen, A.; Tahvanainen, J.; Roininen, H. Chlorogenic acid as an antiherbivore defence of willows against leaf beetles. *Entomol. Exp. Appl.* **2001**, *99*, 47–54. [CrossRef]
61. Ikonen, A.; Tahvanainen, J.; Roininen, H. Phenolic secondary compounds as determinants of the host plant preferences of the leaf beetle, *Agelastica Alni*. *Chemoecology* **2002**, *12*, 125–131. [CrossRef]
62. Malvar, R.A.; Buton, A.; Ordas, B.; Santiago, R. Causes of natural resistance to stem borers in maize. In *Crop Protection Research Advances*; Burton, E.N., Williams, P.V., Eds.; Nova: New York, NY, USA, 2008; pp. 57–100.
63. Cedergreen, N. Quantifying synergy: A systematic review of mixture toxicity studies within environmental toxicology. *PLoS ONE* **2014**, *9*, e96580. [CrossRef] [PubMed]
64. Roell, K.R.; Reif, D.M.; Motsinger-Reif, A.A. An introduction to terminology and methodology of chemical synergy—Perspectives from across Disciplines. *Front. Pharmacol.* **2017**, *8*, 158. [CrossRef] [PubMed]
65. Bauer, L.S.; Miller, D.L.; Maddox, J.V.; McManus, M.L. Interactions between a *Nosema sp.* (Microspora: Nosematidae) and nuclear polyhedrosis virus infecting the gypsy moth, *Lymantria dispar* (Lepidoptera: Lymantriidae). *J. Invertebr. Pathol.* **1998**, *74*, 147–153. [CrossRef]
66. Thomas, M.B.; Watson, E.L.; Valverde-Garcia, P. Mixed infections and insect pathogen interactions. *Ecol. Lett.* **2003**, *6*, 183–188. [CrossRef]

applied biosciences

MDPI

Article

Seed Morpho-Anatomy and Germination Enhancement of the Australian Native Species *Lomandra longifolia* Labill. and *L. hystrix* (R.Br.) L.R. Fraser & Vickery

Fernanda Caro Beveridge *, Alwyn Williams, Robyn Cave, Sundaravelpandian Kalaipandian and Steve W. Adkins

School of Agriculture and Food Sciences, The University of Queensland, Gatton, QLD 4343, Australia
* Correspondence: fernanda.carobeveridge@uq.net.au

Abstract: *Lomandra* species are an important understory component of many Australian native ecosystems, contributing to the floristic richness and stabilizing soils. However, a limited understanding of their germination biology currently hinders their efficient use in seed-based restoration and ornamental plant production. The present study investigated *Lomandra longifolia* and *L. hystrix* diaspore morpho-anatomy and evaluated different mechanical and/or chemical treatments (nicking, leaching, smoke water and gibberellic acid [GA_3]) and under light or dark conditions to enhance germination. Embryos of both species were small and linear with a low embryo to seed ratio (<0.45). Germination rates of both species were significantly hastened by leaching seeds in running water for 36 h as compared to a non-leached seed. The results suggest that pre-treating both *Lomandra* species by leaching could maximize the effectiveness of seed used by resulting in faster, more uniform and, therefore, reliable germination of these species. Finally, seeds of *L. longifolia* had low final germination (<40%), with a high presence of viable but dormant seeds. The ecological cues that promote germination in nature for both species should be further examined.

Keywords: direct seeding; Australian native seed; seed-based restoration; seed dormancy; seed ecology; seed pre-treatments

Citation: Beveridge, F.C.; Williams, A.; Cave, R.; Kalaipandian, S.; Adkins, S.W. Seed Morpho-Anatomy and Germination Enhancement of the Australian Native Species *Lomandra longifolia* Labill. and *L. hystrix* (R.Br.) L.R. Fraser & Vickery. *Appl. Biosci.* **2023**, *2*, 222–235. https://doi.org/10.3390/applbiosci2020016

Academic Editor: Adriana Basile

Received: 16 March 2023
Revised: 28 April 2023
Accepted: 11 May 2023
Published: 16 May 2023

1. Introduction

Lomandra is a genus within the family Asparagaceae [1] that generally consists of small perennial herbs with a rhizomatous growth habit that often form tussocks [2]. *Lomandra longifolia* Labill. (spiny-head mat-rush) is one of the most widely distributed species in Australia [3] (Figure 1) and is particularly common in the south-east region of Australia [4]. It is highly adaptable and can grow in a wide range of habitats, such as on hillsides of dry forests, alongside creeks and coastal headlands. *Lomandra hystrix* (R.Br.) L.R. Fraser and Vickery (creek mat-rush) has a more confined natural distribution than *L. longifolia* and is found in the coastal regions of Queensland and New South Wales. Both species are commonly used as ornamental plants as well as for seed-based restoration projects [5]. These species are important understory components of many Australian ecosystems, providing shelter, breeding sites and food resources for native wildlife [4]. They have multiple benefits for seed-based restoration, which include their contribution to floristic richness and their ability to stabilize soils to prevent erosion due to their fibrous root system [6]. Large quantities of *L. longioflia* seeds are used annually to restore degraded Australian bushlands and disturbed plant communities due to human activity such as mining and construction [6]. Both species can be propagated using freshly produced diaspores (seed encased within the pericarp; hereafter, referred to as seeds) when sown in autumn conditions in Australia.

Even though *Lomandra* spp. are an important ecological, ornamental and restoration component in Australia, their seed biology has been poorly studied, hindering efficient seed use. It is well known that many Australian native species have diverse dormancy mechanisms [7], which manipulates germination through time and space. The most common form of dormancy in dryland Australian natives is physiological dormancy (PD) [7,8]. Physiologically dormant seeds are permeable to water and have fully developed embryos, but the embryo has low growth potential and cannot overcome the mechanical constraints of its covering layers [9]. For this reason, they usually take >28 days to germinate [10]. *Lomandra* seeds often have very slow germination, taking up to 60 days to germinate [11–13]. Grant et al. [13] reported that although germination of *L. longifolia* was very low after 28 days of imbibition, there was a significant increase (63%) in germination after this period. Plummer et al. [2] suggested that this delay in germination is due to dormancy mechanisms, possibly within the pericarp. Further, they found that *Lomandra sonderi* (F.Muell.) Ewart. and *Lomandra drummondii* (Benth.) Ewart. reached full water imbibition within 24 h and concluded that *Lomandra* spp. do not possess water-impermeable seed coats and, therefore, do not present physical dormancy (PY).

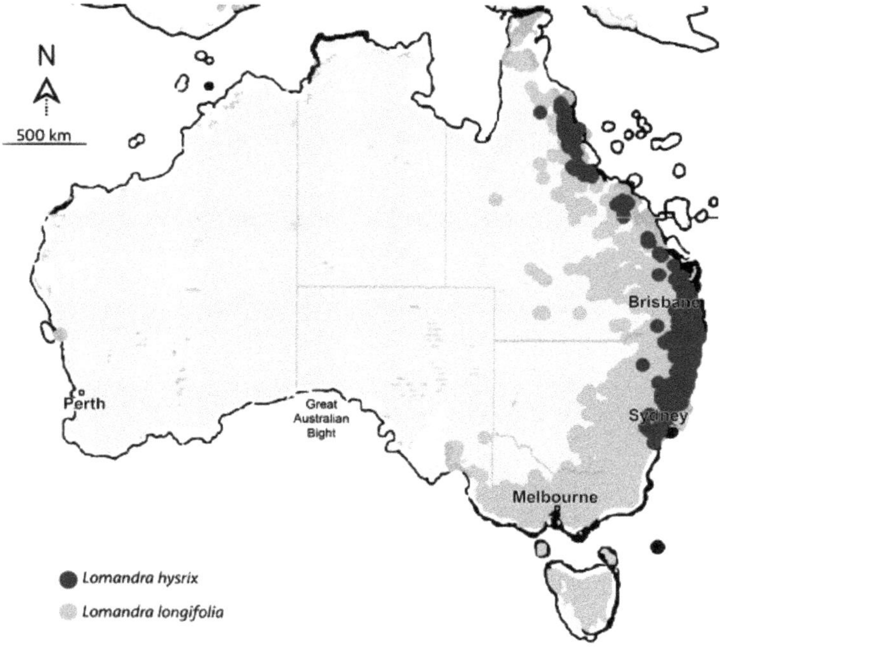

Figure 1. Natural distribution of *Lomandra longifolia* Labill. (light grey) and *Lomandra hystrix* (R.Br.) L.R. Fraser and Vickery (dark grey) in Australia (Source: Atlas of Living Australia).

Identifying the mechanisms that control dormancy and germination, together with finding new ways to hasten seed germination and seedling emergence, could be of great importance to achieve cost-effective usage of *Lomandra* seeds. For example, studies have shown that a combination of smoke water and stratification can increase seed germination to 50% in *Lomandra preisii* (Endl.) Ewart. [7], which could be associated with an overcoming of PD. Moreover, leaching of inhibitors from *L. longifolia* seeds with five cycles of soaking and rinsing combined with warm stratification significantly improved its germination [14]. Similarly, leaching seeds with running tap water or pericarp removal could overcome dormancy in *L. sonderi* and could achieve germination of *ca.* 20% of the seeds. Similar treatments increased germination from 40 to 80% in *L. drummondii* [2]. The authors related the increases in germination with the removal of germination-inhibiting chemicals found in the tissues surrounding the embryo. Gibberellic acid (GA_3) and smoke water have also been used successfully in promoting germination in *L. sonderi* [2]. Gibberellic acid is known to stimulate endosperm weakening and stimulate embryo germination [15]; smoke water has been shown to promote germination in a wide range of Australian native species [7,16,17]. Furthermore, scarification and seed nicking (a small cut through the tissues surrounding the embryo) of the pericarp and seed coat can be effective in overcoming PD or morpho-physiological dormancy (MPD) for several species. These treatments can relieve mechanical restrictions of the fruit tissues and/or seed coat, allowing embryo growth [18,19]. This can be particularly helpful in seeds that have a low embryo growth potential.

Microscopic investigation of seed internal morpho-anatomy is useful to characterize the seed and help elucidate germination biology aspects. Such analysis can determine embryo size, shape and location within the seed to determine the embryo to seed ratio (E:S) and size of the endosperm and/or cotyledons [10]. Documentation of embryo characteristics can also be used in determining the presence of morphological dormancy (MD) or MPD associated with embryo development [20]. Underdeveloped embryos have differentiated organs and tend to have low E:S ratios [21]. Embryos in seeds with MD need to undergo a growth period prior to radicle protrusion [10] (for example, in [22]). There are few studies that focus on morpho-anatomy characterization of *L. longifolia* and *L. hystrix* seed (such as [5,6]). This information could provide helpful insights into explaining the dormancy and slow germination rates observed for both species.

The current study describes the seed morpho-anatomy of both *Lomandra* species and determines methods for elevating seed germination to enable a more cost-efficient use of these seeds in seed-based restoration projects and ornamental plant production. Thus, the objectives were to (1) identify relationships between germination, seed fill and seed morpho-anatomical structures (embryo and seed size); and (2) develop methods to speed up the rate and increase the final seed germination by investigating chemical and mechanical seed treatments to overcome possible PD at optimum alternating temperatures in light or dark conditions. This information will provide an understanding of the dormancy mechanisms that are preventing germination and will help in developing improved seed germination protocols for *L. longifolia* and *L. hystrix*. Improved germination would also lead to increased use of these highly beneficial species in land regeneration projects and reduce the costs associated with seed wastage.

2. Materials and Methods

2.1. Seed Lots

Lomandra longifolia and *L. hystrix* seeds were obtained from a commercial seed supplier (Native Seeds and Land Repair, Maleny, QLD, Australia). *Lomandra longifolia* seeds were collected from the suburb of Redlands, part of the Brisbane metropolitan area in south-east Queensland during December 2017, while *L. hystrix* seeds were collected from the suburb of Caloundra, Sunshine Coast Region in south-east Queensland during January 2017. Both seed batches had >90% viability as determined by the supplier. After delivery, seeds were stored in a seed storage cabinet at 15 $\pm$ 1 °C temperature and 15 $\pm$ 3% relative humidity until used. Seed age was 16 months for *L. longifolia* and 27 months for *L. hystrix* when used.

2.2. Seed Fill, Weight and Morpho-Anatomy

Seeds of both species were examined by using an X-ray machine (Faxitron MX-20 Imaging system, Lincolnshire, IL, USA) to determine seed fill percentage. Seed samples (5 replicates of 25 seeds per species) were exposed to 18 Kv of X-ray tube voltage for 20 s and images were captured using Bioptics software (Olympus, Tokyo, Japan) at 2$\times$ magnification of resolution. The percentage of filled seeds was determined by counting the number of seeds that had a full-sized endosperm and embryo. Filled seeds had a white color, and damaged or unfilled seeds were indicated by black areas inside the seed. The percentage of filled, partially filled (seeds with parts of their endosperm and/or embryo missing) and unfilled seed was determined. Partly filled and unfilled seeds were considered non-viable. Filled seeds with intact and healthy-looking embryos were considered viable. To determine the 100-seed mean weight, 5 samples of 100 seed from each species were randomly selected and weighed.

Seed anatomical structure was determined by photographic analysis by using a light microscope (Olympus SZX7, Mornington, TAS, Australia) with a digital camera attached. Seed and embryo size—specifically, length, width and area—were measured using CellSens software. To determine the E:S ratio, seeds were dissected longitudinally, and the embryo length was divided by seed length [23]. Embryo development was classified based on its anatomy (size and shape) according to Martin [21]. The presence of a developed or underdeveloped embryo was evaluated to identify if MD was present [24].

2.3. Germination Stimulation Using Mechanical and Chemical Treatments

Eight mechanical and/or chemical treatments were used *viz.* seed leaching, seed nicking, chemical treatment with smoke water at three different concentrations (Regen 2000 Smokemaster, batch no. 11957R, Tecnica, Bayswater, VIC, Australia), GA_3 (90% gibberellin A3, Sigma-Aldrich, lot BCBD6798V, St. Louis, MO, USA) or a combination of treatments (Table 1). Prior to treatments, all seeds were surface sterilized in 2% (v/v) sodium hypochlorite (NaOCl) solution for 10 min [25] with two drops of Tween 20 (Labchem, Zelienople, PA, USA) added as a surfactant. Seeds were then washed four times with sterile water and blotted dry. To undertake leaching, seeds of each species were transferred to several mesh ball infusers (diameter of 5 cm) and placed individually into 250 mL glass beakers for 36 h under running, turbulent cold tap water (ambient from main town water supply). To nick seeds, a small cut on the embryo end of the seed was undertaken using a scalpel blade (Figure 2). Chemical treatments (5 mL) were applied to each Petri dish (9 cm diameter) containing two Whatman No. 1 filter papers.

Table 1. The germination stimulant treatments applied to seeds of *Lomandra longifolia* Labill. and *Lomandra hystrix* (R.Br.) L.R. Fraser and Vickery consisted of both mechanical and chemical methods involving leaching, nicking, smoke water (SW) and gibberellic acid (GA_3).

Treatment	Type
Leaching	Mechanical
Nicking	Mechanical
SW1: Smoke water (50 mL L^{-1})	Chemical
SW2: Smoke water (100 mL L^{-1})	Chemical
SW3: Smoke water (200 mL L^{-1})	Chemical
GA3: gibberellic acid (289 μM)	Chemical
Nicking + SW2	Combination
Nicking + GA3	Combination

Previcure® (2% v/v; Bayer Crop Science) was added to the Petri dishes to inhibit fungi growth [19,26]. Then, following the addition of seed, each Petri dish was sealed with Parafilm to prevent evaporation of solutions. This was undertaken in a laminar air flow hood to reduce possible microbial contamination. All treatments were applied under light (with a 12/12 h day/night photoperiod) or dark conditions to simulate the seed being placed on the soil surface (light) or seed burial (dark). For seeds imbibed under light, the 12/12-h photoperiod used had a light intensity of 100 μmol m^{-2} s^{1} (produced by cool white, fluorescent tubes). Dark conditions were achieved by wrapping the Petri dishes with two layers of aluminum foil. Petri dishes were placed in an incubator (TRIL-750 Illuminated Refrigerator Incubator, Thermoline, Wetherill Park, NSW, Australia) using a matching 12/12-h thermoperiod of 20/10 ± 1 °C. The thermoperiod was selected from earlier studies carried out on both *Lomandra* species (unpublished data) and from the published literature (maximum germination for *L. longifolia* was at 20 °C and for *L. hystrix* it was at 15 °C [1]).

Figure 2. Location where seed nicking was applied to seed (pericarp + seed) of *Lomandra hystrix* (R.Br.) L.R. Fraser and Vickery to help relieve mechanical pressure on the embryo by the surrounding tissues. A small nick was formed by applying gentle pressure with a scalpel blade (marked as a black line in the image), resulting in a cut through the pericarp.

2.4. Data Collection and Analysis

For both species, each treatment was replicated 4 times, and each replication had 25 seeds per Petri dish. A completely randomized design was used. All seed germination tests were run for 60 days. Petri dishes were examined for germination twice weekly. Germinated seeds (radicle protrusion $\geq$ 2 mm) were recorded and removed. Seeds germinated under dark conditions were observed under a green safety light (Lion 24 LED Magnetic work lamp, covered with a green plastic sheet) in a dark room. Cumulative germination over time for the different seed treatments for each species was determined using a non-linear regression model fitted with the drm function in package drc [27] using R software, version 3.5.3 [28]. A three-parameter log-logistic model was used [29]. The germination rate index (GRI) was determined according to Maguire [30] (Equation (1)). Germination data (final germination percentage, percentage of dormant seeds and GRI) for each species were analyzed using a two-way factorial analysis of variance (ANOVA). When significant differences were identified, a Tukey's honest significance difference (HSD) test was used as a post-hoc analysis to identify significant differences between treatment means.

Equation (1): Germination rate index (GRI; Maguire [30])

$$\text{GRI}\left(\%\text{d}^{-1}\right) = \frac{\sum(G_i - G_{i-1})}{i} \quad (1)$$

i: Day of germination count
G_i: Percentage of seeds germinated in day i
G_{i-1}: Percentage of seeds germinated the previous count day

3. Results

3.1. Seed Fill and Seed Morpho-Anatomy

Lomandra longifolia had a 100% and *L. hystrix* had a 99% seed fill (Figure 3a,b, respectively). The seed cross-sectional area was 7.5 mm^2 for *L. longifolia* and 11.8 mm^2 for *L. hystrix* (Table 2). Embryos of both species were fully differentiated and seemed to be fully developed. They were small and linear and located in the basal part of the seed (Figure 4). Both species had a large proportion (>70%) of the seed consisting of endosperm tissue surrounding the embryo (Figures 3c and 4). The E:S ratio was 0.4 for both *L. hystrix* and *L. longifolia* (Table 2). The 100-seed weight was 900 mg for *L. hystrix* and 860 mg for *L. longifolia*.

Table 2. Seed characteristics as determined from 10 randomly selected seeds of *Lomandra longifolia* Labill and *Lomandra hystrix* (R.Br.) L.R. Fraser and Vickery. An endosperm was present in both species, and the embryo type was linear. The E:S ratio, seed and embryo length, area, perimeter and width were measured. Mean ± SE.

Tissue	*Lomandra longifolia*	*Lomandra hystrix*
E:S ratio	0.4 ± 0.03	0.4 ± 0.04
Seed length (mm)	3.9 ± 0.1	4.9 ± 0.8
Seed width (mm)	2.3 ± 0.2	2.5 ± 0.4
Seed area (mm^2)	7.5 ± 0.6	11.8 ± 5.2
Embryo length (mm)	1.4 ± 0.1	2.2 ± 0.5
Embryo area (mm^2)	0.3 ± 0.0	0.6 ± 0.1
100 seed weight (mg)	860 ± 20	900 ± 20

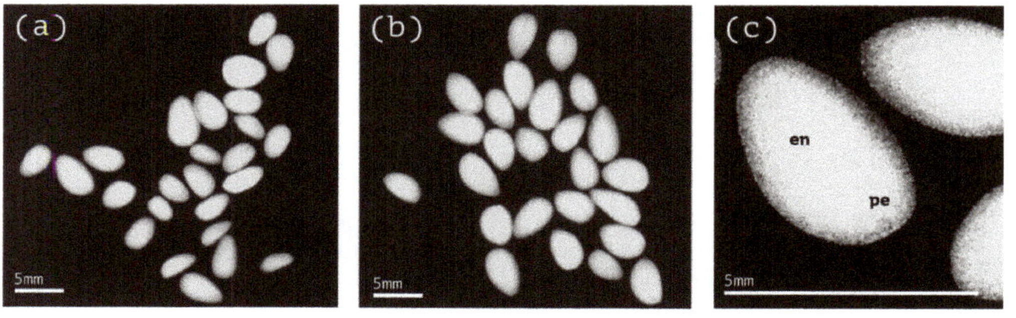

Figure 3. X-ray images (Faxitron MX-20) showing seed fill of (**a**) *Lomandra longifolia* Labill. (100 ± 0% seed fill); (**b**) *Lomandra hystrix* (R.Br.) L.R. Fraser and Vickery (99 ± 1% seed fill); (**c**) seed endosperm (en; white tissue) can be distinguished from the seed pericarp (pe; grey border area) surrounding the seed in *L. hystrix*. Seed fill was calculated by averaging the results of 5 replicates of 25 seeds from each species. All seeds shown are considered filled.

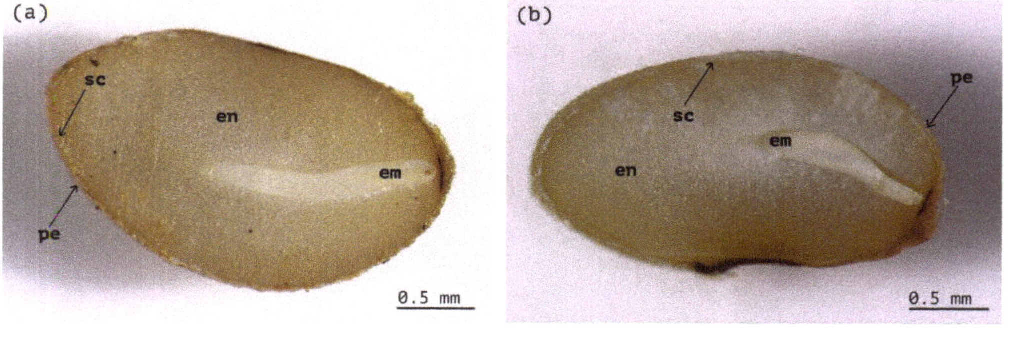

Figure 4. Light microscope images of longitudinal sections of seed (pericarp + seed) of (**a**) *Lomandra longifolia* Labill; and (**b**) *Lomandra hystrix* (R.Br.) L.R. Fraser and Vickery. In image (**b**), the embryo of *L. hystryx* is not shown in full as it sinks into the endosperm. Seeds had a small, linear basal embryo (em). The endosperm (en) filled a large proportion of the seed. Seed coat (sc) and pericarp (pe) are indicated with arrows. Magnification × 160.

3.2. Germination Enhancement Using Mechanical and Chemical Treatments

Leaching significantly increased ($p \leq 0.001$) the GRI for both species in comparison to untreated seeds and other pre-treatments regardless of light conditions (Table 3). Leached seeds were also the first to start germination (Figure 5). For *L. longifolia*, seeds leached in darkness had a higher GRI (1.0 ± 0.1% day^{-1}) as compared to leached seeds under light (0.5 ± 0.1% day^{-1}). Leached seeds for *L. hystrix* incubated under light conditions had GRI of 3.7 ± 0.1% day^{-1} as compared to control of 1.7 ± 0.2% day^{-1} (Table 3). However, the GRI for leached seeds incubated under light was significantly higher (p < 0.0001) than in darkness (2.4 ± 0.2% day^{-1}).

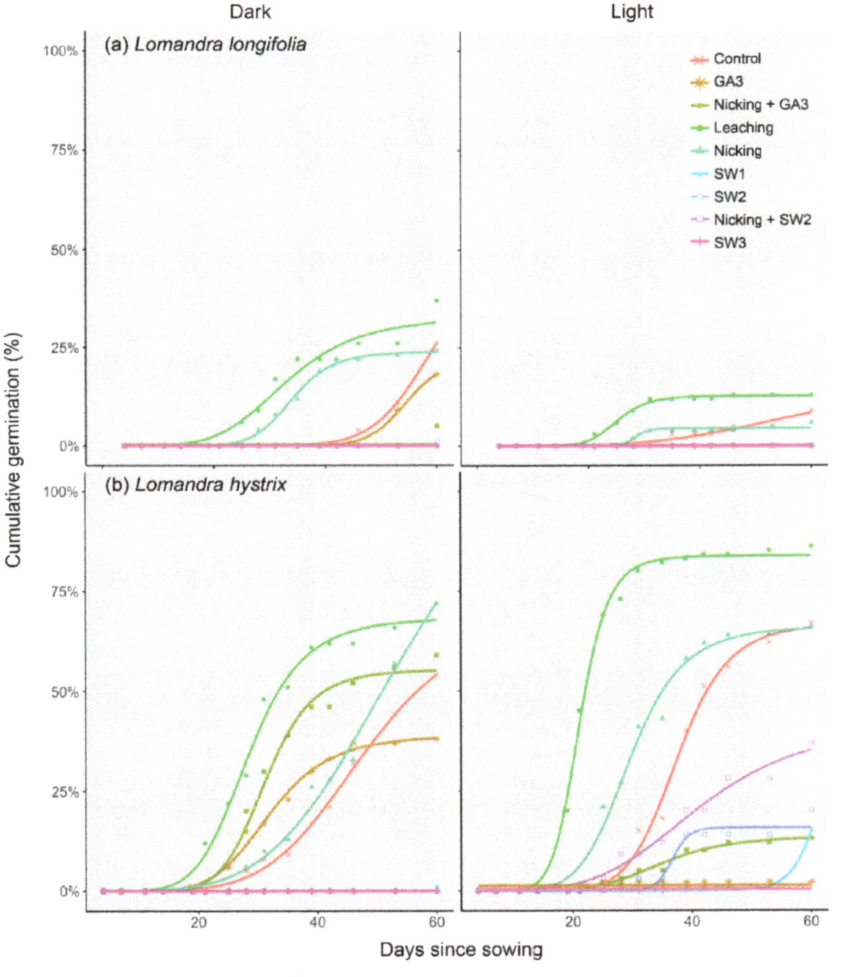

Figure 5. Mean germination percentage during 60 days of incubation (20/10 °C, 12/12-h thermoperiod, light [12/12-h photoperiod] or dark) incorporating chemical and/or mechanical treatments applied to (**a**) *Lomandra longifolia* Labill; and (**b**) *Lomandra hystrix* (R.Br.) L.R. Fraser and Vickery. Four replications per treatment, with twenty-five seeds per replicate per species were used.

Leaching or nicking (mechanical treatments) did not improve the final germination percentage for either species when compared to the untreated seeds (Figure 5). *Lomandra longifolia* had a maximum germination of 37.0 ± 1.9% (Figure 5a; leached seeds incubated in darkness). Furthermore, untreated *L. longifolia* seeds under darkness had a significantly higher final germination percentage (26.0%) as compared to seeds exposed to light (9%; $p \leq 0.01$). In *L. longifolia*, the germination percentages for leached seeds were significantly higher as compared to nicked seeds (7% higher under light and 13% higher under darkness; $p \leq 0.01$). *Lomandra hystrix* had a maximum of 86.0 ± 2.6% germination (Figure 5b; leached seeds incubated under illuminated conditions), and unlike *L. longifolia*, no significant differences ($p > 0.05$) were observed for untreated seeds of *L. hystrix* incubated under light or dark conditions. In *L. hystrix*, untreated seeds, leached seeds and nicked seeds had significantly higher final germination percentages (>53%) than for the remainder of the treatments (<40%; $p \leq 0.05$).

Table 3. Germination rate index (GRI) for *Lomandra hystrix* (R.Br.) L.R. Fraser and Vickery and *Lomandra longifolia* Labill. incubated under complete darkness (24 h) or light/dark (12/12-h photoperiod) conditions following chemical and/or mechanical treatments. All seeds were incubated for 60 days at 20/10 °C with a 12/12-h matching thermoperiod. Treatments were as follows: leaching with running tap water for 36 h, nicking (small cut through pericarp and seed testa), smoke water (SW)—SW1:50, SW2:100, SW3:200 mL L^{-1}; gibberellic acid—GA_3: 289 µM; SW2 + nicking, GA_3 + nicking and control (untreated seeds). Means ± SEM were calculated using 4 replications, 25 seeds per replication. Treatments that had a significantly higher GRI than the control are in bold, and treatments that resulted in zero germination are denoted by a dash.

Treatment	*Lomandra longifolia*		*Lomandra hystrix*	
	Light	**Dark**	**Light**	**Dark**
Leaching	**0.5 ± 0.1**	**1.0 ± 0.1**	**3.7 ± 0.1**	**2.4 ± 0.2**
Nicking	0.2 ± 0.1	0.7 ± 0.0	2.2 ± 0.3	1.6 ± 0.4
SW1	-	-	0.3 ± 0.0	-
SW2	-	-	0.5 ± 0.1	-
SW3	-	-	-	-
GA_3	-	0.3 ± 0.1	-	1.2 ± 0.2
SW2 + nicking	-	-	0.9 ± 0.3	-
GA_3 + nicking	-	0.1 ± 0.0	0.4 ± 0.1	1.7 ± 0.1
Control	0.2 ± 0.1	0.5 ± 0.1	1.7 ± 0.2	1.2 ± 0.1

Treatment with smoke water, at all three concentrations (SW1:50, SW2:100 and SW3:200 mL L^{-1}) and in combination with nicking, gave a significantly lower final germination percentage for both species in light and dark conditions when compared to untreated seeds (≤37% for *L. hystrix* and ≤1% for *L. longifolia*; $p \leq 0.05$). *Lomandra hystrix* and *L. longifolia* seeds treated with a combination of GA_3 and nicking had significantly higher final germination in darkness as compared to light conditions ($p \leq 0.05$; Figure 6). Under light, GA_3 and SW3 gave a significantly lower final germination of <3% than other treatments for *L. hystrix* and *L. longifolia* ($p \leq 0.005$). In *L. longifolia*, significantly higher ($p \leq 0.05$) final germination percentage occurred in dark conditions as compared with light conditions, for leaching, nicking, GA_3 and the control (24, 18, 18, 17% higher, respectively).

After 60 days, ≥50% of *L. longifolia* seeds remained ungerminated (Figure 6a). These seeds were considered dormant as they were filled (determined by X-ray) and firm. No significant differences were observed in dormant seeds of *L. longifolia* between treatments and control ($p \geq 0.05$). For *L. hystrix* (Figure 6b), the number of dormant seeds was significantly higher for seeds treated with smoke water (≥80% dormant seeds in all concentrations), SW2 + nicking (≥63%) and GA_3 + nicking (>40%) (Figure 6). Moreover, for GA_3 and nicking + GA_3 treatments in this species, dormancy was significantly higher in light conditions when compared to darkness (98 vs. 56% and 87 vs. 41% respectively: $p \leq 0.005$) (Figure 6). Seed death occurred mostly in *L. longifolia*, with the highest number observed in SW2 + nicking (51% dead seeds in light conditions) and GA_3 + nicking (>40% for light and dark conditions) (Figure 6). In contrast, a few seeds (≤8%) of *L. hystrix* were dead at the end of the experiment.

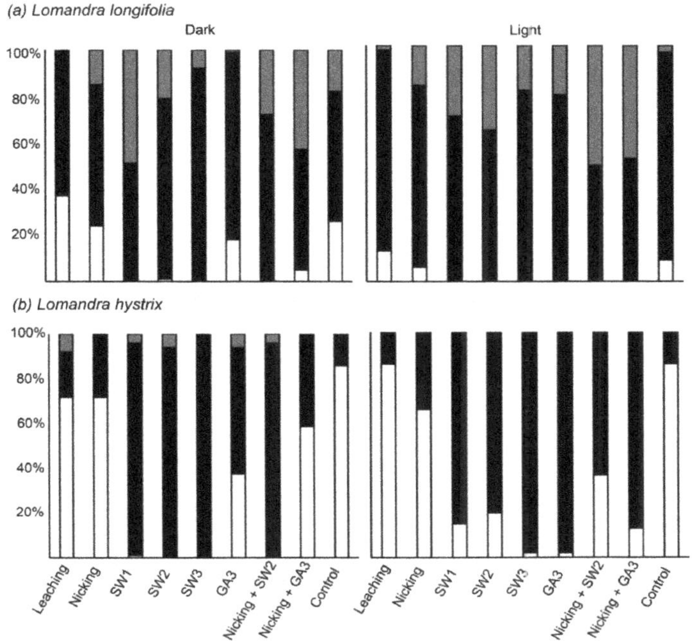

Figure 6. Percentage of germinated (white), dormant (black) and dead (grey) seeds for (**a**) *Lomandra longifolia* Labill. and (**b**) *Lomandra hystrix* (R.Br.) L.R. Fraser and Vickery and *Lomandra longifolia* Labill. germinated under dark or light (12/12-h photoperiod) conditions following chemical and/or mechanical treatments. All seeds were incubated for 60 days at 20/10 °C with a 12/12-h matching thermoperiod. Treatments were as follows: leaching with running tap water for 36 h, nicking (small cut in through the pericarp and seed coat), smoke water—SW1:50; SW2:100; SW3:200 mL L^{-1}; gibberellic acid—GA_3:289 μM, Control: untreated seed. Means of 4 replications with 25 seeds each per species, per treatment are shown.

4. Discussion

This study aimed to characterize seed morpho-anatomy and to determine seed treatments that could enhance the germination of two Australian native *Lomandra* species. The seed traits that influence germination of *L. longifolia* and *L. hystrix* had not been addressed in detail previously and there is a lack of published protocols on how to improve germination for both important seed-based restoration and ornamental species. The current study showed that seeds of both species had PD at the time of sowing, probably due to tissues surrounding the small embryo imposing a mechanical constraint to its protrusion and/or the presence of chemical inhibitors (as shown to occur in other *Lomandra* species). Further results indicate that germination rates of both *Lomandra* species can be significantly improved by leaching seeds under running tap water. Having a deep understanding of seed biology is crucial for seed-based restoration success [31]: it enables seed treatments to be optimized for improving seed performance [32] and can result in faster, more uniform and, therefore, reliable germination. Additionally, understanding seed germination timing will be crucial for seed use under future climate change [33].

4.1. Seed Morpho-Anatomy

The seed fill percentage in both seed lots studied was 100%, suggesting that other factors besides seed viability or seed fill are limiting germination. Both species presented small (E:S = 0.4), linear and basal embryos (Figure 4) surrounded by a thick layer of endosperm.

Even though the embryos were small, microscopic evaluation showed that embryos in this current study were fully differentiated and seemed to be fully developed, consisting of a well-defined primary axis and cotyledon. In contrast, Ruiz-Talonia et al. [14] identified MPD in *L. longifolia* due to the presence of an underdeveloped embryo. In their study, a combination of leaching and warm stratification was needed to overcome dormancy and improve germination. This present study evaluated the use of stored seeds of *L. Longifolia* and *L. hystrix* (for 27 and 16 months, respectively). Seed age can influence seed dormancy status as seeds can undergo after-ripening, dormancy cycling [10] or overcome MD or MPD during storage. Therefore, there is a possibility that embryo growth occurred during storage, overcoming MD. Although, it is important to note that Baskin et al. [22] proposed that a seed with a small embryo in relation to the endosperm does not strictly mean that the embryo is underdeveloped. Some species, such as in the genus *Nymphaea* (Nymphaeaceae [18]) and *Drosera anglica* (Huds.) LePage and W.Bldw. (Droseraceae [22]), have small embryos with low E:S ratios (e.g., 0.24 ± 0.01 for *D. anglica*). However, these embryos did not exhibit growth inside the seed before germination could occur. Therefore, MD was not present. Consequently, to identify the presence of MD or MPD in both of our study species, further studies should be undertaken with freshly collected seeds to determine whether embryo growth occurs within the seed prior to them becoming germinable.

4.2. Seed Germination Ecology and Enhancement

In *L. longifolia*, seeds incubated in the dark had a significantly higher final germination percentage than those imbibed in light conditions (37 vs. 13%; Figure 5). This suggests that *L. longifolia* can germinate to a higher extent if buried in the soil. On the other hand, there was no significant difference between light and dark treatment in *L. hystrix,* although marginal improvements in germination were observed under light. Seeds of many species are sensitive to light intensity and quality, which is a mechanism to avoid plant competition [34]. Therefore, light detection by seeds can be an important germination cue [35]. In *L. longifolia*, germination inhibition by light could be related to the avoidance of germination near or at the soil surface, ensuring seeds are positioned at sufficient depth where moisture is more reliably obtained. This is a common seed adaptation to environments where moisture is limited [36], such as those where *L. longifolia* naturally grows. Moreover, temperature is moderated at greater soil depths [35], which may be an adapdation of *L. longifolia* to enable germination to occur in a wide range of climates (Figure 1). However, burial at depth does not appear to be a requirement of *L. hystrix* seeds, possibly because this species has adapted to growing along watercourses and in rainforests, where water is usually more abundant and temperature fluctuations are less extreme.

Leaching significantly improved the GRI of both species (Figure 5). The positive effects of leaching in this study are consistent with previous observations made for *L. longifolia* [14] and *L. sonderi.* Plummer et al. [2] suggested that water-soluble germination inhibitors located in the pericarp and embryo could inhibit germination in *L. sonderi*; germination inhibition was successfully overcome by removing the pericarp or by leaching seeds for 24 h in running tap water. This may be an adaptation to regions where occasional, but heavy, rainfall can leach out seed germination inhibitors and break down the seed coat tissues [26,37]. This mechanism ensures that conditions are suitable for germination and seedling establishment. On the other hand, Baskin et al. [18] suggested that leaching can also act as a stratification treatment. Periods of warm or cold stratification have been shown to alleviate PD in seeds [38]. For example, stratification at 26/13 °C or 33/18 °C for 4- or 8-weeks alleviated dormancy in *L. preisii* when seeds were germinated at 18/7 °C [7]. Moreover, warm stratification achieved >80% germination in *Acanthocarpus preisii* Lehm. (Asparagaceae) as compared with <20% when seeds were not stratified [39]. Further studies should be directed at identifying if the use of stratification treatments or wet/drying cycles [40] could be involved in overcoming dormancy of *Lomandra* seeds in the soil seedbank.

The improvement in germination of seeds with non-deep PD after scarification has been related to overcoming a mechanical barrier to embryo growth imposed by the tissues surrounding the embryo [10,41]. Likewise, seed nicking has also been shown to relieve embryo growth restrictions [19]. Although in this study, nicking did not significantly promote germination, a positive trend was observed (Figure 5). In seeds presenting PD, the fully developed embryo has a low growth potential; therefore, it cannot overcome the mechanical constraints imposed by its surrounding tissues [10]. Once treatments such as nicking are performed, the embryo can gain sufficient expansive force to protrude through the surrounding tissues. In nature, embryos with low growth potential need cues from the environment to initiate internal chemical signaling which promote certain covering tissues to breakdown and increase the growth potential [10]. However, this resistance can also be weakened over time in the seeds' natural environment by the production of tissue-softening enzymes released by the embryo or by weakening through physical biotic factors such as temperature, fire, animal ingestion, seed burial and saprophytic fungi [37,42]. Embryo germination resistance also varies according to imbibition conditions such as temperature and light/dark interactions [10], and further studies could be undertaken to determine the influence these parameters can have on endosperm resistance.

Constant exposure to smoke water throughout the incubation period significantly inhibited germination of both species, with $\leq$1% final germination for *L. longifolia* and $\leq$20% final germination for *L. hystrix* (Figure 6). This contrasts with the findings of Merritt et al. [43], where smoke water was reported to promote germination in other *Lomandra* species. For example, imbibing seeds in smoke water (1:10 [*v*/*v*]) for 24 h enhanced seed responses to warm stratification providing the highest germination for *L. preisii* seeds (*ca.* 50% [7]). In contrast, Vening et al. [44] reported that smoke water used in an agar-based germination medium at 1:10 (*v:v*) had no significant effect on germination in Australian native forbs from fire prone environments. However, seed sensitivity to smoke water can be a complex process [7], as active constituents of smoke water (such as karrikins) can vary between different stock solutions and species react differently [45]. Furthermore, Adkins et al. [45] found that caryopsis of wild oats (*Avena fatua* L.) had greater germination when exposed to smoke water for 7 days prior to incubation with distilled water as compared to caryopsis that received smoke water before and during incubation.

Considering the above, constant exposure of *Lomandra* seeds to smoke water in this current study could have caused germination inhibition. This may be due to the dual effect of smoke water on the seed germination process reported by Light et al. [46] for 'Grand Rapids' lettuce (*Lactuca sativa* L.). This study proposed that smoke water had an inhibitory component that enters the seed and a promotor component that remains in the seed, inactive until sufficient rainfall has leached out the inhibitor. Additionally, it is important to note that smoke has been proposed to act as a germination enhancer, rather than a dormancy breaker [7,47]. Therefore, smoke water effects might only act to enhance germination after dormancy has been overcome. Further studies on both *Lomandra* species could be undertaken by applying smoke water as a pre-treatment prior to incubation, but after seeds have been treated for dormancy.

Although leaching improved the GRI for *L. longifolia*, the final germination percentage achieved in all treatments was low ($\leq$37% germination). This low germination percentages of *L. longifolia*, even when exposed to GA_3, could suggest the presence of a deeper level of PD. In many species with intermediate or non-deep PD, germination is stimulated by GA_3, while those with deep PD fail to germinate in GA_3 treatments that would normally promote germination [48]. To alleviate PD in *L. sonderi*, the pericarp needed to be removed from the seed (presumably to remove germination inhibition imposed by these tissues); then, GA_3 (145 μM) was applied (to relieve embryo dormancy [2]). Further studies on *L. longifolia* seeds should be undertaken such as excision of the pericarp and seed coat [42] and excision of the embryo [10]; and then, applying GA_3 to identify if the pericarp or endosperm are preventing GA_3 from reaching the embryo. Although GA_3 did not improve germination in both *Lomandra* species, there was a significant interaction between GA_3 and darkness, where seeds treated with GA_3 had higher germination in the dark as compared to light. This could be related to GA_3 interacting with the phytochrome system within seeds [37].

5. Conclusions

This study is one of the first to investigate techniques to enhance germination of the Australian native species *L. longifolia* and *L. hystrix*. The innate slow and initially low germination of both *Lomandra* seeds requires a significantly large number of viable seeds to be sown to achieve the plant density required for ornamental plant production or seed-based restoration projects. Results from this study show that both species had small, linear embryos and a high proportion of endosperm tissue. Slow germination of both species is most likely associated with the presence of at least one mechanism of PD present in the seed. Leaching seeds prior to incubation was the only treatment to significantly hasten seed germination in both *Lomandra* species. To determine the mechanisms by which leaching functioned on hastening germination and to correctly classify dormancy in both species, further studies are now needed on freshly collected seeds. Future areas of research include measuring embryo growth during incubation to test for MPD, undertaking warm stratification prior to germination incubation, and seed treatments with GA_3. Understanding the factors that influence seed germination and pre-treating seeds accordingly, or ensuring these requirements are met in the natural environment (in the case of seed-based restoration), is crucial for the success and cost-efficient use of these seeds. It is also important to consider the scaling-up of treatments for large restoration projects or ornamental plant production and how that could affect seed tissues and the overall cost-effectiveness of the treatment. Moreover, the possibility of providing similar effects naturally in the field by sowing seeds when long periods of rain are forecasted should also be examined.

Author Contributions: Conceptualization, F.C.B. and S.W.A.; original draft preparation, F.C.B., R.C., A.W. and S.W.A.; review and editing, F.C.B., S.K., R.C., A.W. and S.W.A. All authors have read and agreed to the published version of the manuscript.

Funding: This research received no external funding.

Institutional Review Board Statement: Not applicable.

Informed Consent Statement: Not applicable.

Data Availability Statement: Not applicable.

Conflicts of Interest: The authors declare no conflict of interest.

References

1. Seed Information Database (SID), Royal Botanical Gardens Kew. Available online: http://data.kew.org/sid/ (accessed on 14 September 2018).
2. Plummer, J.; Crawford, A.; Taylor, S. Germination of *Lomandra sonderi* (Dasypogonaceae) promoted by pericarp removal and chemical-stimulation of the embryo. *Aust. J. Bot.* **1995**, *43*, 223–230. [CrossRef]
3. Lee, A.; Macfarlane, T. Lomandra. *Fl Aust.* **1986**, *46*, 100–141.
4. Bonney, N.; Miles, A. *What Seed Is That?* N. Bonney: Beverley, Australia, 1994.
5. Ahmad, N.M.; Martin, P.M.; Vella, J.M. Embryology of the dioecious Australian endemic *Lomandra longifolia* (Lomandraceae). *Aust. J. Bot.* **2009**, *56*, 651–665. [CrossRef]
6. Ahmad, N.M.; Martin, P.M.; Vella, J.M. Clonal propagation of *Lomandra longifolia* by somatic embryogenesis. *Sci. Hortic.* **2014**, *180*, 102–110. [CrossRef]
7. Merritt, D.; Turner, S.; Clarke, S.; Dixon, K. Seed dormancy and germination stimulation syndromes for Australian temperate species. *Aust. J. Bot.* **2007**, *55*, 336–344. [CrossRef]
8. Erickson, T.; Barrett, R.; Merritt, D.; Dixon, K. *Pilbara Seed Atlas and Field Guide: Plant Restoration in Australia's Arid Northwest*; CSIRO Publishing: Clayton, Australia, 2016.
9. Baskin, C.; Baskin, J. Types of seeds and kinds of seed dormancy. In *Seeds: Ecology, Biogeography, and Evolution of Dormancy and Germination*; Elsevier: Amsterdam, The Netherlands, 2014; pp. 37–77.
10. Baskin, C.C.; Baskin, J.M. *Seeds: Ecology, Biogeography, and Evolution of Dormancy and Germination*, 2nd ed.; Academic Press: London, UK, 2014.
11. Bonney, N. Understanding botanical pathlines for improved germination of native plant seed. In Proceedings of the Fourth Australian Workshop on Native Seed Biology for Revegetation, Mildura, Australia, 3–4 September 2001; pp. 105–111.
12. Gibson-Roy, P.; Delpratt, J.; Moore, G. Restoring the Victorian Western (Basalt) Plains grassland. 1. Laboratory trials of viability and germination, and the implications for direct seeding. *Ecol. Manag. Restor.* **2007**, *8*, 114–122. [CrossRef]

13. Grant, C.; Campbell, C.; Charnock, N. Selection of species suitable for derelict mine site rehabilitation in New South Wales, Australia. *Water Air Soil Pollut.* **2002**, *139*, 215–235. [CrossRef]
14. Ruiz-Talonia, L.; Whalley, R.; Gross, C.; Carr, D.; Reid, N. Overcoming limitations to propagation from seed of 40 Australian species important for restoration. *New For.* **2022**, 1–20. [CrossRef]
15. Alvarado, V.; Nonogaki, H.; Bradford, K. Expression of endo-β-mannanase and SNF-related protein kinase genes in true potato seeds in relation to dormancy, gibberellin and abscisic acid. In *Dormancy in Plants: From Whole Plant Behaviour to Cellular Control*; Viemont, J., Crabbe, J., Eds.; CABI Publishing: Wallingford, UK, 2000; pp. 347–364.
16. Roche, S. Smoke-a new process for germinating Australian plants. *Aust. Hortic.* **1994**, *92*, 46–47.
17. Dixon, K.W.; Roche, S.; Pate, J.S. The promotive effect of smoke derived from burnt native vegetation on seed germination of Western Australian plants. *Oecologia* **1995**, *101*, 185–192. [CrossRef]
18. Baskin, C.C.; Thompson, K.; Baskin, J.M. Mistakes in germination ecology and how to avoid them. *Seed Sci. Res.* **2006**, *16*, 165–168. [CrossRef]
19. Turner, S.; Merritt, D. Seed germination and dormancy. In *Plant Germplasm Conservation in Australia: Strategies and Guidelines for Developing, Managing and Utilising Ex Situ Collections*; Offord, C.A., Meagher, P.F., Eds.; Australian Network for Plant Conservation Inc.: Canberra, Australia, 2009.
20. Baskin, C.C.; Baskin, J.M. A revision of Martin's seed classification system, with particular reference to his dwarf-seed type. *Seed Sci. Res.* **2007**, *17*, 11–20. [CrossRef]
21. Martin, A. The comparative internal morphology of seeds. *Am. Midl. Nat.* **1946**, *36*, 513–660. [CrossRef]
22. Baskin, C.C.; Baskin, J.M. Underdeveloped embryos in dwarf seeds and implications for assignment to dormancy class. *Seed Sci. Res.* **2005**, *15*, 357. [CrossRef]
23. Commander, L.; Merritt, D.; Rokich, D.; Dixon, K. Seed biology of Australian arid zone species: Germination of 18 species used for rehabilitation. *J. Arid Environ.* **2009**, *73*, 617–625. [CrossRef]
24. Ooi, M.K. Dormancy classification and potential dormancy-breaking cues for shrub species from fire-prone south-eastern Australia. In *Seeds: Biology, Development and Ecology*; Adkins, S., Ashmore, S., Navie, S.C., Eds.; CAB International: Wallingford, UK, 2007; pp. 205–216.
25. Merrit, D. Seed storage and testing. In *Australian Seeds: A Guide to Their Collection, Identification and Biology*; Sweedman, L., Merritt, D., Eds.; CSIRO Publishing: Collingwood, Australia, 2006; pp. 53–60.
26. Cochrane, A.; Kelly, A.; Brown, K.; Cunneen, S. Relationships between seed germination requirements and ecophysiological characteristics aid the recovery of threatened native plant species in Western Australia. *Ecol. Manag. Restor.* **2002**, *3*, 47–60. [CrossRef]
27. Ritz, C.; Baty, F.; Streibig, J.C.; Gerhard, D. Dose-response analysis using R. *PLoS ONE* **2015**, *10*, e0146021. [CrossRef]
28. R Core Team. *R: A Language and Environment for Statistical Computing*; R Foundation for Statistical Computing: Vienna, Austria, 2022.
29. Pedrini, S.; Lewandrowski, W.; Stevens, J.C.; Dixon, K.W. Optimising seed processing techniques to improve germination and sowability of native grasses for ecological restoration. *Plant Biol.* **2018**, *21*, 415–424. [CrossRef]
30. Maguire, J.D. Speed of Germination—Aid in selection and evaluation for seedling emergence and vigor 1. *Crop Sci.* **1962**, *2*, 176–177. [CrossRef]
31. Shaw, N.; Barak, R.S.; Campbell, R.E.; Kirmer, A.; Pedrini, S.; Dixon, K.; Frischie, S. Seed use in the field: Delivering seeds for restoration success. *Restor. Ecol.* **2020**, *28*, S276–S285. [CrossRef]
32. Erickson, T.E.; Muñoz-Rojas, M.; Kildisheva, O.A.; Stokes, B.A.; White, S.A.; Heyes, J.L.; Dalziell, E.L.; Lewandrowski, W.; James, J.J.; Madsen, M.D. Benefits of adopting seed-based technologies for rehabilitation in the mining sector: A Pilbara perspective. *Aust. J. Bot.* **2017**, *65*, 646–660. [CrossRef]
33. Walck, J.L.; Hidayati, S.N.; Dixon, K.W.; Thompson, K.; Poschlod, P. Climate change and plant regeneration from seed. *Glob. Chang. Biol.* **2011**, *17*, 2145–2161. [CrossRef]
34. Yan, A.; Chen, Z. The control of seed dormancy and germination by temperature, light and nitrate. *Bot. Rev.* **2020**, *86*, 39–75. [CrossRef]
35. Fenner, M.; Thompson, K. *The Ecology of Seeds*; Cambridge University Press: Cambridge, UK, 2005.
36. Turner, S.R.; Lewandrowski, W.; Elliott, C.P.; Merino-Martín, L.; Miller, B.P.; Stevens, J.C.; Erickson, T.E.; Merritt, D.J. Seed ecology informs restoration approaches for threatened species in water-limited environments: A case study on the short-range Banded Ironstone endemic *Ricinocarpos brevis* (Euphorbiaceae). *Aust. J. Bot.* **2017**, *65*, 661–677. [CrossRef]
37. Adkins, S.W.; Bellairs, S.M.; Preston, C.; Thompson, L.; Farley, G. Identifying dormancy mechanisms of Australian native plant species. In Proceedings of the Fourth Australian Workshop in Native Seed Biology for Revegetation, Mildura, Australia, 3–4 September 2002; pp. 61–70.
38. Kildisheva, O.A.; Dixon, K.W.; Silveira, F.A.; Chapman, T.; Di Sacco, A.; Mondoni, A.; Turner, S.R.; Cross, A.T. Dormancy and germination: Making every seed count in restoration. *Restor. Ecol.* **2020**, *28*, S256–S265. [CrossRef]
39. Turner, S.; Merritt, D.; Ridley, E.; Commander, L.; Baskin, J.; Baskin, C.; Dixon, K. Ecophysiology of seed dormancy in the Australian endemic species *Acanthocarpus preissii* (Dasypogonaceae). *Ann. Bot.* **2006**, *98*, 1137–1144. [CrossRef] [PubMed]
40. Hoyle, G.; Daws, M.; Steadman, K.; Adkins, S. Mimicking a semi-arid tropical environment achieves dormancy alleviation for seeds of Australian native Goodeniaceae and Asteraceae. *Ann. Bot.* **2008**, *101*, 701–708. [CrossRef]

41. Erickson, T.E.; Shackelford, N.; Dixon, K.W.; Turner, S.R.; Merritt, D.J. Overcoming physiological dormancy in seeds of *Triodia* (Poaceae) to improve restoration in the arid zone. *Restor. Ecol.* **2016**, *24*, S64–S76. [CrossRef]
42. Cochrane, A.; Probert, R. Temperature and dormancy-breaking treatments: Germination of endemic and geographically restricted herbaceous perennials. *Aust. J. Bot.* **2006**, *54*, 349–356. [CrossRef]
43. Merritt, D.; Rokich, D. Seed biology and ecology. In *Australian Seeds: A Guide to Their Collection, Identification and Biology*; Sweedman, L., Merritt, D., Eds.; CSIRO Publishing: Collingwood, Australia, 2006; pp. 19–24.
44. Vening, G.S.; Guja, L.K.; Spooner, P.G.; Price, J.N. Seed dormancy and germination of three grassy woodland forbs required for diverse restoration. *Aust. J. Bot.* **2017**, *65*, 625–637. [CrossRef]
45. Adkins, S.; Peters, N. Smoke derived from burnt vegetation stimulates germination of arable weeds. *Seed Sci. Res.* **2001**, *11*, 213–222.
46. Light, M.; Gardner, M.; Jäger, A.; Van Staden, J. Dual regulation of seed germination by smoke solutions. *Plant Growth Regul.* **2002**, *37*, 135–141. [CrossRef]
47. Thompson, K.; Ooi, M.K. To germinate or not to germinate: More than just a question of dormancy. *Seed Sci. Res.* **2010**, *20*, 209–211. [CrossRef]
48. Baskin, J.M.; Baskin, C.C. A classification system for seed dormancy. *Seed Sci. Res.* **2004**, *14*, 1–16. [CrossRef]

applied biosciences

MDPI

Article

Coupled Biogas and Fiber Production from Agricultural Residues and Energy Crops with Steam Explosion Treatment

Benedikt Hülsemann [1,*], Marian Baumgart [1], Leonhard Lenz [1], Elviliana [2], Marie Föllmer [3], Gregor Sailer [1], Konstantin Dinkler [1] and Hans Oechsner [1]

1 State Institute of Agricultural Engineering and Bioenergy, University of Hohenheim, Garbenstraße 9, 70599 Stuttgart, Germany; leonhard.lenz@uni-hohenheim.de (L.L.)

2 Department of Agro-Industrial Technology, Faculty of Agricultural Technology, Universitas Brawijaya, Malang 65145, Indonesia

3 Hochschule der Medien, University of Applied Sciences, Nobelstraße 10, 70569 Stuttgart, Germany

* Correspondence: benedikt.huelsemann@uni-hohenheim.de

Abstract: The global demand for packaging materials and energy is constantly increasing, requiring the exploration of new concepts. In this work, we presented a bioeconomic concept that uses steam explosion and phase separation to simultaneously generate fibers for the packaging industry and biogas substrate for the energy sector. The concept focused on fiber-rich residues and fiber-rich ecological energy crops from agriculture. Feasibility of the concept in the laboratory using feedstocks, including Sylvatic silphia silage, Nettle silage, Miscanthus, Apple pomace, Alfalfa stalks, and Flax shives was confirmed. Our results showed that we were able to separate up to 26.2% of the methane potential while always extracting a smaller percentage of up to 17.3% of organic dry matter (ODM). Specific methane yields of 297–486 $L_{CH4}\ kg_{ODM}^{-1}$ in the liquid and 100–286 $L_{CH4}\ kg_{ODM}^{-1}$ in the solid phase were obtained. The solid phases had high water absorption capacities of 216–504% due to the steam explosion, while the particle size was not significantly affected. The concept showed high potential, especially for undried feedstock.

Keywords: renewable energy; bioeconomy; biowaste; residuals; silage; liquid–solid separation

Citation: Hülsemann, B.; Baumgart, M.; Lenz, L.; Elviliana; Föllmer, M.; Sailer, G.; Dinkler, K.; Oechsner, H. Coupled Biogas and Fiber Production from Agricultural Residues and Energy Crops with Steam Explosion Treatment. *Appl. Biosci.* **2023**, 2, 278–291. https://doi.org/10.3390/applbiosci2020019

Academic Editor: Robert Henry

Received: 28 February 2023
Revised: 19 May 2023
Accepted: 25 May 2023
Published: 1 June 2023

1. Introduction

The United Nations Development Programme identified 17 Sustainable Development Goals (SDGs) that outline a blueprint for a sustainable future. Five of these SDGs are particularly relevant to agriculture and biogas production, namely Zero Hunger, Affordable and Clean Energy, Industry, Innovation and Infrastructure and Climate Action [1]. Biogas is a renewable energy carrier produced through the anaerobic digestion of organic matter. The biogas technology was proven to be relevant for the reduction in greenhouse gases while simultaneously producing clean energy. Despite its potential, fossil fuels still account to approximately 79% of the global energy consumption [2].

Germany is one of the worldwide market leader for biogas and the sector grew over the last twenty years to roughly 9600 plants in Germany today [3]. The number of biogas plants was growing mainly because the Renewable Energy Sources Act (EEG), in its 2004 and 2009 versions, guaranteed high remunerations for electricity for twenty years [4]. Biogas plants in Germany generated an average electricity revenue of 23.4 EUR -ct/kWhel [5]. However, operation without high legal remuneration is not economically viable. The current version of the EEG (2023) limits the guaranteed remuneration for electricity in a tender procedure to a maximum of 18.03 EUR -ct/kWh for existing biogas plants and 16.07 EUR -ct/kWh for new biogas plants [6]. At the same time, the production costs are 18.9 EUR -ct/kWhel in Germany [5] and the cost increased according to higher costs expected due to stricter regulations, higher feedstock prices, and the advanced age of existing plants. Therefore, operation modes based on only continuous energy generation is not economically feasible for future business models of biogas plants [5].

Alternative products such as the electricity generation on demand, heat sale, production of biomethane or chemicals as well as the generation of fibers must be considered.

Another approach is the reduction in production costs through the use of other energy crops and residues that are cheaper and more ecological than the commonly used maize silage (70% of the energy crops used in German biogas plants [7]).

The use of alternative energy crops that can grow on marginal land also has a positive impact on public acceptance of the biogas technology because these energy crops are not in direct conflict with food production and are characterized with lower CO_2 emission [4,8,9]. Nevertheless, most alternative feedstocks have lower methane potentials and slower anaerobic digestion rates compared to maize silage due to their high fiber content, which has high resistance to microbes during anaerobic digestion [10]. For this reason, the use of these feedstocks is still more expensive than the use of maize silage.

To solve this problem of expensive biogas plant feedstocks, a bioeconomy concept was developed in which energy crops and residual materials are thermo-chemically pretreated via steam explosion (SE) followed by a solid–liquid separation. The solid fraction mainly consists of fibers (hardly digestible in the biogas process), while the liquid fraction contains components that are rapidly and easily degradable in an anaerobic digestion process. Thus, the liquid will be used in a biogas plant for electricity and heat production, while the solid fraction will be used as fiber material in various processes, e.g., for packaging material production.

Lignocellulose is the most abundant organic biomass in the world and an important feedstock for bioenergy technologies, biodegradable materials, and bio-based chemicals in biorefineries [11]. Its complex structure comprises hemicellulose linked cellulose microfibrils embedded in a matrix of lignin, cross-linked polysaccharide networks, and glycosylated proteins [12]. However, the resistance of lignocellulosic substrates to biological degradation in anaerobic digestion, known as biomass recalcitrance, hinders the conversion of the structural polysaccharides of the cell wall into fermentable sugars for use as fuel or chemicals [13]. To tackle this issue, various pretreatment methods were developed, including physical, physicochemical, chemical, and biological methods [14]. Among them, SE is a promising physicochemical approach that simultaneously modifies the biomass chemically, fractures the cell wall, removes hemicellulose, and increases the accessible surface area of cellulose without significant cellulose degradation [15]. Additionally, SE is effective in autocatalytically removing acetic- and uronic-groups forming their respective acids and depolymerizing hemicellulose, making it an attractive method for pretreating lignocellulosic materials [16,17]. SE is usually operated at a temperature of 160–220 °C with a pressure of 0.6–1.0 MPa. Boiling and rapid depressurization break down the lignin structure and degrade the hemicellulose to oligomers and sugars. Pressure around 0.5 MPa depolymerize the hemicellulose but not the cellulose [15,18,19]. Furthermore, after SE, a separation of a liquid and solid phase is easily possible. The pretreatment of fiber-rich substrate to increase biogas production was already investigated in several studies [20–22].

The fibers produced as part of the bioeconomic concept can be used to produce several different products such as paper, packaging, or flower pots. Because of this, the fibers can close the production cycle for these products locally and significantly reduce transport distances and, thus, CO_2 emissions, as they replace wood from *Eucalyptus* spp. or *Pinus* spp. in Germany, which is mainly produced in South America and especially in Brazil [23]. In the case of flower or plant pots, conventional plastic- or peat-based products can be substituted [24]. The production of paper and packaging also holds great potential. The growing popularity of e-commerce led to an increased demand for packaging materials, exacerbating the environmental impact of the packaging industry in terms of CO_2 emissions and energy consumption [25–27]. Kim et al. (2022) reported that due to increasing online trade, the amount of packaging in Germany is 4.8 times higher compared to offline trade and Järvinen et al. (2020) predicted that the demand for paper will almost be doubled by 2050 [28,29].

This study investigates the suitability of various residual materials and ecological energy crops with variable properties (e.g., dry matter (DM)) for the use in the bioeconomy concept with coupled fiber and biogas production by processing the fiber-rich biomass in a SE followed by solid–liquid separation (Figure 1). To investigate the potential of the concept, a wide range of residual materials and ecological energy crops were investigated. The following crops were considered: the silage of the whole plant of *Silphium perfoliatum* L. (sylvatic silphia silage), *Urtica dioica* (nettle silage), and *Miscanthus sinensis* (miscanthus whole plant), as well as the hop grubbing chaff of *Humulus lupulus* (hop bine chaff), the straw of *Miscanthus sinensis* (miscanthus straw), the stalks of *Medicago sativa* (alfalfa stalks), the shives of *Linum usitatissimum* (flax shives), and the pomace of *Malus* (apple pomace). The aim was to evaluate the developed bioeconomy concept and to check what kind of substrate can be used for this concept.

Figure 1. The investigated feedstocks sylvatic silphia silage (**a**), nettle silage (**b**), hop grubbing chaff (**c**), miscanthus whole plant (**d**), miscanthus straw (**e**), apple pomace (**f**), alfalfa stalk (**g**), flax shives (**h**).

2. Materials and Methods

2.1. Substrates and Sampling

Silvatic silphy silage was taken from stored silage (Ostrach, Germany) on 3 September 2021 (see Figure 1). Nettle silage (Ostrach, Germany) was harvested, pressed, and stored as silo bales on 6 September 2021. The silo was open on the day of further treatment steps. Hop grubbing chaff was taken from stored silage (Hallertau, Germany) in January 2021. Miscanthus was harvested as whole plant on 20 October 2021, and as straw on 2 February 2021 (Unterer Lindenhof, Eningen unter Achalm, Germany). Both were chopped before being treated with SE. Apple pomace was taken from juice extractor (Kelterei Widemann, Bermatingen, Germany). Alfalfa stalks were harvested in September 2021 (Futtertrocknung Lamerdingen eG, Lamerdingen, Germany). Flax was harvested at different locations in France. The flax shives were sorted during the production process of flax and were then sent to the authors (Terre de Lin, Saint-Pierre-le-Viger, France). All fresh samples were immediately after harvesting stored in compressed condition in 20 L barrels and at 4 °C for several days prior to testing. Compression was performed by hand.

2.2. Steam Explosion (SE)

All substrates were pretreated using the SE unit at the Department of Yeast Genetics and Fermentation Technology of Hohenheim University (Stuttgart, Germany). The treatment was carried out in a gastight and double-walled laboratory reactor with a volume of 20 L (H & K GmbH Behälter und Edelstahltechnik, Kehl, Germany). The steam was supplied indirectly by heating the substrate–water mixture.

Before SE, the samples were crushed by blades in a Thermomix (Vorwerk, Wuppertal, Germany) for 30 s and mixed with water to obtain a similar water content (Table 1). SE was performed in a gastight and double-walled laboratory reactor with a volume of 20 L (H & K GmbH Behälter und Edelstahltechnik, Kehl, Germany) at 160 °C and 0.5 MPa (see Figure 2). The reactor was heated by steam. A change in temperature and pressure was not possible. All reaction times were set to 10 min, resulting in a severity factor of 2.77, excluding the alfalfa stalks (30 min with a severity factor of 3.54) [30–32].

Table 1. Substrate-to-water ratio for steam explosion and overall water content of different trails.

Sample	Substrate: Water Ratio Substrate: Fresh Water	Water Content %
Sylvatic silpiha silage	3:2	85.8
Nettle silage	2:1	79.7
Hop grubbing chaff	3:2	84.9
Miscanthus whole plant	3:2	74.5
Miscanthus straw	1:2	75.1
Apple pomace	3:1	85.3
Alfalfa stalks	2:1	65.8
Flax shives	1:2	71.1

Figure 2. Overall flow chart of methodology used.

2.3. Solid–Liquid Separation

Solid–liquid separation was carried out using a DPH2/5 hydraulic tincture press (Doninger, Achern, Germany) with a sample volume of 2 L under a pressure of 10 MPa. The separated solid and liquid phases were measured using a Kern PCB precision scale (Kern & Sohn GmbH, Balingen, Germany). The separation process was performed in triplicate for all tests. Untreated flax shives were soaked in water for 24 h prior to the separation process, to ensure sufficient moisture content for successful separation, as low water content would hinder the separation process.

2.4. Biogas Potential Determination

The biogas potential of the liquid and solid phase was determined by Hohenheim biogas yield test (HBT). The HBT is a batch-test performed in 100 mL syringes. These syringes are closed gastight through a hose clamp and silicone. Each syringe was stored in a motor-driven plate to ensure mixing. This plate was stored in a heating cabinet (Memmert, Schwabach, Germany) at 37 °C for 35 days.

The test was carried out with an organic dry matter (ODM)-inoculum-to-substrate ratio of 2.25 and the whole procedure was executed according to VDI 4630 [33]. The substrate was utilized without undergoing additional crushing. The inoculum was taken from a 400 L laboratory reactor that was fed continuously with a broad spectrum of nutrients with a low organic loading rate, to obtain a low specific methane yield of the inoculum. The inoculum was also fermented alone and its methane yield subtracted from the total methane formation. A positive control was performed with concentrated feed and hay as standard feedstock. Further information on the method can be found in the literature [34–36].

2.5. Analytic Parameters Determined for Solid and Liquid Fractions

DM and ODM were measured according to DIN EN 12880 and DIN EN 12879. A DM/ODM correction (with regard to results of the HBT experiments) was performed for all samples according to Weißbach et al. [37,38]. The organic acids were determined by a gas chromatograph GC 2010plus with an AOC-20i autoinjector (Shmiadzu, Jyoto, Japan). The alcohols were determined by Rio detector, the column BioRad Aminex (Hercules, CA, USA), HPLC column HPX-87H (7.8 × 300 mm; part. Size: 5.0 μm) (Sigma-Aldrich, St. Louis, MO, USA), and BioRad–pre-column HPX-87H (Hercules, CA, USA). The concentration of the volatile fatty acids, acetic acid, propionic acid, iso-butyric acid, n-butyric acid, iso-valeric acid, n-valeric acid, and caproic acid was summed up as one cumulative parameter (SUM-VFA). The pH value was determined using a pH meter Type 211 (HANNA Instruments, Woonsocket, RI, USA). All chemical measurements were carried out at least in duplicate.

Due to the low number of replicates, no statistical analysis was performed.

2.6. Characterization of Solid-Phase Specific Properties

The particle size of the solid fraction after SE was determined in triplicate with Analysette 3 SPARTAN (Fritsch GmbH, Idar-Oberstein, Germany). Eight sieves according to ISO 3310-1 (diameter 200 mm, height 50 mm) were used. The mesh sizes used were 0.063 mm, 0.125 mm, 0.25 mm, 0.5 mm, 1 mm, 2 mm, 4 mm, and 8 mm.

Furthermore, water adsorption capacity of the solid fraction after separation was analyzed according to DIN 53923. The samples were first dried at 105 °C for 24 h. Afterwards, 5 g_{DM} of the samples were distributed on a mesh size 0.063 mm sieve (Retsch, Haan, Germany) and placed in a water bath for 6 s at a depth of 20 mm with a temperature of 20 ± 1 °C. To avoid air bubbles, the sieve was inserted at an angle of around 20°. Finally, the sieve was submerged for 120 s and weighed.

3. Results and Discussion

3.1. Mass Balance

The origin and nature of the feedstocks used in this study varied, which resulted in a significant difference in the DM content of the substrates before solid–liquid separation, ranging between 19.6 and 86.8%, as shown in Table 2. However, the DM content of the solid phases after SE and tincture press showed only minor variations, with a value range between 42.6 and 50.0% for most of the substrates investigated. Only the solid phase of apple pomace exhibited a DM content of 37.1%. Zhao et al. (2022) showed that water holding capacity of apple pomace with SE pretreatment is high due to its stable hydrogen bonds [39]. This could explain the lower DM content observed in the solid fraction of apple pomace. Additionally, apple pomace contains a high amount of dietary fiber, e.g., lignocellulosic compounds, which complicate solid/liquid separation [40].

In contrast to the solid phase, the liquid phase of the different biomasses after SE and tincture press showed a huge variation between DM contents (3.5–10.6%), which was due to both the DM variation of the substrate and the changing water to substrate ratio before SE. The DM contents of liquid phases were in the range of manure, and, from a technical perspective, the liquid phases can be considered pumpable with auxiliaries present at biogas plants [41,42]. Nevertheless, low DM concentration of the liquid phase would require a large reactor volume compared to an energy crop fed biogas plant with identic hydraulic retention time and methane production, which increases investment and operating costs. Therefore, dilution should be as low as possible to ensure the function of SE [43] and reduce the water input. Further research to optimize this parameter would be useful.

Table 2. Distribution of ODM in the respective phases after SE, dry matter (DM) and organic dry matter (ODM) of input substrates prior and solid and liquid fractions after steam explosion with solid–liquid separation.

Substrate	Phase	Distribution of ODM after Phase Separation	DM	ODM
		%	%	$\%_{DM}$
Sylvatic silphia silage	Substrate		23.7 ± 0.1	88.4 ± 0.1
	Solid	86.2	47.1 ± 0.9	92.4 ±0.0
	Liquid	13.8	6.0 ± 0.1	58.9 ± 0.1
Settle silage	Substrate		30.5 ± 0.2	79.2 ± 0.0
	Solid	86.9	46.6 ± 1.2	90.3 ±0.4
	Liquid	13.1	7.4 ± 0.5	74.2 ± 0.5
Hop grubbing chaff	Substrate		25.2 ± 1.1	80.1 ± 5.5
	Solid	82.7	48.0 ± 2.6	89.8 ± 1.3
	Liquid	17.3	10.1 ± 0.5	81.8 ± 0.2
Miscanthus whole plant	Substrate		42.6 ± 0.1	96.3 ± 0.1
	Solid	93.9	42.6 ± 1.6	96.1 ± 0.2
	Liquid	6.1	4.5 ± 0.3	77.6 ± 1.3
Miscanthus straw	Substrate		74.6 ± 0.4	96.1 ± 0.1
	Solid	97.3	47.6 ± 0.0	97.1 ± 0.1
	Liquid	2.7	3.5 ± 0.0	77.7 ± 0.5
Apple pomace	Substrate		19.6 ± 0.1	97.8 ± 0.0
	Solid	92.1	37.1 ± 0.4	99.1 ± 0.1
	Liquid	7.9	7.7 ± 0.0	87.9 ± 1.1
aAfalfa stalk	Substrate		51.3 ± 0.8	92.5 ± 0.7
	Solid	88.8	49.8 ± 1.0	95.0 ± 0.2
	Liquid	11.2	11.0 ± 0.1	76.8 ± 0.4
Flax shives	Substrate		86.8 ± 0.7	84.5 ± 1.9
	Solid	95.3	50.0 ± 2.8	96.7 ± 0.9
	Liquid	4.7	5.4 ± 0.0	90.0 ± 0.4

Regarding the distribution of ODM, the majority remained in the solid phase after SE and separation, with a range of 82.7% to 97.2%. Only a small proportion of ODM migrated into the liquid phase, ranging from 2.7% to 17.3%. (Table 2, Figure 3). This proportion was independent from the added water in the observed range of substrate/water ratios. The bioeconomic concept generates new revenue through the recovery of fibers. However, substantial potential is lost for biogas production or as fertilizer if the products are not returned to the field at the end of the lifecycle. In terms of nutrients, it is advantageous to cycle them back to the fields in the form of digestate. Other compounds such as sand or heavy metals could also be accumulated in the liquid phase [44,45]. This may cause problems such as induced phytogenic and/or genotoxic effects in crops and potential health risks for humans and should be further investigated [44,45].

Comparing the ODM ratio of the liquid phase with the fiber (acid detergent lignin (ADL) and neutral detergent fiber (NDF)) content of the feedstock, it was noticeable that a high content of these substances seemed to be associated with a low ODM ratio of the liquid (Figure 4). High fiber content as well as high ADL and NDF content are expected in straw and other late-harvested materials.

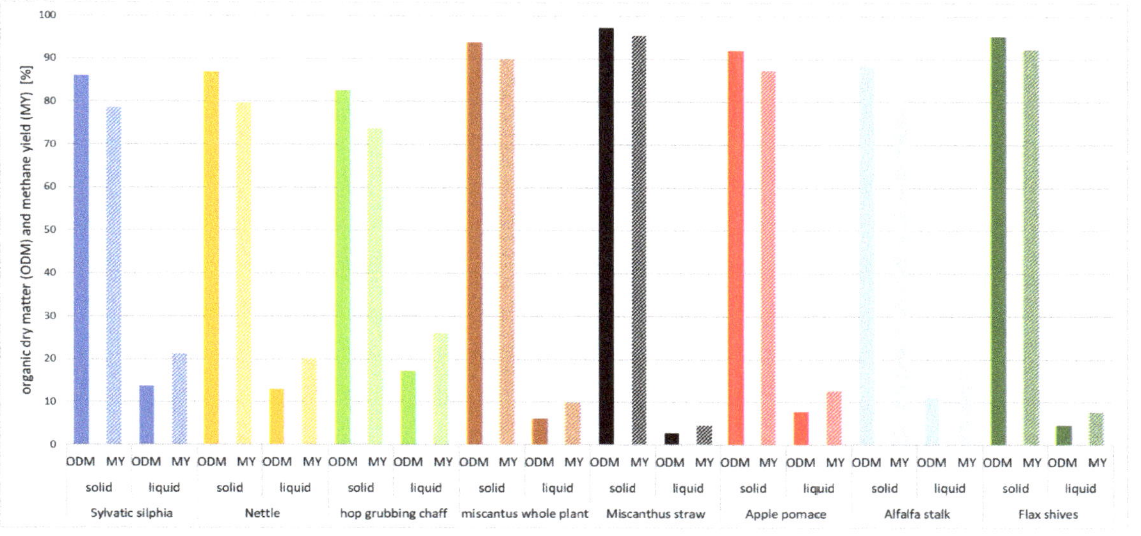

Figure 3. Proportion of organic dry matter (ODM) and methane yield (MY) portion after SE in solid and liquid phase for all investigated substrates.

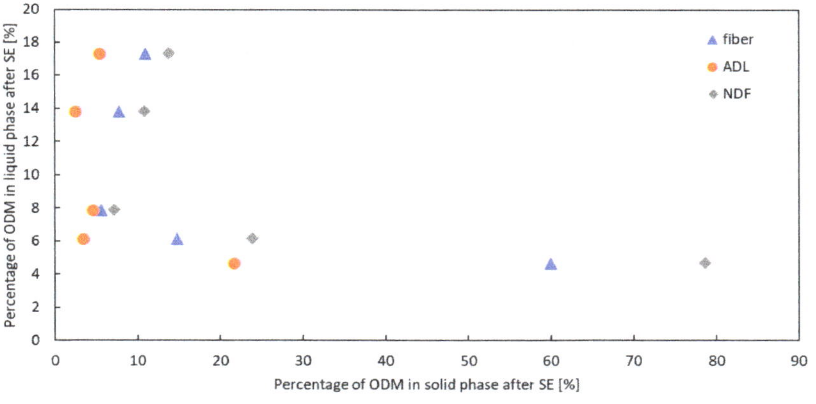

Figure 4. Percentage of organic dry matter (ODM) in liquid phase respective to ODM in the solid fraction (acid detergent lignin (ADL) and neutral detergent fiber (NDF)) after SE.

3.2. Energy Balance

Specific methane yield (SMY) based on ODM and FM were calculated both for solid and liquid phases, as presented in Table 3. The SMY of the solid phases were in the range of 100–287 L_{CH4} $kg_{ODM}{}^{-1}$ and the SMY of the liquid phases were in the range of 297–486 L_{CH4} $kg_{ODM}{}^{-1}$. Based on the results and the kinetics, no inhibition in anaerobic digestion was to be expected.

The SMY of the liquid phases after SE and tincture press were in range or even higher than the SMY of maize silage (355 L $kg_{oTS}{}^{-1}$ [46]). This suggests that there were less anaerobically hard-to-digest or non-digestible components such as fibers in the liquid. The aim of the bioeconomic concept, which is to separate easily degradable cell contents for the biogas process and upcycle poorly degradable fiber constituents, was thus achieved. Furthermore, SE leads to hydrolysis of the substrate, which results in acid generation and, therefore, a reduction in the pH value especially in the liquid phase (Table 4) [17].

High concentrations of SUM-VFAs found in the liquid phase of this study supported this observation (Table 4). Taking into account the mass dependent methane yield, it is observable that the methane yield of the liquid fraction (10–33 L $kg_{FM}{}^{-1}$; Table 3) was much lower than that of maize silage (118 L $kg_{FM}{}^{-1}$) but in the range of cow manure (17 L $kg_{FM}{}^{-1}$) [46]. This suggests an economic use of the liquid phase in a biogas plant, but transport distances should be kept short due to the highwater content. Recirculating the liquid phase instead of adding fresh water could reduce production costs and increase methane yield by enriching the liquid with organics in each circulation step. It is also useful to reduce the required tank volume of the biogas plant through an optimized plant design. Nevertheless, the addition of water is necessary and needs to be adjusted optimally because the moisture content ratio increases the efficiency of SE pretreatment by increasing the mechanical force produced by the expanding gas (water vapor) [31,47].

Table 3. Specific methane yield (SMY) in solid–liquid fractions of the substrates examined in this study. SMY based on organic dry matter (ODM) and fresh matter (FM). Proportion of Methane yield in solid and liquid phase.

Substrate	Phase	SMY	Methane Yield	Proportion of Methane Yield
		L_{CH4} $kg_{ODM}{}^{-1}$	L_{CH4} $kg_{FM}{}^{-1}$	%
Sylvatic silphia silage	Solid	208 ± 6	91 ± 3	78.6
	Liquid	340 ± 10	12 ± 0	21.4
Nettle silage	Solid	217 ± 20	91 ± 8	79.7
	Liquid	334 ± 70	18 ± 4	20.3
Hop grubbing chaff	Solid	179 ± 24	77 ± 10	73.8
	Liquid	336 ± 2	28 ± 0	26.2
Miscanthus whole plant	Solid	280 ± 25	115 ± 10	90.0
	Liquid	391 ± 64	14 ± 2	10.0
Miscanthus straw	Solid	192 ± 10	89 ± 5	95.5
	Liquid	354 ± 4	10 ± 0	4.5
Apple pomace	Solid	287 ± 17	106 ± 6	87.3
	Liquid	486 ± 53	33 ± 4	12.7
Alfalfa stalk	Solid	184 ± 3	85 ± 1	82.3
	Liquid	344 ± 23	28 ± 2	17.7
Flax shives	Solid	100 ± 8	48 ± 4	92.3
	Liquid	297 ± 5	14 ± 0	7.7

As expected, the SMY of the solid phases were lower than that of the liquid phases and, e.g., for sylvatic silphia silage, also lower than the expected methane yield of the substrates according to KTBL (2021) [46]. These results highlight the effective separation of the fiber into the solid phase. Nonetheless, the methane yield of the solid phase (100–287 L $kg_{FM}{}^{-1}$; Table 3) was still high and, for example, in the range of stored solid cow manure (180 L $kg_{oTS}{}^{-1}$ [46]). Based on fresh matter, the SMY of the solid phases were even higher than the literature value of grass silage, a commonly used substrate in German biogas plants (81 L $kg_{FM}{}^{-1}$ [46]). It can be assumed that the solid phase contained other components besides the fibers, which have to be washed out at great expense and are of no benefit to the industry. This assumption is supported by the results of the acid concentrations and the Weender-van-Soest analysis, revealing the presence of SUM-VFA (Table 4), raw protein, and raw fat (1.0–1.9% and 0.5–1.2%) in all samples.

Table 4. pH, volatile fatty acids (SUM-VFA), acetic acid, and butyric acid concentration of solid and liquid phase after SE and separation; standard deviation of all values are <0.0 g kg^{-1}.

Substrate	Phase	pH	SUM-VFA $g\ kg_{FM}{}^{-1}$	Acetic Acid $g\ kg_{FM}{}^{-1}$	Butyric Acid $g\ kg_{FM}{}^{-1}$
Sylvatic silphia silage	solid	5.7	10.6	5.4	0.2
	liquid	5.3	25.0	13.9	0.5
Nettle silage	solid	5.2	7.7	5.7	2.1
	liquid	5.0	16.0	12.2	4.1
Hop grubbing chaff	solid	5.2	16.4	8.6	3.0
	liquid	5.2	29.9	17.4	5.0
Miscanthus whole plant	solid	4.8	1.1	1.0	0.1
	liquid	4.3	4.8	4.8	0.1
Miscanthus straw	solid	6.0	0.4	0.4	0.0
	liquid	5.2	3.2	3.1	0.1
Apple pomace	solid	3.7	1.2	1.1	0.2
	liquid	3.5	1.8	1.6	0.2
Alfalfa stalk	solid	6.0	1.6	1.3	0.3
	liquid	5.0	6.2	5.8	0.4
Flax shives	solid	4.3	3.1	2.9	0.2
	liquid	4.2	7.7	7.4	0.4

The comparison of the methane yield ratio of solid–liquid phase underlines the fact that a high methane yield is not fully exploited, if the solids are not digested (Figure 3). In any case investigated in this study, more than 70% of the methane yield was contained in the solid phase and was, therefore, lost for energy production in the bioeconomy concept. For each kg of substrate, only a maximum of 4.8 $L_{CH4}\ kg_{FM,substrate}{}^{-1}$ was produced, revealing that biogas and energy production were only by-products and the main income must be generated by the fiber utilization. Alternatively, further optimization of the bioeconomy concept may be necessary.

It is important to note that butyric acid has an unpleasant odor for humans and could decrease the quality of the fiber by limiting its potential applications, such as in packaging materials, especially for food, as butyric acid is already detectable at concentrations of 0.06 mg m^{-3}. Even the low concentration of up to 3.0 g $kg_{FM}{}^{-1}$ can, therefore, produce a bad smell, as observed for sylvatic silphia silage, nettle silage, and hop grubbing chaff in this study (Table 4). Based on the data, it was not possible to observe when the butyric acid is generated. Therefore, further research is needed to investigate the inhibition of butyric acid production or the reduction in its concentration especially in the solid phase through optimization of storage, such as in silage, SE parameters, and fiber washing.

The methane yield ratio of the liquid phase was found to be particularly low for flax shives and miscanthus straw, which were dry substrates, with values below 10%, resulting in a maximum production of only 0.7–0.8 L_{CH4} per kg of feedstock. Based on this, it makes sense to harvest miscanthus earlier with higher content of separable cell contents, to get a better separation and to achieve a higher SMY of the liquid phase. According to Tarabanko et al., (2022), flax shives have a similar structure to lignin-rich soft woods [48], in which lignin accounting for up to 25% of the total lignocellulose compound [49,50].

In contrast, sylvatic silphia silage, nettle silage, hop rubbing chaff, and alfalfa stalks (wet substrates) exhibited the best separation of the biogas potential observed in this study, resulting in the highest methane yields in the liquid phases per each kg of feedstock (2–5 $L_{CH4}\ kg_{FM,substrate}{}^{-1}$). This observation is consistent with the finding that these substrates also had the highest SUM-VFA concentrations (Table 4). Despite the possibility of inhibition due to the high SUM-VFA concentration, it was not observed during the experiments because the inoculum used had a high buffer capacity [51–55].

By comparing the combined methane yield per kg of fresh matter of both liquid and solid fractions after SE and separation with the methane yield reported in the literature for the substrates, it was found that the methane yields for sylvatic silphia silage (80 to 59 L_{CH4} kg_{FM}^{-1} [46]) and hop grubbing chaff (69 to 46 L_{CH4} kg_{FM}^{-1} [56]) increased by 36–50% after SE. In a recent study, the utilization of lignin-rich macrophytes by semi-continuous anaerobic digestion with SE pretreatment at a severity factor of 4.4, and, therefore, a bit higher value than used in this study, showed an increase in methane yield of up to 90% [57]. Overall, it can be observed that the proportion of methane yield in the solid fraction was always lower than the proportion of ODM and, therefore, vice versa in the liquid fraction (Figure 3).

In order to achieve a higher methane yield and fiber quality additional pretreatment, separation or anaerobic digestion seems to be reasonable when applying the bioeconomy concept. Furthermore, attention should be given to SE operation conditions, as they were not modified or optimized in this work.

Due to the high SUM-VFA concentration in the liquid fraction, it could be interesting to use this fraction to obtain other value-added products before the energy production [58].

3.3. Characterization of Solid Fraction

The range of water absorption capacities observed for the substrates varied from 216% to 504%. In addition, the calculated average particle sizes for the substrates were found to range from 388 μm to 1857 μm (Table 5).

Table 5. Water absorption capacity and average particle size of investigated substrates.

Substrate	Water Absorption Capacity $\%_{FM}$	Average Particle Size μm
Sylvatic silphia silage	340	1575
Nettle silage	440	1857
Hop grubbing chaff	409	1540
Miscanthus whole plant	401	748
Miscanthus straw	417	756
Apple pomace	216	388
Alfalfa stalk	318	1207
Flax shives	504	520

Both the water capacity and particle size were the lowest for apple pomace. The highest water absorption capacity was observed for flax shives. When the fibers are used in the paper or packaging industry, they are ground by mills to cut the fibers. Low fiber length, as apple pomace fibers in this study, can reduce the additional energy required. Similarly, a low water absorption capacity can reduce the cost of subsequent drying of the paper/packaging material [59]. However, for apple pomace, the dry matter content after separation was the lowest like explained before. The high values of flax shives probably occurred by a multi-porous structure and the high fiber content [60].

The average particle size and water adsorption capacity did not exhibit a dependence. The same behavior was observable on a study on hammer milled palm lignocellulosic by-products [61]. In contrast to the referenced particles (water absorption between 100 and 300%), that were not pretreated with SE, it is observable that the SE pretreated particles showed a higher water absorption captivity with 216–504%. No reason could be found so far.

Particle size was not significantly affected by SE pretreatment for all substrates, as shown in Figure 5 in case of miscanthus (whole plant), because of missing mechanical treatment during the SE pretreatment. According to this, the particle size depends only on the cutting length during harvesting. Particle size should be chosen to optimize storage and SE treatment. The impact on both needs to be considered in future experiments. In the case of SE, several studies suggest large particles for high glucose and xylose concentration. Small particles do not optimize the degradation according to their research [62,63].

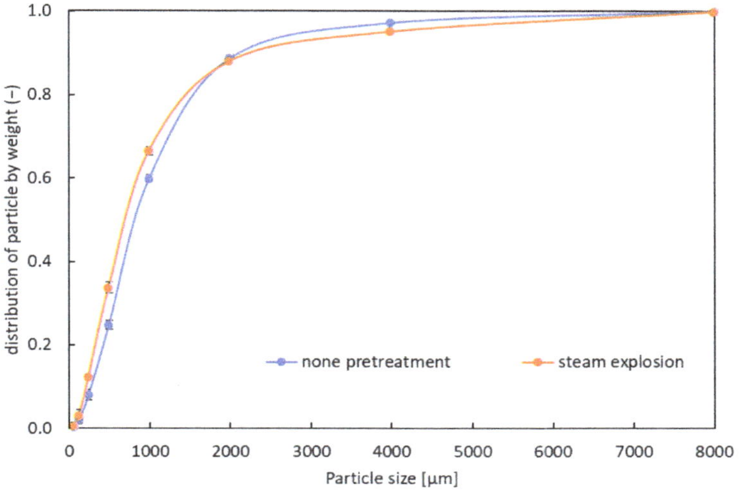

Figure 5. Sum curve of particle size distribution by weight of none pretreated and SE pretreated miscanthus whole plant.

4. Conclusions

In this study, we presented a bioeconomic concept for lignocellulosic feedstock that combined SE, solid–liquid separation, and anaerobic digestion of the liquid fraction. Our results demonstrated that this process effectively separated the easily anaerobically digestible fractions of ODM into a liquid phase while leaving behind fibers with high water adsorption capacity, resulting in a significant separation of SMY. The concept showed particularly high potential for undried feedstocks, with more methane yield being recovered in the liquid phase of undried substrates than for dried ones. However, butyric acid was detected in some of the samples, with levels of up to 3.0 g/kg in the solid phase. This is a major problem for the bioeconomic concept, as it reduces the application range of the fibers. Further investigation into the silage conditions is needed to address this issue. To further develop the bioeconomic concept, it is necessary to optimize the SE operating parameters, the anaerobic digestion process, and search for other biomass types such as biowaste to evaluate the influences on the separation of methane yield and fiber quality, as well as the economic aspects.

Author Contributions: Conceptualization. B.H., M.F. and H.O; methodology. B.H. and H.O.; software. B.H.; validation. B.H., M.B., G.S., K.D., Elviliana and H.O.; formal analysis. B.H. and Elviliana; investigation. B.H.; resources. B.H. and H.O.; data curation. B.H., Elviliana and L.L.; writing—original draft preparation. B.H. and Elviliana; writing—review and editing. B.H., L.L., Elviliana, M.B., G.S., K.D. and H.O; visualization. B.H.; supervision. B.H. and H.O.; project administration. B.H. and H.O.; funding acquisition. B.H. and H.O. All authors have read and agreed to the published version of the manuscript.

Funding: This research was funded by Ministry of Rural Affairs, Food and Consumer Protection Baden-Wuerttemberg, Germany. Project "MLRFaser", grant number 54-8214.07-FP20-106/1.

Institutional Review Board Statement: Not applicable.

Informed Consent Statement: Not applicable.

Data Availability Statement: Not applicable.

Conflicts of Interest: The authors declare no conflict of interest.

References

1. UNDP. Sustainable Development Goals. Available online: https://www.undp.org/sustainable-development-goals (accessed on 21 February 2023).
2. Bilgen, S. Structure and environmental impact of global energy consumption. *Renew. Sustain. Energy Rev.* **2014**, *38*, 890–902. [CrossRef]
3. BMEL. Biogas. Available online: https://www.bmel.de/DE/themen/landwirtschaft/bioeokonomie-nachwachsende-rohstoffe/biogas.html (accessed on 21 February 2023).
4. Theuerl, S.; Herrmann, C.; Heiermann, M.; Grundmann, P.; Landwehr, N.; Kreidenweis, U.; Prochnow, A. The Future Agricultural Biogas Plant in Germany: A Vision. *Energies* **2019**, *12*, 396. [CrossRef]
5. Pohl, M.; Barchmann, T.; Liebetrau, J.; Hülsemann, B.; Oechsner, H.; Zhou, L.; Nägele, H.-J.; Mächtig, T.; Moschner, C.; Kliche, R.; et al. *Biogas-Messprogramm III*; Fachagentur Nachwachsende Rohstoffe: Gülzow, Germany, 2021; ISBN 978-3-942147-42-2.
6. Renewable Energy Source Act (EEG). 2023. Available online: https://www.gesetze-im-internet.de/eeg_2014/ (accessed on 26 May 2023).
7. FNR. Basisdaten Bioenergie Deutschland 2020. 2020. Available online: https://www.fnr.de/fileadmin/Projekte/2020/Mediathek/broschuere_basisdaten_bioenergie_2020_web.pdf (accessed on 9 September 2021).
8. Höller, M.; Lunze, A.; Wever, C.; Deutschle, A.L.; Stücker, A.; Frase, N.; Pestsova, E.; Spiess, A.C.; Westhoff, P.; Pude, R. Meadow hay, Sida hermaphrodita (L.) Rusby and *Silphium perfoliatum* L. as potential non-wood raw materials for the pulp and paper industry. *Ind. Crops Prod.* **2021**, *167*, 113548. [CrossRef]
9. Filippa, F.; Panara, F.; Leonardi, D.; Arcioni, L.; Calderini, O. Life Cycle Assessment Analysis of Alfalfa and Corn for Biogas Production in a Farm Case Study. *Processes* **2020**, *8*, 1285. [CrossRef]
10. Patinvoh, R.J.; Osadolor, O.A.; Chandolias, K.; Sárvári Horváth, I.; Taherzadeh, M.J. Innovative pretreatment strategies for biogas production. *Bioresour. Technol.* **2017**, *224*, 13–24. [CrossRef] [PubMed]
11. Kamm, B.; Kamm, M. Biorefineries--multi product processes. *Adv. Biochem. Eng. Biotechnol.* **2007**, *105*, 175–204. [CrossRef]
12. Sticklen, M.B. Plant genetic engineering for biofuel production: Towards affordable cellulosic ethanol. *Nat. Rev. Genet.* **2008**, *9*, 433–443. [CrossRef] [PubMed]
13. Foston, M.; Ragauskas, A.J. Biomass Characterization: Recent Progress in Understanding Biomass Recalcitrance. *Ind. Biotechnol.* **2012**, *8*, 191–208. [CrossRef]
14. Zhao, X.; Zhang, L.; Liu, D. Biomass recalcitrance. Part II: Fundamentals of different pre-treatments to increase the enzymatic digestibility of lignocellulose. *Biofuels Bioprod. Bioref.* **2012**, *6*, 561–579. [CrossRef]
15. Kabel, M.A.; Bos, G.; Zeevalking, J.; Voragen, A.G.J.; Schols, H.A. Effect of pretreatment severity on xylan solubility and enzymatic breakdown of the remaining cellulose from wheat straw. *Bioresour. Technol.* **2007**, *98*, 2034–2042. [CrossRef]
16. Shaw, M.D.; Karunakaran, C.; Tabil, L.G. Physicochemical characteristics of densified untreated and steam exploded poplar wood and wheat straw grinds. *Biosyst. Eng.* **2009**, *103*, 198–207. [CrossRef]
17. Palmqvist, E.; Hahn-Hägerdal, B. Fermentation of lignocellulosic hydrolysates. II: Inhibitors and mechanisms of inhibition. *Bioresour. Technol.* **2000**, *74*, 25–33. [CrossRef]
18. Garrote, G.; Domínguez, H.; Parajó, J.C. Mild autohydrolysis: An environmentally friendly technology for xylooligosaccharide production from wood. *J. Chem. Technol. Biotechnol.* **1999**, *74*, 1101–1109. [CrossRef]
19. Garrote, G.; Domínguez, H.; Parajó, J.C. Interpretation of deacetylation and hemicellulose hydrolysis during hydrothermal treatments on the basis of the severity factor. *Process Biochem.* **2002**, *37*, 1067–1073. [CrossRef]
20. Bauer, A.; Lizasoain, J.; Theuretzbacher, F.; Agger, J.W.; Rincón, M.; Menardo, S.; Saylor, M.K.; Enguídanos, R.; Nielsen, P.J.; Potthast, A.; et al. Steam explosion pretreatment for enhancing biogas production of late harvested hay. *Bioresour. Technol.* **2014**, *166*, 403–410. [CrossRef]
21. Steinbach, D.; Wüst, D.; Zielonka, S.; Krümpel, J.; Munder, S.; Pagel, M.; Kruse, A. Steam Explosion Conditions Highly Influence the Biogas Yield of Rice Straw. *Molecules* **2019**, *24*, 3492. [CrossRef]
22. Menardo, S.; Bauer, A.; Theuretzbacher, F.; Piringer, G.; Nilsen, P.J.; Balsari, P.; Pavliska, O.; Amon, T. Biogas Production from Steam-Exploded Miscanthus and Utilization of Biogas Energy and CO_2 in Greenhouses. *Bioenerg. Res.* **2013**, *6*, 620–630. [CrossRef]
23. Maximo, Y.I.; Hassegawa, M.; Verkerk, P.J.; Missio, A.L. Forest Bioeconomy in Brazil: Potential Innovative Products from the Forest Sector. *Land* **2022**, *11*, 1297. [CrossRef]
24. Saepoo, T.; Sarak, S.; Mayakun, J.; Eksomtramage, T.; Kaewtatip, K. Thermoplastic starch composite with oil palm mesocarp fiber waste and its application as biodegradable seeding pot. *Carbohydr. Polym.* **2023**, *299*, 120221. [CrossRef]
25. Pålsson, H.; Pettersson, F.; Winslott Hiselius, L. Energy consumption in e-commerce versus conventional trade channels—Insights into packaging, the last mile, unsold products and product returns. *J. Clean. Prod.* **2017**, *164*, 765–778.
26. Kim, R.Y. The Impact of COVID-19 on Consumers: Preparing for Digital Sales. *IEEE Eng. Manag. Rev.* **2020**, *48*, 212–218. [CrossRef]
27. Taylor, K. *The Retail Apocalypse Is Far from over as Analysts Predict 75,000 More Store Closures*; Business Insider: New York, NY, USA, 2019.
28. Kim, Y.; Kang, J.; Chun, H. Is online shopping packaging waste a threat to the environment? *Econ. Lett.* **2022**, *214*, 110398. [CrossRef]

29. Järvinen, J.; Lamberg, J.-A.; Nokelainen, T.; Tikkanen, H. Global Demand for Paper Products: 2006–2050. In *The Evolution of Global Paper Industry 1800–2050*; Lamberg, J.-A., Ojala, J., Peltoniemi, M., Särkkä, T., Eds.; Springer: Dordrecht, The Netherlands, 2012; pp. 307–343. ISBN 978-94-007-5430-0.
30. Jung, J.Y.; Ha, S.Y.; Yang, J.-K. Effect of Water-impregnation on Steam Explosion of Pinus densiflora. *J. Korean Wood Sci. Technol.* **2019**, *47*, 189–199. [CrossRef]
31. Liu, L.-Y.; Qin, J.-C.; Li, K.; Mehmood, M.A.; Liu, C.-G. Impact of moisture content on instant catapult steam explosion pretreatment of sweet potato vine. *Bioresour. Bioprocess.* **2017**, *4*, 675. [CrossRef]
32. Ziegler-Devin, I.; Chrusciel, L.; Brosse, N. Steam Explosion Pretreatment of Lignocellulosic Biomass: A Mini-Review of Theorical and Experimental Approaches. *Front. Chem.* **2021**, *9*, 705358. [CrossRef]
33. VDI-Fachbereich Energietechnik. *Fermentation of Organic Materials—Characterisation of the Substrate, Sampling, Collection of Material Data, Fermentation Tests, VDI 4630*; VDI-Gesellschaft Energie und Umwelt: Düsseldorf, Germany, 2016.
34. Hülsemann, B.; Zhou, L.; Merkle, W.; Hassa, J.; Müller, J.; Oechsner, H. Biomethane Potential Test: Influence of Inoculum and the Digestion System. *Appl. Sci.* **2020**, *10*, 2589. [CrossRef]
35. Mittweg, G.; Oechsner, H.; Hahn, V.; Lemmer, A.; Reinhardt-Hanisch, A. Repeatability of a laboratory batch method to determine the specific biogas and methane yields. *Eng. Life Sci.* **2012**, *12*, 270–278. [CrossRef]
36. Helffrich, D.; Oechsner, H. Hohenheimer Biogasertragstest. *Landtechnik* **2003**, *58*, 148–149. [CrossRef]
37. Weißbach, F. On Assessing the Gas Production Potential of Renewable Primary Products. *Landtechnik* **2008**, *63*, 356–358. [CrossRef]
38. Weißbach, F. Degree of utilization of primary renewable products in biogas production. *Landtechnik* **2009**, *64*, 18–21. [CrossRef]
39. Zhao, Y.; Yu, K.; Tian, X.; Sui, W.; Wu, T.; Wang, S.; Jin, Y.; Zhu, Q.; Meng, J.; Zhang, M. Combined Modification of Soluble Dietary Fibers from Apple Pomace by Steam Explosion and Enzymatic Hydrolysis to Improve its Structural, Physicochemical and Functional Properties. *Waste Biomass Valorization* **2022**, *13*, 4869–4879. [CrossRef]
40. Sobczak, P.; Nadulski, R.; Kobus, Z.; Zawiślak, K. Technology for Apple Pomace Utilization within a Sustainable Development Policy Framework. *Sustainability* **2022**, *14*, 5470. [CrossRef]
41. Liu, Z.; Wang, X. Manure treatment and utilization in production systems. In *Animal Agriculture*; Elsevier: Amsterdam, The Netherlands, 2020; pp. 455–467. ISBN 9780128170526.
42. Irfan Sohail, M.; Arif, M.; Rauf, A.; Rizwan, M.; Ali, S.; Saqib, M.; Zia-ur-Rehman, M. Organic Manures for Cadmium Tolerance and Remediation. In *Cadmium Tolerance in Plants*; Elsevier: Amsterdam, The Netherlands, 2019; pp. 19–67. ISBN 9780128157947.
43. Olugbemide, A.D.; Likozar, B. Assessment of Liquid and Solid Digestates from Anaerobic Digestion of Rice Husk as Potential Biofertilizer and Nutrient Source for Microalgae Cultivation. *Processes* **2022**, *10*, 1007. [CrossRef]
44. Laura Mastellone, M. Exploitation of Digestate in a Fully Integrated Biowaste Treatment Facility: A Case Study. In *Biogas—Recent Advances and Integrated Approaches*; El-Fatah Abomohra, A., Elsayed, M., Qin, Z., Ji, H., Liu, Z., Eds.; IntechOpen: London, UK, 2021; ISBN 978-1-83962-668-5.
45. Kovačić, Đ.; Lončarić, Z.; Jović, J.; Samac, D.; Popović, B.; Tišma, M. Digestate Management and Processing Practices: A Review. *Appl. Sci.* **2022**, *12*, 9216. [CrossRef]
46. KTBL. *Gasausbeute in Landwirtschaftlichen Biogasanlagen: Potenziale, Erträge, Einflussfaktoren*; KTBL: Darmstadt, Germany, 2021; ISBN 978-3-945088-85-2.
47. Wang, W.; Yang, B.; Li, W.; Zhou, Q.; Liu, C.; Zheng, C. Effects of steam explosion pretreatment on the bioactive components and characteristics of rapeseed and rapeseed products. *LWT* **2021**, *143*, 111172. [CrossRef]
48. Tarabanko, V.E.; Vigul, D.O.; Kaygorodov, K.L.; Chelbina, Y.V.; Mazurova, E.V. Catalytic Oxidation of Flax Shives into Vanillin and Pulp. *Catalysts* **2022**, *12*, 1003. [CrossRef]
49. Kazachenko, A.S.; Miroshnikova, A.V.; Tarabanko, V.E.; Skripnikov, A.M.; Malyar, Y.N.; Borovkova, V.S.; Sychev, V.V.; Taran, O.P. Thermal Conversion of Flax Shives in Sub- and Supercritical Ethanol in the Presence of Ru/C Catalyst. *Catalysts* **2021**, *11*, 970. [CrossRef]
50. Chojnacki, J.; Zdanowicz, A.; Ondruška, J.; Šooš, Ľ.; Smuga-Kogut, M. The Influence of Apple, Carrot and Red Beet Pomace Content on the Properties of Pellet from Barley Straw. *Energies* **2021**, *14*, 405. [CrossRef]
51. Xu, Y.; He, Z. Enhanced volatile fatty acids accumulation in anaerobic digestion through arresting methanogenesis by using hydrogen peroxide. *Water Environ. Res.* **2021**, *93*, 2051–2059. [CrossRef]
52. Park, J.-G.; Lee, B.; Jo, S.-Y.; Lee, J.-S.; Jun, H.-B. Control of accumulated volatile fatty acids by recycling nitrified effluent. *J. Environ. Health Sci. Eng.* **2018**, *16*, 19–25. [CrossRef]
53. Photiou, P.; Vyrides, I. Calcined eggshells in anaerobic digestion: Buffering acidification in AD and evaluating end products from phosphate adsorption as soil conditioners. *J. Environ. Chem. Eng.* **2022**, *10*, 107957. [CrossRef]
54. Zhan, J.; Li, Y.; Huang, M.; Zhao, L.; Zou, J.; Tian, D.; He, J.; Lei, Y.; Shen, F. Improvement of anaerobic digestion of food waste by addition of synthesized allophane. *Bioresour. Technol.* **2022**, *361*, 127653. [CrossRef] [PubMed]
55. Sun, H.; Li, R.; Wang, E.; Guo, J.; Zhou, Y.; Dong, R. Coupling biorefinery and biogas production from maize stover by enhancing the ensiling process: Role of the carbon/nitrogen ratio and buffer capacity. *J. Clean. Prod.* **2022**, *339*, 130770. [CrossRef]
56. Hagemann, M.H.; Born, U.; Sprich, E.; Seigner, L.; Oechsner, H.; Hülsemann, B.; Steinbrenner, J.; Winterhagen, P.; Lehmair, E. Degradation of hop latent viroid during anaerobic digestion of infected hop harvest residues. *Eur. J. Plant Pathol.* **2021**, *23*, i19. [CrossRef]

57. Akizuki, S.; Suzuki, H.; Fujiwara, M.; Toda, T. Impacts of steam explosion pretreatment on semi-continuous anaerobic digestion of lignin-rich submerged macrophyte. *J. Clean. Prod.* **2023**, *385*, 135377. [CrossRef]
58. Valentino, F.; Munarin, G.; Biasiolo, M.; Cavinato, C.; Bolzonella, D.; Pavan, P. Enhancing volatile fatty acids (VFA) production from food waste in a two-phases pilot-scale anaerobic digestion process. *J. Environ. Chem. Eng.* **2021**, *9*, 106062. [CrossRef]
59. Fleiter, T.; Fehrenbach, D.; Worrell, E.; Eichhammer, W. Energy efficiency in the German pulp and paper industry—A model-based assessment of saving potentials. *Energy* **2012**, *40*, 84–99. [CrossRef]
60. Zou, D.; Li, K.; Li, W.; Li, H.; Cao, T. Effects of pore structure and water absorption on internal curing efficiency of porous aggregates. *Constr. Build. Mater.* **2018**, *163*, 949–959. [CrossRef]
61. Saadaoui, N.; Rouilly, A.; Fares, K.; Rigal, L. Characterization of date palm lignocellulosic by-products and self-bonded composite materials obtained thereof. *Mater. Des.* **2013**, *50*, 302–308. [CrossRef]
62. Ballesteros, I.; Oliva, J.M.; Negro, M.J.; Manzanares, P.; Ballesteros, M. Enzymic hydrolysis of steam exploded herbaceous agricultural waste (*Brassica carinata*) at different particule sizes. *Process Biochem.* **2002**, *38*, 187–192. [CrossRef]
63. Liu, Z.-H.; Qin, L.; Pang, F.; Jin, M.-J.; Li, B.-Z.; Kang, Y.; Dale, B.E.; Yuan, Y.-J. Effects of biomass particle size on steam explosion pretreatment performance for improving the enzyme digestibility of corn stover. *Ind. Crops Prod.* **2013**, *44*, 176–184. [CrossRef]

applied biosciences

MDPI

Article

Exploring the Synergistic Impacts of Cover Crops and Fertilization on Soil Microbial Metabolic Diversity in Dryland Soybean Production Systems Using Biolog EcoPlates

Durga P. M. Chinthalapudi [1,2], Sapna Pokhrel [3], William L. Kingery [1], Mark W. Shankle [4] and Shankar Ganapathi Shanmugam [1,2,*]

[1] Department of Plant and Soil Sciences, Mississippi State University, Starkville, MS 39762, USA; dc2484@msstate.edu (D.P.M.C.); wkingery@pss.msstate.edu (W.L.K.)

[2] Institute for Genomics, Biocomputing and Biotechnology, Mississippi State University, Starkville, MS 39762, USA

[3] School of Plant and Environmental Sciences, Virgina Tech, Blacksburg, VA 24061, USA; spokhrel20@vt.edu

[4] Pontotoc Ridge-Flatwoods Branch Experiment Station, North Mississippi Research and Extension Center, 8320, Hwy 15 South, Pontotoc, MS 38863, USA; mark.shankle@msstate.edu

* Correspondence: sg383@igbb.msstate.edu; Tel.: +1-(662)-325-0807

Citation: Chinthalapudi, D.P.M.; Pokhrel, S.; Kingery, W.L.; Shankle, M.W.; Ganapathi Shanmugam, S. Exploring the Synergistic Impacts of Cover Crops and Fertilization on Soil Microbial Metabolic Diversity in Dryland Soybean Production Systems Using Biolog EcoPlates. *Appl. Biosci.* **2023**, *2*, 328–346. https://doi.org/10.3390/applbiosci2030022

Academic Editor: Robert Henry

Received: 11 May 2023
Revised: 29 June 2023
Accepted: 30 June 2023
Published: 6 July 2023

Abstract: The metabolic diversity of soil microbiota embodies diverse functional capabilities that support ecosystem resilience, driving essential biogeochemical processes and facilitating the optimization of sustainable agricultural systems. Integrating cover crops into agricultural systems cultivates a diverse array of metabolic activities among soil microbes, synergistically enhancing ecosystem services and bolstering soil health for sustainable and productive farming practices. In an effort to gain deeper insights and expand our knowledge, we conducted a study examining the effects of cover crops and fertilizer sources, thereby shedding light on their combined impacts on the metabolic activity dynamics of soil microbial communities. In this investigation, we employed a split-plot design with two factors: (a) cover crop with three solo cover crop species—Cereal rye (*Secale cereale*), wheat (*Triticum aestivum*), hairy vetch (*Vicia villosa*), and one mixture of mustard (*Brassica rapa*) and cereal rye (*Secale cereale*) (CC-mix), (b) Fertilizer source includes poultry litter, chemical fertilizer, and no-fertilizer treatments. We assessed the metabolic potential of soil microbiota by using carbon substrates utilizing Biolog EcoPlates. The findings revealed that the plots with CC-mix treatment exhibited greater metabolic diversity compared to the other treatments, while among the fertilizer sources, poultry litter demonstrated higher metabolic activity. Furthermore, both treatment factors predominantly metabolized carbohydrates and polymers compared to other carbon substrate categories. The principal component analysis accounted for 46.4% of the variance, collectively represented by PC1 and PC2, emphasizing the substantial contributions of carbohydrates, amino acids, and carboxylic acids to the observed metabolic diversity. Canonical correspondence analysis revealed that pH had positively correlated with microbial functional diversity, whereas total carbon (TC), total nitrogen (TN), and water-stable aggregates (WSA) showed a negative correlation. In conclusion, cover cropping and type of fertilizer source had a notable impact on soil microbial functional diversity, with the cover crop mixture exhibiting a more pronounced influence than the individual cover crop treatments.

Keywords: microbial communities; soil microbiota; fertilization; poultry litter; Shannon diversity; Simpson diversity; principal component analysis; canonical correspondence analysis; cover cropping; living mulch; microbial functional diversity; catabolic diversity

1. Introduction

Soil is a highly diverse, intricate, and dynamic ecosphere that comprises numerous living and non-living elements persistently communicating with each other. Soil health can undergo changes over time as a result of both natural occurrences and human activities. Soil health is improved through effective management and land-use choices that consider the diverse functions of soil [1]. Unsustainable soil management practices worsened soil erosion, a deterioration in soil microbiome activity, and a consequent reduction in nutrient availability for plant uptake [2]. Nature-based solutions (NBS) strategies, such as reduced soil tillage, soil covering with crop residuals or cover crops, soil amendments, and crop rotation to achieve microbial diversity, have been suggested as a substitution for conventional agriculture for improving soil health characteristics [3]. The association between soil biology and agricultural sustainability is paramount, as microbes play a pivotal role in facilitating the decomposition of intricate organic compounds into plant-available forms [4].

Over the past few years, our comprehension of soil microbial communities in dryland ecosystems has broadened, as these organisms play a vital role in maintaining soil health and ecosystem functions [5]. For instance, the soil microbiome mediates various essential ecosystem services, such as nitrogen incorporation [6], organic carbon decomposition, carbon capture [7], bio-remediation of environmental pollutants [3], and resistance to biotic and abiotic stresses [8,9]. Nonetheless, such ecosystem services are compromised by intensive farming systems and tillage practices [10] due to excessive synthetic fertilizers and monoculture cropping systems [10]. This lowers soil microbial functional diversity, which significantly influences crop performance [11]. Expanding diversified cropping systems promotes sustainable agriculture, potentially improving ecosystem functioning by enhancing soil microbial diversity as observed in natural ecosystems [12,13]. The objective is to enhance the functional diversity of soil microbiota by promoting microbial diversity, thereby boosting its capacity to metabolize a broad range of organic substances. Introducing diversity in organic matter residues in agricultural systems, e.g., cover crop mixtures, is anticipated to enhance functional diversity [14,15]. Integration of cover crops and organic amendments alters the abundance, diversity, and metabolic activities of microoganisms present in the agricultural soils, and these altered microbial groups have various effects on their soil ecosystem services [16], which require further exploration.

A large body of literature specifies that, different cover crops with varying crop rotations and periods have diverse impacts and roles on soil health properties and microorganisms [17,18]. The legume cover crops are potent nitrogen fixers (i.e., symbiotic association with *Rhizobium*) and can transform atmospheric nitrogen into available forms (NH_3 and NH_4) [19]. Legumes generate residue with a comparatively higher nitrogen content that is readily degradable and accessible to crops and soil microbes. Grass species with their root system have strong nutrient scavenging capability, specifically for N, and hence have been used to reduce N leaching [20]. Brassicaceous species cover crops have a deep root system, can provide an allelopathic effect to control weeds, and have bio fumigation properties for the suppression of soil pathogens [21,22]. Cover crops are characterized by their rapid growth rate and exhibit a range of C:N ratios, leading to the production of residues that possess a stoichiometry that differs from microbial biomass [23]. Winter cover crops such as cereal rye and hairy vetch are widely utilized due to their ability to withstand low temperatures, incorporate substantial biomass [24], N fixing [25], and stimulate soil biological activity [26]. Three-year research conducted by Buyer et al. (2010) reported that incorporating vetch and rye into crop rotations significantly enhanced soil microbial biomass and metabolic activity [27]. In addition, cover crops, compost, and manure are recognized as vital factors in sustainable farming as they supply vital nutrients to plants, enhance soil structure and quality, and add organic matter [28].

Single-cover crop species cannot deliver all the benefits for soil health [29]. Therefore, multifunctional cover crops, such as combining legumes and grasses or integrating multispecies mixture into crop production systems, can be more advantageous for improving soil health since they can expand the variety of substrates available to the ecosystem and provide a number of ecosystem services [29,30]. The inclusion of plant mixtures with varying C:N ratios can provide diverse substrate options for the soil system, which might expand the niche breadth of the soil microbiota, thus enhancing its metabolic potential [31]. For instance, a combination of oats, radish, and vetch showed significant enhancement in the abundance of gram-positive bacteria, and phosphatase function using phospholipid fatty acid analysis more effectively than a combination of oats and radish [32]. Given that different cover crop species have chemically different root exudates, Housman et al. (2021) reported that higher microbial biomass and improved enzymatic activity in multiple cover crops compared to solo species also differed effects between species [33]. Moreover, elevated soil microbial functional activity was observed in cover crop mixture plots compared to solo treatments [30].

The biological variation of microbial community composition is being studied from three perspectives, viz., species diversity, genetic diversity, and functional diversity [34,35]. The functional diversity of soil microbiome is a component of biodiversity that comprises a broad range of metabolic behaviors that can affect many elements of ecosystem functions, such as ecological stability, nutrient cycling, ecosystem dynamics, and so on [36]. Identifying the functional microbial diversity in soils entails selecting the most appropriate approaches for examining the maximum diversity present in the samples.

Cultivation-dependent approaches require the preparation of cell suspensions, dilution, inoculation on solid media, and phenotypic examination of microbial species. The majority of research on the structure of heterotrophic microbial communities has used isolate-based techniques [37]. The isolation of microbial samples varies among species, and cell type (cell, spore, or mycelium) generates bias. Furthermore, designing nutrient media and appropriate conditions for all members of microbial communities presents a significant challenge, which are major disadvantages of these approaches [36]. Microbial lipid-based techniques without actual culturing of the microorganisms, namely, PLFA and fatty acid methyl ester (FAME), have been extensively used for the examination of microbiome metabolic and compositional changes due to agricultural practices [38]. These techniques, however, unquestionably have drawbacks, including complicated operational protocols, time-consuming assays, poor precision, and inconsistent repeatability [39,40].

Community-Level Physiological Profiling (CLPP) offers an alternative approach, comparing and categorizing microbial communities based on their utilization of carbon substrates. The Biolog® EcoPlates™ (Biolog Inc., Hayward, CA, USA) has been specifically designed to examine the functional diversity of microbial communities. This plate has 31 separate carbon source wells, each with three replications, representing six different types of carbon compound families: carbohydrates, carboxylic acids, amino acids, polymers, amines, and phenolic compounds, with a blank well as a control [41]. The principle behind this assay involves the reduction reaction of water-soluble colorless triphenyl tetrazolium chloride to purple triphenyl formazan [41]. This technique is straightforward to implement, as it does not necessitate the use of isolated cultures and preserves the metabolic traits of microbial communities at optimal levels [42].

Soybean (*Glycine max* L.) is a legume oil seed crop known to fix atmospheric nitrogen through symbiotic interactions with soil bacteria [43]. It has numerous applications, including the production of soy milk, soy sauces, tofu, edible oil, economically recoverable phytohormones, and biodiesel. Traditional soybean monoculture has been coupled with numerous ecological concerns, especially regarding soil health parameters, such as loss of organic matter [44] and available phosphorus [45]. Soil conservation and sustainable agriculture approaches such as cover cropping, crop rotations, and organic fertilizers have widely been recognized as substitutes for soybean monocropping with positive influences on soil organic carbon, aggregate soil stability, nutrient cycling, N fixation, and high-water holding capacity [46–48].

The ecosystem services of different cover crops with fertilizer sources vary, which allows us to evaluate cover crop-mediated soil health promotion processes specific to microbial functional diversity and to correlate soil physicochemical characteristics such as pH, organic carbon, active carbon (POXC), total nitrogen, and glomalin in dryland soybean production systems. Overall, the objective of this experiment was to examine the effect of cover crops and different types of fertilizer sources (organic and inorganic) on microbial community composition, functional diversity, and soil health parameters.

In this experiment, we executed of two-year field study in which cover crop solo and mixture species were integrated into dryland soybean production systems to test the following hypotheses:

- Soil microbial functional diversity and soil health characteristics were positively affected by cover crop mixtures and organic amendments in dryland soybean production systems.
- Cover crop mixture leads to higher microbial functional diversity compared to solo cover crop treatments.

2. Materials and Methods

2.1. Description of the Experimental Sites

The study area was situated at the Pontotoc Ridge-Flatwoods Branch Experimental Station in Pontotoc, MS, USA (34°09′ N, 88°58′ W). The soil type at the research location belongs to the Atwood silt loam series, characterized as a semi-active, mixed soil with fine-silty texture, thermic Typic Paleudalf, situated on a moderate 3% slope. The soil here consists of 13.9% clay, 17.6% sand, and 68.5% silt (Soil Survey, Natural Resources Conservation Service (NRCS)). The investigation was carried out under rainfed and zero tillage conditions, with initial soil characterization at the experimental site revealing 1.57% organic matter and a pH of 6.67. Prior to this study, the experimental site had been planted with corn and soybeans for 2016 and 2017, respectively. Cover crops were established in 2016 and maintained until 2019, while soybeans were cultivated from 2017–2019. The study site has an average January temperature of 5.4 °C, an average July temperature of 23 °C, and an annual average rainfall of 1483 mm [49].

2.2. Experimental Design and Field Methods

The study was arranged in a split-plot arrangement with a randomized block design and three replications each. In this experiment, two factors—cover crops as a primary factor and fertilizer source as a secondary factor were taken into consideration. The experimental design comprised 15 whole plots of 167.2 m^2 each, with 45 sub-plots of 74.3 m^2 designated for the different cover crop and fertilizer source treatments.

Cover crops included cereal rye (*Secale cereale*) (CR), wheat (*Triticum aestivum*) (WT), vetch (*Vicia villosa*) (VT), mustard (*Brassica juncia*) + cereal rye (CC-mix), and native vegetation (NV) (no cover crop with natural weeds) as a control. Sub-factor fertilizer sources contained poultry litter (PL) as an organic amendment, inorganic fertilizer (P, K, and S; CL), and no fertilizer (NO) as a control. Detailed information regarding cover crop planting dates, management practices, fertilizer source chemical composition, and application dosages was described by Pokhrel et al. (2021) [50].

2.3. Soil Sampling and Analysis of Physicochemical Properties

At each plot, soil samples were collected at rhizosphere region 1–15 cm depth after the termination of cover crops. Roughly 500 g of soil was collected into a Ziplock bag and preserved at a −20 °C freezer used for estimation of the metabolic diversity of microbial species. Soil physicochemical characters such as total C, active carbon (POXC), total nitrogen, glomalin (EE-GRSP), water stable aggregate (WSA), and soil pH were estimated from samples that were air-dried overnight, grounded, and passed through a 2.0 mm sieve. Detailed protocols for estimating these soil parameters were described in our previous studies [16,50].

2.4. Biolog™ EcoPlates

The community-level physiological profiling of soil microbiota was analyzed using Biolog™ EcoPlates. Each plate consists of 31 wells with different carbon sources and one control well. All these are replicated three times to control variation in the sample inoculum. The consumption of carbon substrates in wells by microbial communities results in the color development of their respective wells. This color development is quantified spectrophotometrically using a plate reader. Firstly, 10 g of soil was measured and placed in a 250 mL conical flask, then added 100 mL of sterilized NaCl solution (0.85%), later agitated at 150 rpm for 30 min, and then left undisturbed in a refrigerator for 30 min. The clear suspension was transferred to a petri dish, and with the help of a multi-channel pipette, samples were loaded into a Biolog EcoPlate. The inoculated sample plates were placed in the dark during incubation at 25 °C, and absorbance was measured using a BioTek 800 TS (Agilent™, Santa Clara, CA, USA) microplate reader every 24 h for five days.

2.5. Soil Microbial Communities Catabolic Profiling

2.5.1. Quantification of Average Well Color Development in Biolog EcoPlates Wells

The metabolic process of soil microbial populations in individual wells of Biolog EcoPlate leads to the formation of formazan, thus change in color of the tetrazolium dye is induced [42]. The absorbance values of individual wells at 590 nm wavelength represent metabolic activity and are expressed as average well color development (***AWCD***) [37].

$$AWCD = \sum_{i}^{31} \frac{C_i - R}{n} \tag{1}$$

Here, C_i represents the absorbance value (OD) of each well containing a carbon source; R represents the absorbance value of a blank well (without carbon substrate); and n represents the total number of wells with carbon substrates.

2.5.2. Determination of Diversity Indices of Microbial Populations:

Average well-color development estimated from Biolog EcoPlates can be utilized to estimate microbial metabolic diversity indices proposed by Yan (2011) [51]. Absorbance measurements and optical density (OD) at 96 h were employed to estimate the average well-color development and metabolic diversity parameters. This time point was selected as it exhibited an ideal range of absorbance values for the analysis. The following functional diversity parameters were calculated:

(1) *Shannon Diversity Index* (H'):

$$H' = \sum_{i=1}^{31} P_i(\ln P_i) \tag{2}$$

Here, P_i represents the ratio between $(C_i - R)$ and the total absorbance of the total plate wells. ln denotes the natural logarithm with respect to P_i. The Shannon diversity index (H') is used to measure the microbial communities' heterogeneousness based on the concept of uncertainty. Higher uncertainty refers to greater diversity present in communities, and it gives microbial species diversity and evenness within the community.

(2) *Shannon Evenness Index* (E):

$$E = \frac{H'}{\ln S} \tag{3}$$

This method centers on the uniformity of absorbance values at each well for all utilized carbon substrates [41]. In this equation, S represents the total absorbance value of 31 wells. The Shannon evenness index (E) identifies the evenness of microbial-type abundance in the communities.

(3) *Inverse Simpson Diversity Index* ($1/D$):

$$D = \sum \frac{n_i \times (n_i - 1)}{N \times (N - 1)} \tag{4}$$

The Simpson Diversity Index accounts for both the overall number of species and the proportional representation of each individual species [52]. In this formula, n_i denotes the relative absorbance value in each ith well; N is the sum of the absorbance values of 31 wells [23]. The inverse Simpson diversity index ($1/D$) defines the species diversity by estimating the probability of one species encountering another [53].

(4) *McInthosh Index (U):*

$$U = \sqrt{\sum(n_i^2)} \quad (5)$$

The McInthosh index was calculated according to [54]. Where, n_i is corrected absorbance value with blank, i.e., $(C_i - R)$. It measures the microbial communities' homogeneity [55].

(5) *Substrate richness (SR):*

Substrate richness (R) refers to the count of carbon substrates that are utilized by soil microorganisms. It was estimated as the sum of oxidized wells, which had to be at least 0.5 after 96 h incubation [56].

(6) *Statistical analysis:*

Diversity parameters were calculated using Microsoft Excel Version 16.74 (Microsoft Corporation 2023, Redmond, WA, USA). Analysis of variance (ANOVA) and LSD (least significant differences) for two factors was performed for the cover crop and fertilizer source factors using JMP®, Version 17.0. SAS Institute Inc., Cary, NC, 1989–2021. The plots were generated using the 'ggplot2' package Wickham (2016) within the R programming language (R Core Team, 2023) [57] and Sigma Plot, Version 15 (Systat Software Inc., San Jose, CA, USA).

To assess the functional composition of the bacterial community, we conducted a Principal Component Analysis (PCA) on the corrected values of Biolog absorbance. The Biolog absorbance values were adjusted by subtracting the absorbance values from the control well of the Biolog EcoPlate. PCA is a widely used statistical technique that simplifies complex multidimensional data into a smaller set of interpretable variables called principal components [37]. This analysis enabled us to examine whether the measured variables could effectively differentiate the soil management treatments based on their principal components. PERMANOVA is a multivariate statistical method that allows for the analysis of dissimilarities in microbial community composition based on categorical or continuous variables. It helps to determine whether there are significant differences in community structure between different treatments or groups [58]. In addition to PCA, Canonical Correspondence Analysis (CCA) was applied to explore and visualize the relationships between microbial communities and environmental variables. CCA helped us identify the key environmental factors that influenced the composition of microbial communities, providing valuable insights into the ecological processes that shape the dynamics of the microbial community [59].

The PERMANOVA (Permutational Multivariate Analysis of Variance) test, CCA (Canonical Correspondence Analysis), and PCA (Principal Component Analysis) were performed using a vegan package in R programming language [60].

3. Results

3.1. *Differences in AWCD over Time in Soils with Cover Crops and Fertilizer Source Treatment*

Generally, the amount of carbon substrate utilization is directly proportional to the metabolic ability of the respective microbial communities, as determined by *AWCD* [37]. Our results showed that *AWCD* values increased over time, indicating that soil microbiome has a greater ability to utilize carbon substrates when there is a longer interaction with the substrate (Figure 1A,B). The results revealed that the *AWCD* values from 0 to 24 h for all treatments were small, indicating a noticeable lag phase in the first 24 h. After 24 h, the metabolic activity of microbial populations rapidly increased for all treatments, indicaing soil microbes were at their growth phase and capable of utilizing carbon substrates in plates.

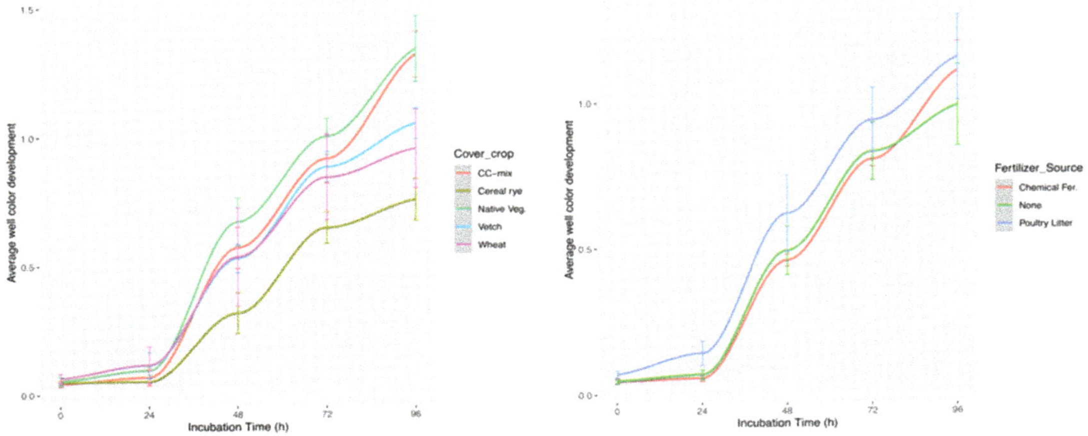

(A) (B)

Figure 1. Changes in average well color development (AWCD) of the soil microbial communities in cover crop (**A**) and fertilizer source (**B**) treatments with incubation time.

Cereal rye treatment consistently exhibited the lowest *AWCD* at all incubation periods. Native vegetation treatment had significantly higher *AWCD* after 24, 48 h, and 72 h than other cover crop treatments followed by CC-mix ($p < 0.05$) (Figure 1A). After 96 h of incubation, both native vegetation and CC-mix had similar *AWCD*, which was higher than the remaining cover crop treatments. In regard to fertilizer source treatments, *AWCD* was highest with PL (Poultry litter) treatment, followed by chemical fertilizer. The fertilizer treatment exhibited the lowest *AWCD* after a 96-h incubation period (Figure 1B). During the entire incubation period, the *AWCD* values for each cover crop followed the following pattern NV > CC-mix > VT > WT > CR. These results suggest that the metabolic activity of soil microbial communities in utilizing carbon substrates differs among cover crops and fertilizer sources.

3.2. Influence of Cover Cropping and Fertilizer Source Treatments on Microbial Metabolic Diversity Indices

Absorbance measured as OD values at 96 h were used for the estimation of functional diversity parameters and average well-color development since these indicated the optimal range of absorbance. The influence of cover crops and fertilization on functional diversity indices such as the Shannon diversity index (H'), Shannon evenness index (E), McIntosh index (U), substrate richness (R), and inverse Simpson index ($1/D$) of soil microbial communities is shown in Table 1.

All the functional diversity indices except $1/D$ demonstrated significant differences ($p < 0.05$). The H', E, R, and $1/D$ indices were most pronounced in the CC-mix treatment, succeeded by native vegetation and vetch treatments, whereas the values observed in the cereal rye treatment were considerably lower. Additionally, the native vegetation treatment demonstrated the highest *AWCD* and U values.

In the poultry litter treatment, the highest values were observed across all functional diversity indices, including *AWCD*, and were closely followed by the chemical fertilizer treatment (Table 1). Notably, significant differences were observed between fertilizer source treatments for diversity indices, with the exception of $1/D$.

Table 1. Main effects of cover crop and fertilizer source on soil microbial community diversity indices.

	AWCD [1]	*H′*	*E*	*U*	*R*	*1/D*
			Cover Crop			
Cereal rye	0.76 (0.04) d	3.14 (0.02) d	0.91 (0.006) d	5.15 (0.27) d	16.55 (1.00) d	67.79 (3.60)
CC-mix [2]	1.32 (0.04) a	3.31 (0.009) a	0.96 (0.002) a	8.03 (0.26) a	25.6 (0.37) a	71.88 (3.70)
Native vegetation	1.34 (0.07) a	3.29 (0.016) a	0.96 (0.004) a	8.25 (0.40) a	25.33 (0.78) a	66.30 (2.72)
Vetch	1.05 (0.03) b	3.24 (0.004) b	0.94 (0.001) b	6.77 (0.19) b	21.44 (0.92) b	84.41 (7.46)
Wheat	0.95 (0.07) c	3.20 (0.01) c	0.93 (0.003) c	6.23 (0.44) c	19.88 (1.67) c	84.00 (8.83)
CC*FT [3]	s*	s*	s*	s*	s*	ns
			Fertilizer Source			
CL	1.11 (0.054) a	3.21 (0.019) b	0.93 (0.005) b	7.13 (0.26) a	22.46 (0.58) a	72.02 (2.12)
None	0.99 (0.07) b	3.22 (0.024) b	0.94 (0.007) b	6.32 (0.39) b	20.46 (1.54) b	77.67 (6.10)
PL	1.15 (0.08) a	3.27 (0.011) a	0.95 (0.003) a	7.21 (0.47) a	22.4 (1.21) a	74.97 (5.53)

[1] Variable in column with no letters is not significant at the 0.05 level using Fisher's protected LSD; Standard error in parenthesis. [2] CC-mix =Cereal rye and mustard; [3] CC*FT = Interaction between cover crop and fertilizer source (ns = non-significant, s* = significant. *AWCD* = Average color development; *H′* = Shannon Diversity Index; *E* = Shannon evenness index; *U* = Mcinthosh Index; *R* = Substrate richness; *1/D* = Inverse Simpson Diversity, CL = Chemical fertilizer, PL = Poultry litter.

The microbial functional diversity indices also exhibited significant variations, which were influenced by the interaction between cover crop and fertilizer source treatments. Notably, the CC-mix + PL treatment demonstrated the highest values for all the diversity indices, followed by CC-mix + CL treatment for *AWCD*, *U*, and *R* indices and NV + NO for *H′* and *E*. The soil samples under CR + NO treatment, on the other hand, displayed lower values for all the diversity indices. These results generally align with the patterns observed for the main effects of cover crop, except for the WT + PL treatment, which displayed higher values compared to the vetch treatment in combination with other fertilizer sources (Table 2).

Table 2. Combined effects of cover crop and fertilizer source on soil microbial community diversity indices.

Treatment [2]	*AWCD* [1]	*H′*	*E*	*U*	*R*
MX +PL	1.57 (0.036) a	3.33 (0.018) a	0.97 (0.0032) a	9.47 (0.28) a	27 (0.83) a
MX + CL	1.47 (0.038) ab	3.32 (0.012) a	0.967 (0.0032) a	8.87 (0.29) ab	26.33 (1.03) ab
NV + NO	1.34 (0.047) bc	3.32 (0.0134) a	0.968 (0.0023) a	8.09 (0.28) bc	26 (0.93) ab
NV + PL	1.32 (0.074) bc	3.31 (0.0324) ab	0.964 (0.0093) ab	8.05 (0.23) bc	26.33 (1.22) ab
WT + PL	1.23 (0.048) cd	3.26 (0.0284) bcd	0.95 (0.0042) bcd	7.76 (0.31) c	25 (0.73) abc
MX + NO	1.17 (0.083) cd	3.30 (0.0173) abc	0.963 (0.0024) abc	7.17 (0.32) cd	24.67 (0.69) abc
NV + CL	1.12 (0.037) de	3.23 (0.0182) d	0.943 (0.0043) d	7.18 (0.35) cd	22.67 (0.73) cde
VT + NO	1.12 (0.028) de	3.24 (0.0284) cd	0.943 (0.0032) cd	7.15 (0.29) cd	23.67 (0.83) bcd
VT + CL	1.10 (0.047) de	3.24 (0.0482) cd	0.944 (0.0024) cd	7.12 (0.23) cd	22.67 (0.93) cde
WT + CL	0.94 (0.058) ef	3.17 (0.0284) ef	0.923 (0.0032) ef	6.26 (0.32) de	21 (1.02) de
VT + PL	0.93 (0.057) ef	3.24 (0.0138) cd	0.944 (0.0023) cd	6.04 (0.29) e	18 (1.10) fg
CR + CL	0.91 (0.075) fg	3.11 (0.0324) fg	0.905 (0.0024) fg	6.22 (0.28) de	20 (1.12) ef
CR + PL	0.73 (0.036) gh	3.23 (0.0231) de	0.940 (0.043) de	4.76 (0.23) f	16.33 (1.11) gh
WT + NO	0.70 (0.037) h	3.19 (0.2842) de	0.931 (0.0024) de	4.69 (0.24) f	13.67 (1.03) hi
CR + NO	0.63 (0.048) h	3.07 (0.0124) g	0.894 (0.0036) g	4.47 (0.25) f	13.33 (1.20) i

[1] Variable in column with no letters are not significant at the 0.05 level using Fisher's protected LSD; Standard Error in parenthesis. [2] Treatment = CR + CL = Cereal rye + Chemical fertilizer; CR + NO = Cereal rye + No fertilizer; CR + PL = Cereal rye + Poultry litter; MX + CL = CC-mix + Chemical fertilizer; MX + NO = CC-mix + No fertilizer; MX + PL = CC-mix + Poultry litter; NV + CL = Native vegetation + Chemical fertilizer; NV + NO = Native vegetation + No fertilizer; NV + PL = Native vegetation + Poultry litter; VT + CL = Vetch + Chemical fertilizer; VT + NO = Vetch + No fertilizer; VT + PL = Vetch + Poultry litter; WT + CL = Wheat + Chemical fertilizer; WT + NO = Wheat + No fertilizer; WT+PL = Wheat + Poultry litter.

3.3. *Classification of Carbon Substrate Utilization Categories in Biolog EcoPlate*

The Biolog EcoPlate comprises 31 carbon substrates, which are classified into six distinct categories: carbohydrates, carboxylic acids, amino acids, polymers, amines, and phenolic acids (Table 3).

Table 3. Carbon substrates present in Biolog EcoPlates related to PC1 and PC2.

Carbon Substrate	PC1	PC2
	Polymers	
Tween 40	0.96179268	−0.03904354
Tween 80	0.34211498	0.03092762
Alpha-Cyclodextrin	−0.83462371	−0.41750008
Glycogen	−0.63170406	−0.23706666
	Carbohydrates	
Glucose-1-Phosphate	−0.25453698	0.58790260
D-L-Alpha-Glycerol Phosphate	0.19979821	−0.74993661
D-Cellubiose	−0.45806275	0.15852001
Alpha-D-Lactose	−0.50881668	0.77378681
Beta-Methyl-D-Glucoside	−0.70499935	0.02685358
D-Xylose	−0.74269362	0.35992147
I-Erythritol	−0.56968692	0.59688337
D-Mannitol	0.85824297	−0.38592539
N-Acetyl-D-Glucosamine	0.58653954	−0.02068943
	Carboxylic acids	
Pyruvic Acid Methyl Ester	0.42378445	−0.26569367
D-Glucosaminic acid	0.52295150	−0.39975288
D-Galactonic Acid-Gamma-Lactone	0.63099447	−0.06814916
D-Galacturonic acid	0.92169910	−0.40273613
Gamma-Amino-Butyric Acid	0.75517974	−0.05258175
Itaconic Acid	−0.23717004	−0.45819712
Beta-Keto Butyric Acid	−0.65735782	−0.34116388
D-Malic Acid	0.93858412	−0.04106865
	Amino acids	
L-Arginine	0.60169520	0.39097528
L-Asparagine	1.00149965	0.01168636
L-Phenylalanine	−0.58543371	−0.02388333
L-Serine	0.08844583	0.36381245
L-Threonine	−0.62958879	−0.42939563
Glycyl-L-Glutamic Acid	−0.78420635	−0.36112807
	Amines	
Phenythyl-Amine	−0.52359729	0.15857889
Putrescine	0.84357073	0.29691745
	Phenolic acids	
2-Hydroxy Benzonic Acid	−0.18237857	−0.16476193
4-Hydroxy Benzonic Acid	0.23406065	0.47218892

The metabolic activity of soil microbial populations on these six different categories of carbon substrates distinctly varies among cover crops and fertilizer source treatments. Figure 2A,B showed that the relative utilization of different carbon group substrates significantly differs with treatments ($p < 0.001$).

In the CC-mix treatment, higher utilization of carbon substrates in the polymer and carboxylic acid groups was observed, while the microbial communities from native vegetation treatment showed the highest consumption of carbohydrates, amino acids, amines, and phenolic compounds. Among fertilizer source treatments, the chemical fertilizer treatment showed greater carbon substrate utilization across all carbon groups except polymers, while the highest consumption of polymer group carbon substrates was evident in the poultry litter treatment. Collectively, both treatments indicated that carbohydrates and polymers were the primary carbon substrate groups harnessed by soil microbial communities residing within their respective treatments.

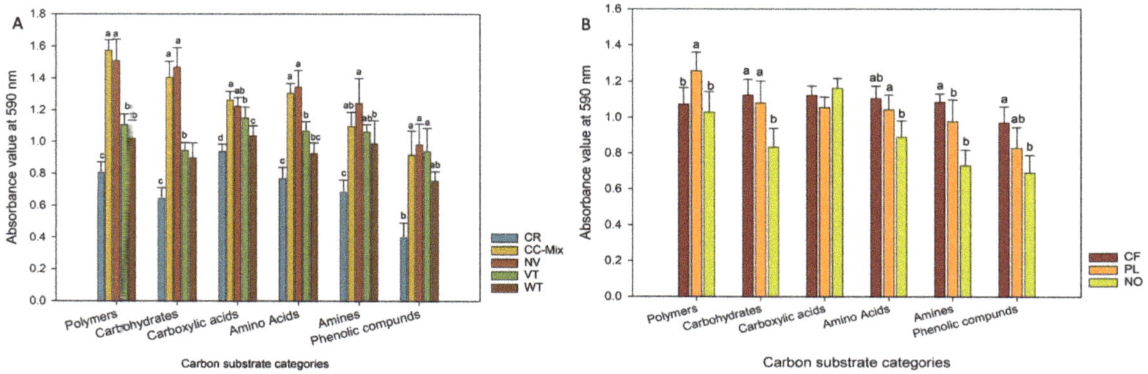

Figure 2. Utilization of carbon substrate categories by soil microbial communities from the cover crop (**A**) and fertilizer source (**B**) treatments. Bars with letters denotes significance at the level 0.05 level using Fisher's protected LSD. (CR = Cereal rye; CC-mix = Cereal rye and mustard; NV = Native vegetation; VT = Vetch; WT = Wheat, CF = Chemical fertilizer; PL = Poultry litter; and NO = No fertilizer).

3.4. Analysis of Carbon Substrate Utilization by Soil Microbial Communities Using Principal Component Analysis (PCA)

To establish the degree of separation of treatments in relation to carbon substrate utilization, each of the cover crop and fertilizer treatments was subjected to principal component analysis and Bray–Curtis ordination. Principal components—PC1 and PC2 scores for 31 carbon substrates are given in Table 3. The higher loading scores represent larger effects of carbon substrate on the principal components. The first and second principal components described 34.4% and 12% variation, respectively. As shown in Figure 3A, there were distinct differences in the utilization of soil microbial communities from different cover crop treatments (PERMANOVA, $p = 0.009$).

The results from the principal component analysis showed that PC1 was positively correlated to carbon sources, including L-Asparagine, Tween 40, D- Malic acid, D- Galacturonic acid, D-Mannitol, Putrescine, Gamma-Amino-Butyric acid, N-Acetyl-D-Glucosamine, L-Arginine, and D-Glucosaminic acid, as evidenced by their high loading scores exceeding 0.5 (Table 3). On the other hand, PC2 showed a positive correlation with Glucose-1-Phosphate, Alpha-D-Lactose, and I-Erythritol. Collectively, these results highlight the significant contribution of carbohydrates, amino acids, and carboxylic acids to the two principal components.

3.5. Association between Soil Microbial Metabolic Diversity and Soil Physicochemical Characteristics

The effects of cover crops and fertilizer source treatments on soil physicochemical characteristics are given in Table 4.

A canonical correspondence analysis (CCA) was carried out on the soil physicochemical properties and carbon substrate groups (Figure 4). The initial two CCA axes accounted for 79.5% and 12.1% of the variability in data, respectively. The cumulative amount of interpretation of the first 2 CCA axes reached 92.06 %, which could reflect the association between microbial community species based on carbon group utilization and soil parameters. The soil pH = 0.7367 indicated a substantial positive correlation with the major axis of the CCA, whereas total carbon TC = −0.28704, total nitrogen TN = −0.31267, and POXC = −0.25727 displayed a negative association.

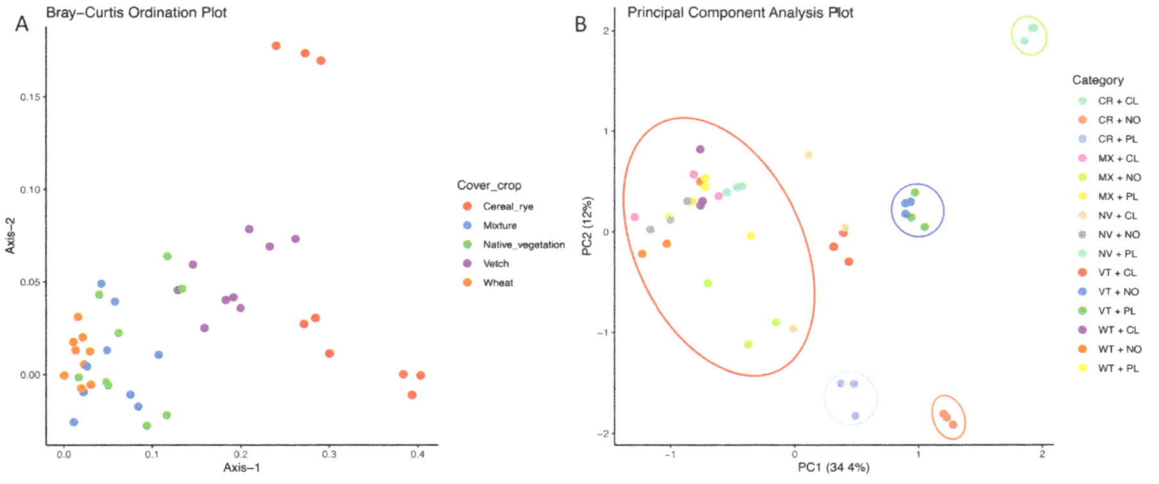

Figure 3. (**A**) Bray Curtis ordination and (**B**) Principal component analysis (PCA) of carbon substrate utilization of soil microbial communities from cover crops and fertilizer source treatments. (Category—CR + CL = Cereal rye + Chemical fertilizer; CR + NO = Cereal rye + No fertilizer; CR + PL = Cereal rye + Poultry litter; MX + CL = CC-mix + Chemical fertilizer; MX + NO = CC-mix + No fertilizer; MX + PL = CC-mix + Poultry litter; NV + CL = Native vegetation + Chemical fertilizer; NV + NO = Native vegetation + No fertilizer; NV + PL = Native vegetation + Poultry litter; VT + CL = Vetch + Chemical fertilizer; VT + NO = Vetch + No fertilizer; VT + PL = Vetch + Poultry litter; WT + CL = Wheat + Chemical fertilizer; WT + NO = Wheat + No fertilizer; WT + PL = Wheat + Poultry litter.).

Table 4. Main effects of cover crops and fertilizer source treatments on soil physicochemical characteristics.

	TC [1] (%)	TN (%)	WSA (%)	EEGSP (mg/kg)	POXC (mg/kg)	pH
			Cover crop			
Cereal rye	1.66 (0.104)	0.160 (0.006) ab	52.11 (2.67)	82.55 (2.86)	550.77 (21.39)	5.64 (0.12) ab
CC-mix [2]	1.70 (0.07)	0.173 (0.009) ab	49.11 (2.38)	89.88 (3.42)	569.33 (28.61)	5.56 (0.08) ab
Native vegetation	1.58 (0.08)	0.156 (0.006) ab	56.77 (2.89)	85.77 (2.43)	521.11 (18.94)	5.75 (0.09) a
Vetch	1.72 (0.10)	0.178 (0.01) a	55 (3.08)	83.11 (3.34)	561 (40.02)	5.48 (0.07) b
Wheat	1.52 (0.07)	0.153 (0.005) b	46.44 (2.63)	79.88 (0.44)	503.22 (1.67)	5.76 (0.06) a
CC*FT [3]	ns	ns	ns	ns	ns	ns
			Fertilizer source			
Mineral	1.58 (0.06)	0.15 (0.005)	49.73 (2.12)	81.2 (2.57)	537.13 (22.53)	5.48 (0.063) b
None	1.60 (0.06)	0.16 (0.006)	52.2 (2.84)	82.4 (2.52)	530 (21.60)	5.766 (0.08) a
Poultry litter	1.72 (0.07)	0.17 (0.007)	53.73 (1.66)	89.13 (2.20)	555.46 (21.34)	5.68 (0.04) a

[1] Variable in column with no letters are not significant at the 0.05 level using Fisher's protected LSD; Standard Error in parenthesis. [2] CC-mix =Cereal rye and mustard; [3] CC*FT = Interaction between cover crop and fertilizer source (ns = non-significant). TC = Total carbon; TN = Total nitrogen; WSA = Water-stable aggregate; EEGSP = Easily extractable soil glomalin; POXC = Permanganate oxidizable carbon.

The CCA also revealed a strong positive correlation between carbon substrate groups and primary axes, such as carbohydrates (0.94), polymers (0.67), and a bit less to carboxylic acids (0.39) and amino acids (0.23).

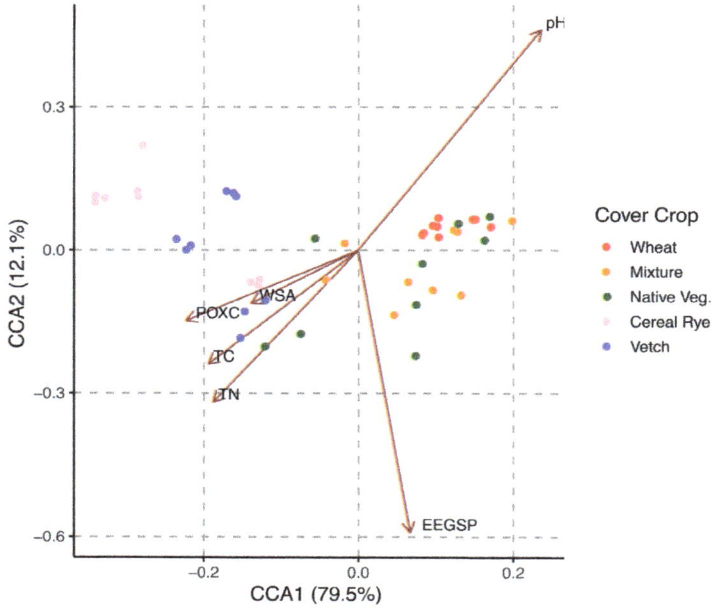

Figure 4. Canonical Correspondence Analysis (CCA) ordination plot revealing the relationship between cover crop treatments and soil physicochemical properties.

4. Discussion

This is one of the few experiments investigating the synergistic effects of cover cropping and fertilizer sources on both the physicochemical properties of the soil and the functional diversity of the microbial community. We hypothesized that the synergistic effects of cover crops and fertilizer sources would cause significant differences in the metabolic diversity and distribution of soil microbial populations. The results showed that different cover crop and fertilizer source treatments exerted different impacts on the metabolic activities of soil microorganisms in soybean fields.

AWCD is proportional to the number and species diversity of soil microbes that can metabolize carbon substrates, and it indicates the carbon source catabolic activity of microbial communities to utilize the carbon substrates [37,61]. In the present investigation, *AWCD* of soil microbial populations from cover crop and fertilizer source treatments exhibited classic microbial growth over time as the duration of incubation increased (Figure 1A,B), and these results were in accordance with other findings [40,53]. In this experiment, during the entire incubation period, native vegetation and cover crop mix (cereal rye + mustard) exhibited higher *AWCD* values than wheat, vetch, and cereal rye. Elevated values of average well-color development (*AWCD*) indicate a higher level of metabolic activity exhibited by soil microbial populations [62]. Additionally, plots with cover crop mix treatment showed high Shannon diversity and substrate richness, whereas native vegetation showed high *U* and *E*. Drost et al. (2020) reported multiple cover crop species results in elevated amounts of substrate niches that increased the functional diversity of soil microbial populations [30].

Incorporating a mix of two or more species in cover cropping can provide additional benefits by promoting both the abundance of beneficial soil microbial populations and the diversity of the soil microbiota [63]. The quality of the residue from cover crops, such as cereal rye, mustard, vetch, and wheat, differed significantly. Additionally, different types of fertilizer sources were used in our two-year study. Despite these variations, our findings showed that the community-level physiological profiling using the Biolog EcoPlates exhibited convergence in diversity indices between cover crop mix and native vegetation treatments.

Studies have shown that mixed cover crops favor the release of certain allelochemicals, primary metabolites, and secondary metabolites significantly alter the soil microbial abundance and their functions [53]. However, cover crop biomass, quality, different soil types, C:N ratio of residues, environmental circumstances, and their interactions significantly influence the abundance and functional diversity of soil microbial populations [64]. Consistently, the cereal rye treatment exhibited the lowest average well-color development (*AWCD*) values, as well as lower diversity index values, which is consistent with previous findings [28]. The *AWCD* of the vetch cover crop was moderate relative to other treatments, as were the indexes of microbial functional diversity. Vetch, being a leguminous cover crop, has a low C: N ratio and decomposes rapidly, leading to the availability of more nitrogen in the soil. This is reflected in the significant differences observed in the soil's physicochemical properties compared to other treatments ($p = 0.0018$, Table 4).

The *AWCD* values of fertilizer source treatments were also significantly different. Plots with poultry litter treatment showed higher *AWCD* than chemical fertilizer and no fertilizer (Table 1); these findings are similar to the other discoveries. Many researchers reported application of organic amendments results in elevated soil microbiological activity as determined by dehydrogenase function, which reflects the complete range of oxidative activity of microbial populations [65,66]. The utilization of organic amendments has been widely recognized for its beneficial impact on the structure and functioning of soil microbial communities [67–70]. A study conducted by Marinari et al. (2000) [67], it was found that the application of organic fertilizers resulted in enhanced soil microbial activity. The researchers attributed this effect to a synergistic interaction between the microbial populations present in the soil and the organic materials used as fertilizers.

The combined effects of fertilizer source and cover crop demonstrated significant differences in functional diversity parameters. Specifically, the soil samples from the CC-mix + poultry litter treatment plot exhibited higher values across all functional diversity indices. This outcome is likely attributed to the synergistic benefits derived from the combination of mixed cover crop species and the application of poultry litter as a fertilizer. Furthermore, it is interesting to note that the soil samples from the wheat + poultry litter treatment plot also displayed higher values compared to the main effects of wheat and even outperformed the main effects of the vetch treatment This suggests that the addition of poultry litter to the wheat plot might have a greater contribution to the microbial communities in that specific plot.

Additionally, our observations revealed that cereal rye, as a cover crop in the main effects analysis, displayed lower values in the functional diversity indices. Interestingly, this pattern persisted when examining the combined effects of cereal rye with different fertilizer sources. The consistently lower values suggest that the influence of cover crops, as a collective factor, exerts a stronger impact on the functional diversity indices compared to individual cover crop effects alone. This finding highlights the significance of considering the combined effects of cover crops and fertilizer sources in shaping microbial functional diversity in the soil.

A total of 31 carbon substrates are classified into six categories, and the metabolism patterns of these six categories were significantly different among treatments ($p < 0.001$). Cover crop mix had a higher metabolism rate of polymers and carboxylic acids, whereas microbial communities under native vegetation treatment showed higher utilization of carbohydrates, amines, and phenolic compounds. Overall, both cover crop and fertilizer source treatments showed that carbohydrates and polymers were mainly used by the microbial populations in the soil. These results were in accordance with Lan et al. (2019) [53].

Nivelle et al. (2016) reported higher metabolism of phenolic compounds and carbohydrates under no-tillage with oats, phacelia, and flax [71]. Many researchers have reported a high degree of carbohydrate utilization by soil microbial communities in the uppermost soil layer (0–10 cm) [28,72], owing to the presence of sugars and organic matter generated by the decomposition process and rhizodeposits. This abundance of carbohydrates in the topsoil may offer greater accessibility to soil microbial communities. The higher utilization of polymers (Glycogen, alpha-cyclodextrin, tween-80, and tween-40) was observed by CC-mix treatment. Generally, these carbon substrates are more stable or recalcitrant; specifically, tween-40 and tween-80 don't represent plant-derived products [73]; furthermore, they are often found in processed organic matter [74]. The microbial communities also exhibited better utilization of amines which includes phenylethylamine and putrescine. Additionally, phenylethylamine was discovered in fermented soybean, whereas putrescine was produced in a variety of soybean plant material, including immature roots [75,76]. These results infer that the major crop also has a role in structuring metabolic patterns of soil microbial populations.

Fertilizer source treatments also exhibited significant differences among the utilization of carbon groups. Poultry litter had higher consumption of polymers, whereas chemical fertilizer had higher utilization of carbohydrates, carboxylic acids, amino acids, amines, and phenolic compounds. Guanghua et al. (2008) reported chemical fertilizer treatment metabolized more polymers, whereas chemical fertilizer + FYM showed higher utilization rates in carboxylic acids, which contradicted our findings [77].

Through CCA analysis, it was determined that total carbon, total nitrogen, and POXC had a negative correlation with carbon substrate categories, indicating that these parameters have an effect on the metabolism of different carbon substrate categories in different cover crops and fertilizer sources. The soil characters' pH and EEGSP positively correlated with the primary axis. Similar results were reported by Lan et al. (2019) [53]; however, they conducted an investigation in forest soils of Eucalyptus trees.

The present investigation revealed varying utilization patterns of the six categories of carbon substrates among the soil microbial communities subjected to different treatments (Figure 2A,B); this may be attributed to the varied structures of soil microbial communities resulting from the different treatments involving cover crop and fertilizer source. However, carbohydrates were found to be the dominant carbon substrate utilized across all treatments, as indicated by the results of principal component analysis (PCA). The PCA results further indicated that carbohydrates made the largest contribution to both PC1 and PC2, followed by amino acids and carboxylic acids. Polymers are also highly metabolized by soil microbial populations in CC-mix treatment; however, PCA didn't reveal in the first two components.

Treatments strongly influenced microbial functional diversity distribution in both ordination plots; the differences in utilization of carbon substrates across cover crop and fertilizer source plots significantly influenced microbial functional diversity distribution (Figure 3A,B). The PCA separation suggests the significant combined effects of cover cropping and fertilizer sources on soil microbial metabolic activity and functional diversity (Figure 3B). The Bray–Curtis ordination plot clearly shows the grouping of wheat, cover crop mix, and native vegetation, whereas vetch and cereal rye formed separate groups. These results infer that grass species cover crops harboring soil microorganisms with similar catabolic profiles, whereas vetch, which is leguminous species distributed separately. The effect of fertilizer sources on vetch and cereal rye is also evident in the PCA ordination plot (Figure 3B). The use of different fertilizer sources with cereal rye resulted in the formation of three separate groups, indicating that the application of poultry litter, chemical fertilizer, or no fertilizer creates distinct microbial communities with unique catabolic profiles. Similarly, vetch with poultry litter and no fertilizer formed a single group, while vetch with chemical fertilizer formed a separate group. These results infer that combined impacts of cover cropping and fertilizer source significantly influence the metabolic activity and community structure of soil microbiota. To further confirm our findings, we conducted PERMANOVA, which revealed that both cover crops and fertilizer sources significantly influenced the metabolic patterns of soil microbial communities ($p = 0.0012$ (PCA), $p = 0.009$ (Bray–Curtis)). Our results are consistent with previous studies by Nair and Ngouajio (2012), Bucher and Lanyon (2005), and Gomez et al. (2005) that investigated the impact of organic manures such as compost, dairy manure, and FYM on the metabolic activity of soil microbiota [28,78,79].

Recent breakthroughs in plant-microbe interaction studies have demonstrated that plants may alter their rhizosphere microbial communities [80,81]. In addition to this, several primary metabolites, including sugars and organic acids, are secreted by soybean into the rhizosphere [82,83]. In addition to soybean root exudates, the root system of cover crops may supply alternate carbon pools that impact soil microorganisms' metabolic activity and community structure in soybean fields.

5. Conclusions

Our study demonstrated that the Biolog EcoPlates approach utilizing community-level physiological profiling methodology was capable of detecting short-term changes resulting from management techniques. Our results showed that organic amendment considerably enhanced microbial functional diversity. We also demonstrated how the implementation of cover cropping impacted the functional diversity and metabolic capacity of the soil microbiota. Soil microbes occur in the active/labile portion of soil organic matter participating in various soil ecological services; thus, soil microbial metabolic activities may respond more rapidly than physicochemical properties of soils to changes in management techniques such as the use of cover crops, organic amendments application, and tillage operations or environmental circumstances [78]. Our results also demonstrated that cover cropping with multispecies can enhance and alter the metabolic diversity of soil microbial communities and the association between cover crop species and microbial communities' metabolic patterns. Our findings suggest that the incorporation of cover crops, especially muti-species, could potentially benefit dryland soybean production through enhanced soil microbial metabolic diversity. However, it is important to note that these results only reflect a portion of the entire microbial communities in the soil to a limited extent; some microbial species don't metabolize carbon substrates in the Biolog EcoPlates, and some might be in a dormant state and couldn't be accounted for the overall metabolic diversity of microbial communities. Novel approaches like soil metagenomics (amplicon sequencing and shotgun metagenomics) should be utilized to further explore how carbon substrate associated with management practices affect the abundance, species diversity, and in-depth metabolic patterns of soil microbial populations.

Author Contributions: Conceptualization, M.W.S.; methodology, S.G.S. and W.L.K.; investigation, M.W.S., S.P., W.L.K. and S.G.S.; writing—original draft, D.P.M.C.; writing—review and editing, S.G.S. and M.W.S.; supervision, M.W.S. and S.G.S.; data curation and visualization, S.G.S. and D.P.M.C.; formatting and submission, S.G.S.; funding acquisition, M.W.S. All authors have read and agreed to the published version of the manuscript.

Funding: The research presented in this study was made possible through funding provided by the Mississippi Soybean Promotion Board (MSPB) under project number 21-2020, which was awarded to Mark W. Shankle.

Institutional Review Board Statement: Not Applicable.

Informed Consent Statement: Not Applicable.

Data Availability Statement: The corresponding author can be contacted for access to the data presented in this study upon request.

Acknowledgments: We extend our sincere gratitude to the dedicated individuals at the Pontotoc Ridge-Flatwoods Branch Experiment Station, located in Pontotoc, MS, USA. Their invaluable assistance in setting up the experimental site and conducting field operations is greatly appreciated. Additionally, we would like to express our deep appreciation to the lab members, colleagues, and friends from the Department of Plant and Soil Sciences and the Institute of Genomics Biocomputing and Biotechnology, Mississippi State University, who provided support and assistance throughout the duration of this study. Their contributions and guidance were instrumental in the successful execution of this research. We are immensely grateful for the collaborative efforts and contributions of all those involved, as their collective expertise and support were vital in the completion of this project.

Conflicts of Interest: The authors declare no conflict of interest.

References

1. Doran, J.W Soil Health and Global Sustainability: Translating Science into Practice. *Agric. Ecosyst. Environ.* **2002**, *88*, 119–127. [CrossRef]
2. Bogužas, V.; Skinulienė, L.; Butkevičienė, L.M.; Steponavičienė, V.; Petrauskas, E.; Maršalkienė, N. The Effect of Monoculture, Crop Rotation Combinations, and Continuous Bare Fallow on Soil CO2 Emissions, Earthworms, and Productivity of Winter Rye after a 50-Year Period. *Plants* **2022**, *11*, 431. [CrossRef] [PubMed]
3. Norris, C.E.; Congreves, K.A. Alternative Management Practices Improve Soil Health Indices in Intensive Vegetable Cropping Systems: A Review. *Front. Environ. Sci.* **2018**, *6*, 50. [CrossRef]
4. Friedel, J.K.; Gabel, D.; Stahr, K. Nitrogen Pools and Turnover in Arable Soils under Different Durations of Organic Farming: II: Source-and-Sink Function of the Soil Microbial Biomass or Competition with Growing Plants? *J. Plant. Nutr. Soil Sci.* **2001**, *164*, 421–429. [CrossRef]
5. Lüneberg, K.; Schneider, D.; Siebe, C.; Daniel, R. Drylands Soil Bacterial Community Is Affected by Land Use Change and Different Irrigation Practices in the Mezquital Valley, Mexico. *Sci. Rep.* **2018**, *8*, 1413. [CrossRef] [PubMed]
6. Mahmud, K.; Makaju, S.; Ibrahim, R.; Missaoui, A. Current Progress in Nitrogen Fixing Plants and Microbiome Research. *Plants* **2020**, *9*, 97. [CrossRef]
7. Six, J.; Frey, S.D.; Thiet, R.K.; Batten, K.M. Bacterial and Fungal Contributions to Carbon Sequestration in Agroecosystems. *Soil Sci. Soc. Am. J.* **2006**, *70*, 555–569. [CrossRef]
8. Enebe, M.C.; Babalola, O.O. The Impact of Microbes in the Orchestration of Plants' Resistance to Biotic Stress: A Disease Management Approach. *Appl. Microbiol. Biotechnol.* **2019**, *103*, 9–25. [CrossRef]
9. Selvakumar, G.; Panneerselvam, P.; Ganeshamurthy, A.N. Bacterial Mediated Alleviation of Abiotic Stress in Crops. In *Bacteria in Agrobiology: Stress Management*; Maheshwari, D.K., Ed.; Springer: Berlin/Heidelberg, Germany, 2012; pp. 205–224, ISBN 978-3-642-23465-1.
10. Dubey, A.; Malla, M.A.; Khan, F.; Chowdhary, K.; Yadav, S.; Kumar, A.; Sharma, S.; Khare, P.K.; Khan, M.L. Soil Microbiome: A Key Player for Conservation of Soil Health under Changing Climate. *Biodivers Conserv.* **2019**, *28*, 2405–2429. [CrossRef]
11. Tsiafouli, M.A.; Thébault, E.; Sgardelis, S.P.; de Ruiter, P.C.; van der Putten, W.H.; Birkhofer, K.; Hemerik, L.; de Vries, F.T.; Bardgett, R.D.; Brady, M.V.; et al. Intensive Agriculture Reduces Soil Biodiversity across Europe. *Glob. Chang. Biol.* **2015**, *21*, 973–985. [CrossRef]
12. Fierer, N.; Strickland, M.S.; Liptzin, D.; Bradford, M.A.; Cleveland, C.C. Global Patterns in Belowground Communities. *Ecol. Lett.* **2009**, *12*, 1238–1249. [CrossRef]
13. Wagg, C.; Bender, S.F.; Widmer, F.; van der Heijden, M.G.A. Soil Biodiversity and Soil Community Composition Determine Ecosystem Multifunctionality. *Proc. Natl. Acad. Sci. USA* **2014**, *111*, 5266–5270. [CrossRef]
14. Baumann, K.; Marschner, P.; Smernik, R.J.; Baldock, J.A. Residue Chemistry and Microbial Community Structure during Decomposition of Eucalypt, Wheat and Vetch Residues. *Soil Biol. Biochem.* **2009**, *41*, 1966–1975. [CrossRef]
15. Hättenschwiler, S.; Tiunov, A.V.; Scheu, S. Biodiversity and Litter Decomposition in Terrestrial Ecosystems. *Annu. Rev. Ecol. Evol. Syst.* **2005**, *36*, 191–218. [CrossRef]
16. Kodadinne Narayana, N.; Kingery, W.L.; Shankle, M.W.; Ganapathi Shanmugam, S. Differential Response of Soil Microbial Diversity and Community Composition Influenced by Cover Crops and Fertilizer Treatments in a Dryland Soybean Production System. *Agronomy* **2022**, *12*, 618. [CrossRef]
17. Fageria, N.K.; Baligar, V.C.; Bailey, B.A. Role of Cover Crops in Improving Soil and Row Crop Productivity. *Commun. Soil Sci. Plant Anal.* **2005**, *36*, 2733–2757. [CrossRef]
18. Balota, E.; Calegari, A.; Nakatani, A.; Coyne, M. Benefits of Winter Cover Crops and No-Tillage for Microbial Parameters in a Brazilian Oxisol: A Long-Term Study. *Agri. Ecosyst. Environ* **2014**, *197*, 31–40. [CrossRef]
19. Somenahally, A.; DuPont, J.I.; Brady, J.; McLawrence, J.; Northup, B.; Gowda, P. Microbial Communities in Soil Profile Are More Responsive to Legacy Effects of Wheat-Cover Crop Rotations than Tillage Systems. *Soil Biol. Biochem.* **2018**, *123*, 126–135. [CrossRef]
20. Kaspar, T.C.; Jaynes, D.B.; Parkin, T.B.; Moorman, T.B.; Singer, J.W. Effectiveness of Oat and Rye Cover Crops in Reducing Nitrate Losses in Drainage Water. *Agric. Water Manag.* **2012**, *110*, 25–33. [CrossRef]
21. Haramoto, E.R.; Gallandt, E.R. Brassica Cover Cropping for Weed Management: A Review. *Renew. Agric. Food Syst.* **2004**, *19*, 187–198. [CrossRef]
22. Kruger, D.H.M.; Fourie, J.C.; Malan, A.P. Cover Crops with Biofumigation Properties for the Suppression of Plant-Parasitic Nematodes: A Review. *S. Afr. J. Enol. Vi.* **2013**, *34*, 287–295. [CrossRef]
23. Xu, W.; Ge, Z.; Poudel, D.R. Application and Optimization of Biolog EcoPlates in Functional Diversity Studies of Soil Microbial Communities. *MATEC Web Conf.* **2015**, *22*, 04015. [CrossRef]
24. Lundquist, E.J.; Jackson, L.E.; Scow, K.M.; Hsu, C. Changes in Microbial Biomass and Community Composition, and Soil Carbon and Nitrogen Pools after Incorporation of Rye into Three California Agricultural Soils. *Soil Biol. Biochem.* **1999**, *31*, 221–236. [CrossRef]
25. Abdul-Baki, A.A.; Teasdale, J.R.; Korcak, R.; Chitwood, D.J.; Huettel, R.N. Fresh-Market Tomato Production in a Low-Input Alternative System Using Cover-Crop Mulch. *HortScience* **1996**, *31*, 65–69. [CrossRef]
26. Carrera, L.M.; Buyer, J.S.; Vinyard, B.; Abdul-Baki, A.A.; Sikora, L.J.; Teasdale, J.R. Effects of Cover Crops, Compost, and Manure Amendments on Soil Microbial Community Structure in Tomato Production Systems. *Appl. Soil Ecol.* **2007**, *37*, 247–255. [CrossRef]

27. Buyer, J.S.; Teasdale, J.R.; Roberts, D.P.; Zasada, I.A.; Maul, J.E. Factors Affecting Soil Microbial Community Structure in Tomato Cropping Systems. *Soil Biol. Biochem.* **2010**, *42*, 831–841. [CrossRef]
28. Nair, A.; Ngouajio, M. Soil Microbial Biomass, Functional Microbial Diversity, and Nematode Community Structure as Affected by Cover Crops and Compost in an Organic Vegetable Production System. *Appl. Soil Ecol.* **2012**, *58*, 45–55. [CrossRef]
29. Chu, M.; Jagadamma, S.; Walker, F.R.; Eash, N.S.; Buschermohle, M.J.; Duncan, L.A. Effect of Multispecies Cover Crop Mixture on Soil Properties and Crop Yield. *Agric. Environ. Lett.* **2017**, *2*, 170030. [CrossRef]
30. Sytske, M.D.; Rutgers, M.; Wouterse, M.; Wietse de Boer, P.; Bodelier, L.E. Decomposition of Mixtures of Cover Crop Residues Increases Microbial Functional Diversity. *Geoderma* **2020**, *361*, 114060. [CrossRef]
31. Redin, M.; Recous, S.; Aita, C.; Dietrich, G.; Skolaude, A.C.; Ludke, W.H.; Schmatz, R.; Giacomini, S.J. How the Chemical Composition and Heterogeneity of Crop Residue Mixtures Decomposing at the Soil Surface Affects C and N Mineralization. *Soil Biol. Biochem.* **2014**, *78*, 65–75. [CrossRef]
32. Chavarría, D.N.; Verdenelli, R.A.; Serri, D.L.; Restovich, S.B.; Andriulo, A.E.; Meriles, J.M.; Vargas-Gil, S. Effect of Cover Crops on Microbial Community Structure and Related Enzyme Activities and Macronutrient Availability. *Eur. J. Soil Biol.* **2016**, *76*, 74–82. [CrossRef]
33. Housman, M.; Tallman, S.; Jones, C.; Miller, P.; Zabinski, C. Soil Biological Response to Multi-Species Cover Crops in the Northern Great Plains. *Agric. Ecosyst. Environ.* **2021**, *313*, 107373. [CrossRef]
34. Bowles, T.M.; Acosta-Martínez, V.; Calderón, F.; Jackson, L.E. Soil Enzyme Activities, Microbial Communities, and Carbon and Nitrogen Availability in Organic Agroecosystems across an Intensively-Managed Agricultural Landscape. *Soil Biol. Biochem.* **2014**, *68*, 252–262. [CrossRef]
35. Bundy, J.G.; Davey, M.P.; Viant, M.R. Environmental Metabolomics: A Critical Review and Future Perspectives. *Metabolomics* **2009**, *5*, 3–21. [CrossRef]
36. Prosser, J.I. Molecular and Functional Diversity in Soil Micro-Organisms. In *Diversity and Integration in Mycorrhizas*; Smith, S.E., Smith, F.A., Eds.; Springer: Dordrecht, The Netherlands, 2002; pp. 9–17, ISBN 978-90-481-5933-8.
37. Garland, J.L.; Mills, A.L. Classification and Characterization of Heterotrophic Microbial Communities on the Basis of Patterns of Community-Level Sole-Carbon-Source Utilization. *Appl. Environ. Microbiol.* **1991**, *57*, 2351–2359. [CrossRef]
38. Miura, T.; Makoto, K.; Niwa, S.; Kaneko, N.; Sakamoto, K. Comparison of Fatty Acid Methyl Ester Methods for Characterization of Microbial Communities in Forest and Arable Soil: Phospholipid Fraction (PLFA) versus Total Ester Linked Fatty Acids (EL-FAME). *Pedobiologia* **2017**, *63*, 14–18. [CrossRef]
39. Kim, J.-W.; Rehmann, L.; Ray, M.B. Development of Microalgal Bioassay Based on the Community Level Physiological Profiling (CLPP). *Algal Res.* **2017**, *25*, 47–53. [CrossRef]
40. Ge, Z.; Du, H.; Gao, Y.; Qiu, W. Analysis on Metabolic Functions of Stored Rice Microbial Communities by BIOLOG ECO Microplates. *Front. Microbiol.* **2018**, *9*, 1375. [CrossRef]
41. Sofo, A.; Ricciuti, P. A Standardized Method for Estimating the Functional Diversity of Soil Bacterial Community by Biolog®EcoPlatesTM Assay—The Case Study of a Sustainable Olive Orchard. *Appl. Sci.* **2019**, *9*, 4035. [CrossRef]
42. Preston-Mafham, J.; Boddy, L.; Randerson, P.F. Analysis of Microbial Community Functional Diversity Using Sole-Carbon-Source Utilisation Profiles—A Critique. *FEMS Microbiol. Ecol.* **2002**, *42*, 1–14. [CrossRef]
43. Gresshoff, P.M.; Hayashi, S.; Biswas, B.; Mirzaei, S.; Indrasumunar, A.; Reid, D.; Samuel, S.; Tollenaere, A.; van Hameren, B.; Hastwell, A.; et al. The Value of Biodiversity in Legume Symbiotic Nitrogen Fixation and Nodulation for Biofuel and Food Production. *J. Plant Physiol.* **2015**, *172*, 128–136. [CrossRef] [PubMed]
44. Crop Rotations and Nitrogen Fertilization to Manage Soil Organic Carbon Dynamics—Studdert—2000—Soil Science Society of America Journal—Wiley Online Library. Available online: https://acsess.onlinelibrary.wiley.com/doi/full/10.2136/sssaj2000.6441496x?casa_token=1K9nk6ndXnMAAAAA%3Ah-zsB8QZf6qSdNskQBnm21ylLKSzeoTDor686VzO4EwZS6cHQX78UbjNq95WtJli_P927aW69lnldo (accessed on 24 November 2022).
45. Fernandez-Gnecco, G.; Smalla, K.; Maccario, L.; Sørensen, S.J.; Barbieri, P.; Consolo, V.F.; Covacevich, F.; Babin, D. Microbial Community Analysis of Soils under Different Soybean Cropping Regimes in the Argentinean South-Eastern Humid Pampas. *FEMS Microb. Ecol.* **2021**, *97*, fiab007. [CrossRef] [PubMed]
46. McDaniel, M.D.; Grandy, A.S. Soil Microbial Biomass and Function Are Altered by 12 Years of Crop Rotation. *SOIL* **2016**, *2*, 583–599. [CrossRef]
47. Venter, Z.S.; Jacobs, K.; Hawkins, H.-J. The Impact of Crop Rotation on Soil Microbial Diversity: A Meta-Analysis. *Pedobiologia* **2016**, *59*, 215–223. [CrossRef]
48. Tamburini, G.; Bommarco, R.; Wanger, T.C.; Kremen, C.; van der Heijden, M.G.A.; Liebman, M.; Hallin, S. Agricultural Diversification Promotes Multiple Ecosystem Services without Compromising Yield. *Sci. Adv.* **2020**, *6*, eaba1715. [CrossRef] [PubMed]
49. IPS—Record of Climatological Observations—Select Month | IPS | National Climatic Data Center (NCDC). Available online: https://www.ncdc.noaa.gov/IPS/coop/coop.html?_page=2&state=MS&foreign=false&stationID=227111&_target3=Next+%3E (accessed on 18 April 2023).
50. Pokhrel, S.; Kingery, W.L.; Cox, M.S.; Shankle, M.W.; Shanmugam, S.G. Impact of Cover Crops and Poultry Litter on Selected Soil Properties and Yield in Dryland Soybean Production. *Agronomy* **2021**, *11*, 119. [CrossRef]
51. Zak, J.C.; Willig, M.R.; Moorhead, D.L.; Wildman, H.G. Functional Diversity of Microbial Communities: A Quantitative Approach. *Soil Biol. Biochem.* **1994**, *26*, 1101–1108. [CrossRef]

52. Simpson's Diversity Index. Available online: https://geographyfieldwork.com/Simpson%27sDiversityIndex.htm (accessed on 25 November 2022).
53. Lan, X.; Du, H.; Peng, W.; Liu, Y.; Fang, Z.; Song, T. Functional Diversity of the Soil Culturable Microbial Community in Eucalyptus Plantations of Different Ages in Guangxi, South China. *Forests* **2019**, *10*, 1083. [CrossRef]
54. Zhen, T.; Fan, W.; Wang, H.; Cao, X.; Xu, X. Monitoring Soil Microorganisms with Community-Level Physiological Profiles Using Biolog EcoPlates^{TM} in Chaohu Lakeside Wetland, East China. *Eurasian Soil Sc.* **2020**, *53*, 1142–1153. [CrossRef]
55. YaNan, T. HongQi, W. Application of biolog to study of environmental microbial function diversity. *Environ. Sci. Technol.* **2011**, *34*, 50–57.
56. Németh, I.; Molnár, S.; Vaszita, E.; Molnár, M. The Biolog EcoPlate^{TM} Technique for Assessing the Effect of Metal Oxide Nanoparticles on Freshwater Microbial Communities. *Nanomaterials* **2021**, *11*, 1777. [CrossRef]
57. Wickham, H. *Ggplot2: Elegant Graphics for Data Analysis*, 2nd ed; Use R! Springer: Cham, Switzerland, 2016; ISBN 978-3-319-24277-4.
58. Anderson, M.J.; Walsh, D.C.I. PERMANOVA, ANOSIM, and the Mantel Test in the Face of Heterogeneous Dispersions: What Null Hypothesis Are You Testing? *Ecol. Monogr.* **2013**, *83*, 557–574. [CrossRef]
59. Ter Braak, C.J.F. The Analysis of Vegetation-Environment Relationships by Canonical Correspondence Analysis. *Vegetatio* **1987**, *69*, 69–77. [CrossRef]
60. Oksanen, J. Vegan: Community Ecology Package. 2010. Available online: http://CRAN.R-project.org/package=vegan (accessed on 6 June 2023).
61. Choi, K.-H.; Dobbs, F.C. Comparison of Two Kinds of Biolog Microplates (GN and ECO) in Their Ability to Distinguish among Aquatic Microbial Communities. *J. Microbiol. Methods* **1999**, *36*, 203–213. [CrossRef]
62. Wang, Y.; Ouyang, Z.; Zheng, H.; Wang, X.; Chen, F.; Zeng, J. Carbon Metabolism of Soil Microbial Communities of Restored Forests in Southern China. *J. Soils Sediments* **2011**, *11*, 789–799. [CrossRef]
63. Wortman, S.E.; Francis, C.A.; Bernards, M.L.; Drijber, R.A.; Lindquist, J.L. Optimizing Cover Crop Benefits with Diverse Mixtures and an Alternative Termination Method. *Agron. J.* **2012**, *104*, 1425–1435. [CrossRef]
64. Garbeva, P.; van Veen, J.A.; van Elsas, J.D. Microbial Diversity in Soil: Selection Microbial Populations by Plant and Soil Type and Implications for Disease Suppressiveness. *Annu. Rev. Phytopathol.* **2004**, *42*, 243–270. [CrossRef]
65. Perucci, P. Enzyme Activity and Microbial Biomass in a Field Soil Amended with Municipal Refuse. *Biol. Fertil. Soils* **1992**, *14*, 54–60. [CrossRef]
66. Marinari, S.; Masciandaro, G.; Ceccanti, B.; Grego, S. Influence of Organic and Mineral Fertilisers on Soil Biological and Physical Properties. *Bioresour. Technol.* **2000**, *72*, 9–17. [CrossRef]
67. Bulluck, L.R.; Brosius, M.; Evanylo, G.K.; Ristaino, J.B. Organic and Synthetic Fertility Amendments Influence Soil Microbial, Physical and Chemical Properties on Organic and Conventional Farms. *Appl. Soil Ecol.* **2002**, *19*, 147–160. [CrossRef]
68. Naidu, R. Influence of Lime, Fertilizer and Manure Applications on Soil Organic Matter Content and Soil Physical Conditions: A Review. *Nutr. Cycl. Agroecosyst.* **1998**, *51*, 123–137.
69. Melero, S.; Porras, J.C.R.; Herencia, J.F.; Madejon, E. Chemical and Biochemical Properties in a Silty Loam Soil under Conventional and Organic Management. *Soil Tillage Res.* **2006**, *90*, 162–170. [CrossRef]
70. Marschner, H. Ion Uptake Mechanisms of Individual Cells and Roots. In *Mineral Nutrition of Higher Plants*; Elsevier: Amsterdam, The Netherlands, 1995; pp. 6–78, ISBN 978-0-12-473542-2.
71. Nivelle, E.; Verzeaux, J.; Habbib, H.; Kuzyakov, Y.; Decocq, G.; Roger, D.; Lacoux, J.; Duclercq, J.; Spicher, F.; Nava-Saucedo, J.-E.; et al. Functional Response of Soil Microbial Communities to Tillage, Cover Crops and Nitrogen Fertilization. *Appl. Soil Ecol.* **2016**, *108*, 147–155. [CrossRef]
72. Wu, Y.; Lyu, T.; Yue, B.; Tonoli, E.; Verderio, E.A.M.; Ma, Y.; Pan, G. Enhancement of Tomato Plant Growth and Productivity in Organic Farming by Agri-Nanotechnology Using Nanobubble Oxygation. *J. Agric. Food Chem.* **2019**, *67*, 10823–10831. [CrossRef] [PubMed]
73. Nunan, N.; Lerch, T.Z.; Pouteau, V.; Mora, P.; Changey, F.; Kätterer, T.; Giusti-Miller, S.; Herrmann, A.M. Metabolising Old Soil Carbon: Simply a Matter of Simple Organic Matter? *Soil Biol. Biochem.* **2015**, *88*, 128–136. [CrossRef]
74. Grandy, A.S.; Neff, J.C. Molecular C Dynamics Downstream: The Biochemical Decomposition Sequence and Its Impact on Soil Organic Matter Structure and Function. *Sci. Total Environ.* **2008**, *404*, 297–307. [CrossRef]
75. Chen, Y.-H.; Liu, X.-W.; Huang, J.-L.; Baloch, S.; Xu, X.; Pei, X.-F. Microbial Diversity and Chemical Analysis of Shuidouchi, Traditional Chinese Fermented Soybean. *Food Res. Int.* **2019**, *116*, 1289–1297. [CrossRef]
76. Eller, M.H.; Warner, A.L.; Knap, H.T. Genomic Organization and Expression Analyses of Putrescine Pathway Genes in Soybean. *Plant Physiol. Biochem.* **2006**, *44*, 49–57. [CrossRef]
77. Guanghua, W.; Junjie, L.; Xiaoning, Q.; Jian, J.; Yang, W.; Xiaobing, L. Effects of Fertilization on Bacterial Community Structure and Function in a Black Soil of Dehui Region Estimated by Biolog and PCR-DGGE Methods. *Acta Ecol. Sin.* **2008**, *28*, 220–226. [CrossRef]
78. Gomez, E.; Ferreras, L.; Toresani, S. Soil Bacterial Functional Diversity as Influenced by Organic Amendment Application. *Bioresour. Technol.* **2006**, *97*, 1484–1489. [CrossRef]
79. Bucher, A.E.; Lanyon, L.E. Evaluating Soil Management with Microbial Community-Level Physiological Profiles. *Appl. Soil Ecol.* **2005**, *29*, 59–71. [CrossRef]
80. Berendsen, R.L.; Pieterse, C.M.J.; Bakker, P.A.H.M. The Rhizosphere Microbiome and Plant Health. *Trends Plant Sci.* **2012**, *17*, 478–486. [CrossRef]

81. Kanasugi, M.; Sarkodee-Addo, E.; Ansong Omari, R.; Mohammad Golam Dastogeer, K.; Fujii, Y.; Oppong Abebrese, S.; Bam, R.; Asuming-Brempong, S.; Okazaki, S. Exploring Rice Root Microbiome; The Variation, Specialization and Interaction of Bacteria and Fungi In Six Tropic Savanna Regions in Ghana. *Sustainability* **2020**, *12*, 5835. [CrossRef]
82. Tawaraya, K.; Horie, R.; Shinano, T.; Wagatsuma, T.; Saito, K.; Oikawa, A. Metabolite Profiling of Soybean Root Exudates under Phosphorus Deficiency. *Soil Sci. Plant Nutr.* **2014**, *60*, 679–694. [CrossRef]
83. Timotiwu, P.B.; Sakurai, N. Identification of Mono-, Oligo-, and Polysaccharides Secreted from Soybean Roots. *J. Plant Res.* **2002**, *115*, 0077–0085. [CrossRef] [PubMed]

applied biosciences

MDPI

Communication

Review of Partial Hybrids between Herbaceous *Medicago sativa* and Woody *Medicago arborea* and Their Potential Role in Alfalfa Improvement

John Irwin [1,*] and Edwin Bingham [2]

[1] School of Agriculture and Food Sciences, University of Queensland, Brisbane QLD 4072, Australia
[2] Agronomy Department, University of Wisconsin, Madison, WI 53706, USA; ebingham@wisc.edu
* Correspondence: j.irwin@uq.edu.au

Simple Summary: Partial hybrids between *Medicago sativa* (alfalfa) and *Medicago arborea*, named Alborea, are the result of reproductive abnormalities in the *M. sativa* seed parent. These hybrids have the potential, through the addition of traits outside the alfalfa gene pool, to increase the productivity and utilisation of alfalfa.

Abstract: *Medicago sativa* (2n = 4x = 32) and *M. arborea* (2n = 4x = 32) were thought to be reproductively isolated until hybrids (Alborea) were produced by sexual reproduction for the first time in 2003 in Wisconsin. The hybrids were asymmetric, at or near 2n = 4x = 32, and with a predominance of the alfalfa genome. Only *M. sativa* seed parents with reproductive abnormalities, including unreduced eggs, have produced hybrids; where *M. arborea* has been used as the seed parent, no hybrids have resulted. Pedigree selection within derivatives of the two original *M. sativa* seed parents (MB and M8) has been successful in increasing the frequency of hybrids produced. While Alborea individuals more closely resemble *M. sativa*, a number of *M. arborea*-specific traits have been observed across different hybrid individuals. These include single-coil flat pods, large seeds, yellow flowers, indeterminate growth, a minimal crown, lodging, frost resistance, and anthracnose resistance. These *M. arborea* traits have the potential to restructure alfalfa to increase its versatility and utilisation. There is emerging evidence from North and South America and Australia that some Alborea selections have the capacity to complement adapted alfalfa cultivars for yield. Work is continuing to introgress *M. arborea* traits of value into alfalfa.

Keywords: partial hybrids; interspecific hybrids; unreduced gametes; reproductive abnormalities; plant breeding

Citation: Irwin, J.; Bingham, E. Review of Partial Hybrids between Herbaceous *Medicago sativa* and Woody *Medicago arborea* and Their Potential Role in Alfalfa Improvement. *Appl. Biosci.* **2023**, 2, 373–383. https://doi.org/10.3390/applbiosci2030024

Academic Editor: Jakub Sawicki

Received: 4 May 2023
Revised: 3 July 2023
Accepted: 7 July 2023
Published: 13 July 2023

1. Introduction

Alfalfa (*Medicago sativa* L.) is the world's oldest known cultivated forage species; historical records date back to 1300 BC in Turkey [1]. Alfalfa is grown in over 80 countries, and the world area is in the range of 30–35 m ha, of which 70% is collectively located in the USA, former USSR, and Argentina [2]. Alfalfa's wide usage is driven by it possessing valuable agronomic traits, including a high feeding value, its perenniality, and its ability to fix atmospheric nitrogen [3]. Also, the deep rootedness of alfalfa makes it suitable for reducing nitrate leaching and preventing the expansion of salt-degraded soils. Cultivated alfalfa is autotetraploid (2n = 4x = 32) [4] as well as being subject to outbreeding and very subject to inbreeding depression. Because of this, alfalfa breeding methodologies leading to commercialisation as synthetic cultivars have remained largely unchanged over several decades [5]. This has contributed to a yield stagnation, which has developed in both North America [6,7] and Australia [8].

In alfalfa breeding, two subspecies have been extensively exploited; the purple-flowered *M. sativa* subspecies sativa and the yellow-flowered *M. sativa* subspecies *falcata*.

Other taxa in the above subspecies complex include *M. sativa* subspecies *glomerata*, *M. sativa* subspecies *varia*, *M. sativa* subspecies *caerulea*, and *M. sativa* subspecies × *hemicycla*. Both 2n = 4x = 32 and 2n = 2x = 16 exist, except for *M. sativa* subspecies × *hemicycla*, subspecies *caerulea*, and subspecies *glutinosa*; all of which are 2n = 2x = 16 [9]. Until now, the above subspecies within the *M. sativa* complex have provided the gene pool for alfalfa cultivar improvement. However, despite the breeding of multiple disease- and pest-resistant cultivars over the last 100 years as well as winter hardiness, alfalfa yields and persistence levels appear to have plateaued.

High levels of genetic diversity exist within the *M. sativa* subspp. Intermating between multiple alfalfa germplasms over the entire subspp. complex has been extensively practised since 1950 in an attempt to ingress multiple agronomically valuable traits [10]. However, there are traits not present in the *M. sativa* subspecies complex that would increase the utilisation of alfalfa; these include a larger seed size, increased drought tolerance, and increased tolerance to salinity [11]. All of these traits and others are known to exist in *Medicago arborea*. In this paper, we review research undertaken to generate hybrids between *M. sativa* and *M. arborea*, with the aim of introgressing traits unique to *M. arborea* such as a larger seed size into the alfalfa gene pool.

2. Phylogenetic Relationships between *M. sativa* subspp. and *M. arborea*

The genus *Medicago* has received extensive revision since it was first comprehensively monographed by Urban [12], where 46 species were recognised. Small and Jomphe [13] provided a comprehensive revision of the genus based on morphology, and defined 83 species spread across 12 sections of the genus. The perennial species *M. arborea* and *M. strasseri* were placed in section *Dendrotelis* and *Medicago sativa* subspp. in section *Medicago*; the distinction is based on section *Dendrotelis* having ligneous branches and section *Medicago* having herbaceous to lignescent branches. Within *Dendrotelis*, three species have been recognized; *M. arborea*, *M. strasseri* (both 2n = 4x = 32), and *M. citrina* (2n = 6x = 48). The main distinction between *M. arborea* and *M. strasseri* was the fruit size, with the fruit and seed diameter twice as large in *M. arborea*.

Molecular Phylogenies

Steele et al. [14] used plastid and nuclear DNA sequences to reassess phylogeny and character evolution in *Medicago*. *Medicago sativa* subspp. and relatives in section *Medicago* formed a weakly to strongly supported monophyletic group, which was referred to as the section *Medicago* clade [14]. *Medicago arborea* and its close relatives (previously placed in section *Dendrotelis*) always resolved within the *Medicago* clade. Steele et al. [14], based on an analysis of all molecular markers employed, concluded that *M. arborea* was part of a group with other species of section *Medicago*, although it often resolved as part of a basal polytomy within that group. They also proposed that the common ancestor of these shrubby polyploid species was a herbaceous perennial in section *Medicago*; woodiness is a derived character state of these species. It was also concluded that hybridisation is difficult to accomplish between species in section *Medicago*.

3. Early Attempts to Hybridise *M. sativa* and *M. arborea*

Several previous attempts have been made to generate hybrids between *M. sativa* and *Medicago* spp. outside the *M. sativa* subspecies complex using sexual reproductive processes [15,16]. In the work of Fridriksson and Bolton [15], pollination of *M. sativa* stigmas from highly self-incompatible clones with a range of annual and perennial *Medicago* spp., including *M. arborea*, failed to produce mature embryos. However, following pollination with *M. arborea*, the early stages of embryonic growth were initiated, indicating fertilisation had taken place. Interspecific hybrids were obtained by crossing 2x *M. sativa* with a range of diploid perennial *Medicago* spp., including *M. papillosa* [17]. Ovule embryo culture was necessary for the recovery of hybrids with balanced genomes of the two species. Only tetraploid hybrids were fertile in *M. sativa* × *M. papillosa*. These showed only disomic segregation patterns, indicating little or no exchange between the homologous genomes [18].

Somatic hybrids were generated between *M. sativa* and *M. arborea* through the symmetrical fusion of mesophyll protoplasts of *M. sativa* with callus protoplasts of *M. arborea* [11]. While considerable genomic rearrangements were evident in the resulting hybrids, their morphology was generally intermediate to the parents, and about half of the species-specific RFLP bands of both parents was present in them. These somatic hybrids did not flower during the first 2 years after generation, even though several floral buds were observed. The authors noted that flowering may only have been delayed because *M. arborea* takes about 2 years to flower after seeding and the paper was written only 2 years after generating the hybrids. Mizukami et al. [19] used the electrofusion of protoplasts of the annual species *M. rugosa* and *M. scutellata* with 4x *M. sativa* to produce somatic hybrids. These hybrids were unstable and chromosomes were rapidly lost during vegetative growth.

4. Generation of Alborea and Its Characteristics

Alfalfa and *M. arborea* were considered to be reproductively isolated until relatively recently. McCoy and Echt [20] stated that it was possible to obtain hybrids between alfalfa and all other species of the subgenus *Medicago*, with the exception of *M. arborea*. The paper by Nenz et al. [11], where sterile somatic hybrids were reported between *M. sativa* and *M. arborea*, rekindled interest in hybrids between the two species in Wisconsin, and screening of male sterile alfalfa seed parents commenced in 1998.

Over the period 1998–2002 in Wisconsin, 5 alfalfa seed parents yielded no hybrids after pollinating at least 200 florets per plant per year. In 2003, the first hybrids were produced after pollinating clone MB (obtained from Magnum 111 X Blaser XL). Over the period 2002–2013, MB produced a hybrid for every 250 florets pollinated. A progeny of MB, WA 2071, yielded a hybrid at the rate of one for every 420 florets pollinated [21]. MB and its derivatives represented the first level of weakening the hybridisation barrier between *M. sativa* and *M. arborea*. The hybrids were described as a new cultigen, Alborea [22]. Another alfalfa genotype, M8, was discovered in Wisconsin, which was a more efficient hybridiser with *M. arborea* than MB and its derivatives [21]. M8 was derived from crossing *M. sativa* subspecies *sativa*, *caerulea*, and *falcata* in a three-way x three-way cross. M8 produced a hybrid for every 85 florets pollinated [21]. An S1 plant of M8 tested in Queensland, WA2570, yielded a hybrid for every 38 florets pollinated [23]. Both MB and M8 produce small amounts of pollen; both have been shown to possess the reproductive abnormalities of unreduced eggs and pollen [23].

The reciprocal cross *M. arborea* × *M. sativa* has failed to produce hybrids despite pollinating hundreds of *M. arborea* florets over 5- and 4-year periods in Wisconsin and Queensland, respectively [21]. This indicates that in the *M. arborea* seed parents, the hybridisation barrier is complete as it appears to be in several alfalfa plants from commercial cultivars pollinated with *M. arborea*.

4.1. Morphological Characterization

Morphologically, the hybrids exhibit traits of both parents (Table 1 and Figure 1), with the greatest resemblance being to the *M. sativa* parent. Either purple- or cream-flowered *M. sativa* have always been used as the seed parent. The hybrids can be identified by variegated flowers, with the yellow pigment coming from *M. arborea*. Another diagnostic character of hybrids is a 1–1.5-coil flat seed pod; alfalfa parents had 3-coil round pods and *M. arborea* had a 1–1.5-coil flat pod. Other *M. arborea* traits found in some hybrids include winter activity, no crown, a branched root system, an erect growth habit, and large dark-green leaves [24]. *M. arborea* has seeds four times the size of alfalfa, and some hybrids exhibit seeds approaching the size of *M. arborea*. Complementation for forage yield has also been observed in other Alborea × alfalfa crosses [24–26]. Anthracnose (*Colletotrichum trifolii*) resistance has been transferred from *M. arborea* to Alborea [27] and used in alfalfa breeding.

Figure 1. (**a**) Yellow/orange flowers of *M. arborea*; (**b**) co-expression of purple and yellow pigments in Alborea; (**c**) purple flower of *M. sativa* cv Sequel; (**d**) 1–1.5-coil immature flat pods of *M. arborea* ca. 1 cm in diameter; (**e**) immature 3-coil pods of *M. sativa* cv Sequel with tight coiling; (**f**) mature pods and seeds, generated in Wisconsin, of (left to right) alfalfa, Alborea, and *M. arborea*. Reproduced with permission from Ref. [24]. Copyright 2016 Australian Institute of Agricultural Science and Technology.

Table 1. *Medicago arborea* traits and other traits observed in Alborea in Wisconsin and Queensland.

Trait	Observations
	Observed in greenhouses and fields
Flower colour	Yellow flowers per se and variegated flowers
Indeterminate growth	Plants grow up to 4 m in height
Minimal crown	Observed in ca. 20% of plants
Large leaves	Leaves larger than in both parents in some plants
M. arborea pod shape and size	1- to 1.5-coil flat pods ca. 1 cm in diameter observed in ca. 20% of plants
Large seeds	Seeds twice the size of alfalfa and half the size of *M. arborea*
Short racemes	5–6 florets versus 15–25 in alfalfa
Fewer seeds per pod	0–50% of alfalfa, although 8–9 per pod observed in one plant
Pollen quantity	0–25% of alfalfa
Autogamy	Full seed set not observed; 10–25% of cross-fertility level
Crossability	Low frequency of Alborea plants that do not cross well with alfalfa but cross with Alborea
	Observed in fields
Lodging resistance	About 25% of Alborea plants resist lodging after rain and wind
Frost resistance	Low frequency of Alborea plants (ca. 5–10%) stay green down to −8 °C
Solid stem	About 5% of Alborea plants have a solid stem above the base like *M. arborea*; alfalfa has hollow stems
Heterosis in crosses	About 25% have heterosis for vigour in crosses with alfalfa
Branching roots	Some plants show absence of a tap root like *M. arborea*
Winter activity	Around 65% of plants with a group 9 dormancy level
Late flowering	May take some plants 2 years to flower, as for *M. arborea*

4.2. Genomic Characterisation

The root-tip chromosome number has been determined for Alborea and was found to be near tetraploid (2n = 4x = 30–32) [21,27]. An AFLP analysis was conducted on five hybrids and parents in Australia [27]. An MB derivative, WA2071, was used as the seed parent. The work confirmed the asymmetric genomic composition of the hybrids, with up to only 4% of the *M. arborea*-specific AFLP bands present in any hybrid. Of the bands, 27% was monomorphic in both the parents and the hybrid, so more than 4% of the *M. arborea* genome was possibly transferred. In total, 7% of the total number of bands was unique to the hybrid, indicating possible chromosomal rearrangements due to the introgression of fragments of *M. arborea* chromosome/s contributing to new alleles at the sites of introgression [27].

The presence of around 4% of *M. arborea*-specific alleles in the hybrid could indicate the transfer of a whole chromosome or the introgression of several smaller parts as, theoretically, each *M. arborea* chromosome constitutes an average of 1/32, or 3.1%, of the genome.

M. arborea introgression into 7 hybrids generated in Wisconsin [28] was investigated using the marker profiles generated by 46 SSR primers of known *M. sativa* genomic locations covering all 8 linkage groups [29–31]. Introgression of *M. arborea*-specific alleles was found from linkage groups 1, 6, and 7, with two hybrids showing introgression from all three linkage groups [32]. Test crosses were made between the hybrids and *M. sativa*, and the transmission of *M. arborea*-specific alleles was demonstrated in the test cross-progeny. Whether the introgressed *M. arborea* genome represents a chromosome, an arm, or several small pieces of chromosome/s remains unresolved, with the latter most likely.

5. Similarities between Alborea and Other Asymmetric Plant Hybrids

De Wet et al. [33] reported what they described as "counterfeit" hybrids between *Tripsacum austral* (2n = 2x = 36) and *Zea mays* (2n = 2x = 20). Hybrids were characterised by 18 or 36 *Tripsacum* and 10 or rarely 20 *Zea* chromosomes, with those containing 36 *Tripsacum* and 20 *Zea* chromosomes most likely deriving from the fertilisation of unreduced eggs and pollen. However, some hybrids resembled true hybrids in phenotype, but lacked *Zea* chromosomes and were characterised by 36 *Tripsacum* chromosomes. These individuals were named counterfeit hybrids. To explain the origin of counterfeit hybrids, it was suggested that genes had been transferred from the genome of *Zea* sperm to the genome of an unreduced egg during the fertilisation process. The situation described above is similar in many ways to what was observed with F1 Alborea hybrids. They are near tetraploid, phenotypically more closely resemble the alfalfa egg parent, but demonstrate some *M. arborea* phenotypic traits and only contain around 5% of *M. arborea*-specific alleles [27]. Unreduced eggs and pollen have been reported in Alborea parents [23,34], which were also produced in *Tripsacum* [33].

The example described above was referred to as introgressive hybridisation by Allard [35]. Here, one species is enriched to a small degree with genes derived from another, thus broadening its base of variability and increasing the variety of recombination products that may be secured from it. Allard used as an example the introgression of *Tripsacum* genes into corn by repeated natural backcrossing. As noted by Allard [35], the issue was highly speculative at that time, but there is now unequivocal evidence for introgressive hybridization [36].

Li and Heneen [37] generated intergeneric hybrids between diploid *Brassica campestris* (2n = 20) and *Orychophragmus violaceus* (2n = 24). The hybrid was mixoploid (2n = 23–42) and cells 2n = 34 were frequent. The partial separation of parental genomes during mitosis, leading to the addition of *O. violaceus* chromosomes to the *B. campestris* complement, was proposed to explain the partial hybrids.

Tu et al. [38] studied intertribal hybrids between *Brassica rapa* (2n = 2x = 20) and *Isatis indigotica* (2n = 2x = 14). One hybrid closely resembled the seed parent, *B. rapa*, and was 2n = 22. Genomic in situ hybridisation (GISH) of this F1 individual revealed two chromosomes from *I. indigotica*; the remainder were from *B. rapa*. An AFLP analysis confirmed *I. indigotica*-specific bands in the hybrid. Some of the introgressions were mitotically stable. While Alborea has not been subjected to GISH analysis, the process of formation of the partial hybrids appears to be similar to what is described above.

6. Importance of Reproductive Abnormalities in Generating Interspecific Hybrids

In the generation of Alborea, it is when only certain alfalfa genotypes are used as the seed parent that hybrids result. Repeated pollination of *M. arborea* with *M. sativa* has failed to produce hybrids [21]. It has been established that alfalfa seed parents and their derivatives that produce hybrid also produce unreduced male and female gametes [23,34]. Alfalfa parent MB had a lower frequency of 2n eggs than the other parent M8; MB is a less efficient hybridiser than M8 [34]. The production of 2n eggs, arising from second division restitution [39] by the MB or M8 parents, would appear to be integral to hybrid generation. Fertilisation of the 2n egg by a 2x gamete of *M. arborea*, followed by the subsequent loss of some *M. arborea* chromosomes due to differences in the timing of mitotic events between the two species, provides an explanation for the near 4x hybrids with a predominance of alfalfa DNA. Unreduced eggs are produced by M8 and MB at a frequency similar to that with which they produce hybrids [34]. Unreduced gamete formation in plants and their use in breeding was reviewed by Brownfield and Kohler [40]. Cheng et al. [41] reported that unreduced female gametes may have been involved in the generation of partial hybrids between *B. napus* (2n = 38) and *Orychophragmus violaceous* (2n = 24). The suspected hybrids were 2n = 38 and contained chromosome fragments of *O. violaceous*. Asynchronous mitotic cell cycles were thought to be responsible for the loss of *O. violaceous* chromosomes after fertilisation.

In the production of counterfeit hybrids between *Tripsacum* and *Zea*, it was reported that the parthenogenetic development of non-reduced female gametes was involved in all cases [33] and that *Zea* sperm had successfully reached the embryo sac at the time of normal fertilisation. The non-random transfer of *Zea* genes or the non-random incorporation of selected *Zea* DNA fragments into specific positions of the *Tripsacum* genome were suggested. For such events to occur, reproductive abnormalities in the egg parent would seem to be essential. Seed parents of F1 Alborea show much-reduced pollen and produce 2n pollen [21]. The genetic control of the absence of cytokinesis after restitutional meiosis in 2x alfalfa eggs was studied by Baraccia et al. [42] and five genes were proposed to control the production of an unreduced gamete among members of the *M. sativa* subspp. complex. Chromosome doubling by unreduced gametes has undoubtedly played an important role in the evolution of autoploidy in alfalfa, but its role in facilitating hybrid embryo development between different species in the genus *Medicago* has been undetermined until the work with Alborea [23,34].

Another cultivated autotetraploid with a large number of related wild and cultivated species at different ploidy levels is the potato, *Solanum tuberosum* (2n = 4x = 48). Sexual polyploidisation through unreduced gametes has been a significant factor in its evolution [43]. Boyes and Thompson [44] and Valentine [45] established that as well as 2n gametes, the need for a balance of chromosome sets between maternal tissue, the embryo, and endosperm was required for successful interploidy crosses. It would appear that the endosperm of M8 and MB Alborea hybrids are able to tolerate an unbalanced genome dosage. The Endosperm Balance Number (EBN) hypothesis, developed in the early 1980s and reviewed by Carputo et al. [46], postulates that each species has a genome-specific ploidy, the EBN, which must be in a 2:1 maternal-to-paternal ratio in the hybrid endosperm for normal endosperm development. With unreduced 4x eggs in M8 and MB and normal reduced 2x *M. arborea* pollen, it could be anticipated that the Alborea endosperm could be 10x (4:1 EBN ratio) compared with 6x (2:1 EBN ratio) from a normal 4x × 4x cross. These EBN values are only theoretical; the actual EBN of the Alborea hybrids remains undetermined. However, the postulated higher maternal-to-paternal genome ratio may have facilitated endosperm development, which is a necessary precursor to embryo development.

7. Use of Alborea in Alfalfa Breeding

Nine distinct germplasm sources have been recognised and analysed within the *M. sativa* subspp. complex [47]. Since 1971, the alfalfa cultivars released in North America typically include genetic variations from all nine germplasm sources [47]. Kidwell et al. [48] conducted an RFLP evaluation of these nine germplasm sources and only two, *M. sativa* subsp. falcata cv WISFAL and Peruvian, formed distinct clusters; the remaining seven sources were not clearly discriminated by the analysis. Recent molecular marker studies showed that 4x alfalfa populations maintained high levels of within-cultivar genetic diversity [49]; however, alfalfa yields have stagnated in North America and Australia [6–8]. This has provided the impetus to attempt to introduce new traits from other *Medicago* species into the *M. sativa* subspp. complex, which will enhance alfalfa productivity and persistence. *M. arborea* is one such species that has many traits that would increase the versatility of alfalfa (Table 1). It can grow to 4 m in height, has a shrubby growth habit, is drought- and salt-tolerant, and is the longest-lived of all the *Medicago* species. In addition, its seeds are four times larger than alfalfa (100 per gram versus 400 per gram) [3].

In a test of transmission of a single *M. arborea* trait from a yellow-flowered Alborea plant, yellow pigment was backcrossed into the alfalfa cultivar Pegesis (Bingham, unpublished data). Pegesis alfalfa has a purple flower colour; the addition of the yellow pigment resulted in variegated flowers due to the co-expression of purple and yellow. The variegated plants that were selected backcrossed three generations and did not express other *M. arborea* traits, indicating the gene for yellow pigment was likely carried on a short segment of *M. arborea* chromatin. The pollen and seed production of variegated plants was normal. The behaviour of yellow in the backcross was consistent with yellow incorporated in an alfalfa chromosome.

Some *M. arborea* accessions were shown to be resistant to races 1 and 2 of *Colletotrichum trifolii* [50] and to *Phytophthora medicaginis* (Irwin, unpublished data). Armour et al. [27] reported a transfer of resistance to *C. trifolii* race 2 from *M. arborea* to *M. sativa*. The resistance in the Alborea hybrid was successfully transferred to testcross progeny (Alborea x susceptible *M. sativa* clone), with 35% of the testcross progeny resistant. The resistant testcross plants were polycrossed and the resistance was again transferred to the polycross progeny. This indicated that the anthracnose resistance from *M. arborea* was stable and transmissible through at least two generations. It has subsequently been utilised in developing experimental alfalfa synthetics.

Alborea and Alborea × alfalfa crosses have been agronomically evaluated in a range of experiments conducted across locations in different continents. Irwin et al. [32] reported yield increases of up to 42% over adapted cultivars in the subtropics at Gatton, Queensland. Humphries et al. [25] tested selections of Alborea 101 in a Mediterranean environment in South Australia. One selection, CTA 033, had an estimated 23–30% higher yield than the commercial control cultivars. In field evaluations in Mediterranean environments in South Australia and Chile, an Alborea hybrid derived from the cross *M. sativa* cv Genesis x Alborea showed one of the highest dry matter yields and persistence levels [26]. Based on the above, evidence is emerging that Alborea has the capacity to enhance alfalfa productivity and persistence, at least in subtropical and Mediterranean environments.

Tani et al. [51] researched responses to salt shock and stress in seedlings of alfalfa, *M. arborea*, and their hybrid (Alborea). The two parental species appeared to regulate different components of the salt-tolerance mechanism. The Alborea population studied was more sensitive to all salt treatments than its parents, except the low dose (50 mM NaCl). This was most likely due to no selection for salt tolerance being undertaken during the development of the Alborea population. However, it should be possible to pyramid salt-tolerance genes in individual plants by cycles of selection for salt tolerance.

8. The Future

While each Alborea plant contains less than half of the *M. arborea* genome, it has been shown that introgressions have occurred from at least three of the *M. arborea* chromosomes [32,52]. Thus, so long as Alborea populations generated from several different hybrid individuals are used in breeding, it should be possible to transfer most of the *M. arborea* traits to alfalfa. The use of DNA marker technology, as described in Reference [53] and specific to the *M. arborea* genome, would facilitate the identification of the partial hybrids. Currently, only two alfalfa seed parents (MB and M8) and their progeny have been successfully used to generate hybrids [21,23]. The genome of MB codes for winter dormancy and low yields; its derivatives were relatively low yielding in Australia [32]. However, it should be possible to breed higher-yielding alfalfa seed parents by crossing M8 or MB, for example, with adapted genotypes and selecting for the reproductive abnormalities of both unreduced eggs and pollen. Unreduced pollen, due to its much-increased size, can be readily identified, and it provides an effective screening methodology to identify seed parents that also produce unreduced eggs, which appears to be essential for hybrid development. MB, M8, and all Alboreas examined produce unreduced eggs and pollen, and it is easier to screen for unreduced pollen than for unreduced eggs [23].

Having being derived from alfalfa parents with reproductive abnormalities, Alborea plants vary in the amount of pollen they produce, from a trace to about 25% of normal alfalfa [23,27]. Female fertility is expressed at a high level when pollinated with 4x normal *M. sativa*. This indicates that Alborea lines, when used as seed parents, could be useful in semi-hybrid development, as proposed by [54,55]. In Wisconsin and Queensland, work continues to identify adapted alfalfa cultivars/lines that complement Alborea selections for yield and persistence. Recent work has demonstrated the capacity for Alborea crosses with alfalfa to enhance alfalfa yield and persistence in Mediterranean environments, which are characterised by drought in summer [25,26]. Kang et al. [56] identified traits in alfalfa that provided tolerance to drought. These traits included lower-leaf wilting, delayed senescence, and leaf tolerance to desiccation under stress, smaller leaves, lower early root and shoot growth, and lower stomata density. Screening Alborea for the above traits, which exist in *M. arborea*, and their use in breeding may facilitate the development of more drought-tolerant alfalfas. Del Pozo [57] recently reported two Alborea selections to be the most drought-tolerant and productive entries in an alfalfa diversity panel tested in a Mediterranean environment in central Chile.

It has been shown that specific *M. arborea* alleles and traits can be transferred from the hybrids to alfalfa [21,25–27,32], indicating a degree of stability for the alleles/traits studied. However, as no genomic in situ hybridisation (GISH) studies have been performed on partial hybrids, it is not known whether chromosome fragments from *M. arborea* become integrated into the alfalfa genome or if a chromosome from *M. arborea* is transferred. It could be expected that a univalent would not be meiotically stable. Stability has been demonstrated for several traits, including anthracnose resistance, winter activity, and flower colour, and work is continuing for the other traits listed in Table 1. The results suggest that fragments of the *M. arborea* chromosomes have introgressed into the chromosomes of the *M. sativa* seed parent, providing meiotic stability. The high proportion (7%) of AFLP bands that were unique to the hybrid [27] was indicative of genome rearrangement, as could be expected with the integration of fragments of *M. arborea* chromosomes, which are more likely to be meiotically stable than univalents. Future research should utilise GISH studies to increase the understanding of the cytogenetic basis of partial hybridity. Pyramiding *M. arborea* traits from Alborea into adapted alfalfa cultivars may provide opportunities to increase the productivity, persistence, and utilisation of this important forage legume.

9. Conclusions

The importance of reproductive abnormalities, including unreduced eggs, in the *M. sativa* seed parents to the generation of partial hybrids between *M. sativa* and *M. arborea*, with a predominance of the *M. sativa* genome, was experimentally established. Although each hybrid contained about 5% of the *M. arborea*-specific alleles, it was possible to transfer a range of *M. arborea*-specific traits to the hybrids, which might have value in alfalfa improvement. Such traits included a larger seed size, indeterminate growth, frost tolerance, and heterosis for productivity and persistence in crosses with alfalfa. It is possible to conclude that the reviewed research conducted over the last 20 years has demonstrated the potential of the interspecific partial hybrids to enhance alfalfa improvement through long-term breeding activities.

Author Contributions: Writing, J.I. and E.B. All authors have read and agreed to the published version of the manuscript.

Funding: J.I. received support from PGG Wrightson Seeds Limited, New Zealand, 2009–present.

Institutional Review Board Statement: Not applicable.

Informed Consent Statement: Not applicable.

Data Availability Statement: Not applicable.

Conflicts of Interest: The authors declare no conflict of interest.

References

1. Hendry G. Alfalfa in history. *J. Am. Soc. Agron.* **1923**, *15*, 171–174. [CrossRef]
2. Michaud, R.; Lehman, W.; Rumbaugh, M. World distribution and development. In *Alfalfa and Alfalfa Improvement*; American Society of Agronomy: Madison, WI, USA, 1988; Volume 29, pp. 26–28.
3. Small, E. *Alfalfa and Relatives: Evolution and Classification of Medicago*; NRC Research Press: Ottawa, ON, Canada, 2011; p. 727.
4. Stanford, E. Tetrasomic inheritance in alfalfa. *Agronomy* **1951**, *43*, 222–225. [CrossRef]
5. Tysdal, H.; Kiesselbach, T.; Westover, H. Alfalfa breeding. Nebraska Agric. Expt. Station Res Bull. 1942 Number 124. Available online: https://digitalcommons.unl.edu/ardhistrb/219/ (accessed on 30 June 2023).
6. Holland, J.; Bingham, E. Genetic improvement for yield and fertility of alfalfa cultivars representing different eras of breeding. *Crop Sci.* **1994**, *34*, 953–957. [CrossRef]

7. Lamb, J.; Shaeffer, C.; Rhodes, L.; Sule, R.; Undersander, D.; Brummer, E. Five decades of alfalfa cultivar improvement: Impact on forage yield, persistence, and nutritive value. *Crop Sci.* **2006**, *46*, 902–909. [CrossRef]
8. Lowe, K.; Bowdler, T.; Casey, N.; Pepper, P. Evaluating temperate species in the subtropics. 3. Irrigated lucerne. *Trop. Grassl.* **2009**, *44*, 1–23.
9. Sakiroglu, M.; Llhan, D. *Medicago sativa* species complex: Revisiting the century old problem in the light of molecular tools. *Crop Sci.* **2020**, *61*, 827–838. [CrossRef]
10. Bingham, E.; Groose, R.; Woodfield, D.; Kidwell, K. Complementary gene interactions in alfalfa are greater in autotetraploids than in diploids. *Crop Sci.* **1994**, *34*, 823–829. [CrossRef]
11. Nenz, E.; Pupilli, F.; Damiani, F.; Arcioni, S. Somatic hybrid plants between the forage legumes *Medicago sativa* L. and *Medicago arborea* L. *Theor. Appl. Genet.* **1994**, *93*, 183–189. [CrossRef]
12. Urban, I. *Prodromus einer Monographie der Gattung. Medicago L.*; Kommissions-Verlag von R Gaertner: Brandenb, Germany, 1873; Volume 15, pp. 1–85.
13. Small, E.; Jomphe, M. A synopsis of the genus *Medicago* (Leguminosae). *Can. J. Bot.* **1988**, *67*, 3260–3294. [CrossRef]
14. Steele, K.; Ickert-Bond, S.; Zarre, S.; Wojciechowski, M. Phylogeny and character evolution in *Medicago* (Leguminosae): Evidence from analyses of plastid TRN/MatK and nuclear GA 30x1 sequences. *Am. J. Bot.* **2010**, *97*, 1142–1155. [CrossRef]
15. Fridriksson, S.; Bolton, J. Development of the embryo of *Medicago sativa* L. after normal fertilisation and after pollination by other species of *Medicago*. *Can. J. Bot.* **1963**, *41*, 23–33. [CrossRef]
16. Lesins, K.; Lesins, I. *Genus Medicago (Leguminosae), a Taxonomic Study*; Dr W Jurk: The Hague, The Netherlands, 1979.
17. McCoy, T.; Smith, L. Interspecific hybrids of perennial *Medicago* species using ovule-embryo culture. *Theor. Appl. Genet.* **1986**, *71*, 772–783. [CrossRef] [PubMed]
18. McCoy, T.; Quarisa, G. Allotetraploid behaviour of hybrids of *Medicago sativa* L. and *Medicago papillosa*. *Bioss Genome* **1989**, *32*, 6–11. [CrossRef]
19. Mizukami, Y.; Kato, M.; Takamizo, T.; Kanbe, M.; Inami, S.; Hattori, K. Interspecific hybrids between *Medicago sativa* L. and annual *Medicago* containing Alfalfa weevil resistance. *Plant Cell Tissue Organ Cult.* **2006**, *84*, 80–89. [CrossRef]
20. McCoy, T.; Echt, C. Potential of trispecies bridge crosses and random amplified polymorphic DNA markers for introgression of *Medicago daghestanica* and *M. pironae* germplasm into alfalfa (*M. sativa*). *Genome* **1993**, *36*, 594–601. [CrossRef]
21. Bingham, E.; Armour, D.; Irwin, J. The hybridization barrier between herbaceous *Medicago sativa* and woody *M. arborea* is weakened by selection of seed parents. *Plants* **2013**, *2*, 343–353. [CrossRef]
22. Bingham, E.; Irwin, J. "Alborea": A new cultigen developed from hybrids of alfalfa X *M. arborea*. Medicago Genetic Reports. 2014. Available online: www.medicago-reports.org (accessed on 30 June 2023).
23. Bingham, E.; Irwin, J. The hybridization barrier between herbaceous *Medicago sativa* and woody *M. arborea* is weakened by reproductive abnormalities in *M. sativa* seed parents. *Plants* **2023**, *12*, 962. [CrossRef]
24. Irwin, J.; Sewell, J.; Woodfield, D.; Bingham, E. Restructuring lucerne (*Medicago sativa*) through introgression of the *Medicago arborea* genome. *Agric. Sci.* **2016**, *28*, 40–47.
25. Humphries, A.; Ovalle, C.; Hughes, S.; del Pozo, A.; Inostroza, L.; Barahona, V.; Yu, L.; Yerzhanova, S.; Rowe, T.; Hill, J.; et al. Characterization, preliminary evaluation and pre-breeding of diverse alfalfa crop wild relatives originating from drought-stressed environments. *Crop Sci.* **2020**, *61*, 69–88. [CrossRef]
26. Inostroza, L.; Espinoza, S.; Barahona, V.; Gerding, M.; Humphries, A.; del Pozo, A.; Ovalle, C. Phenotypic diversity and productivity of *Medicago sativa* subspecies from drought-prone environments in Mediterranean type climates. *Plants* **2021**, *10*, 862. [CrossRef]
27. Armour, D.; Mackie, J.; Musial, J.; Irwin, J. Transfer of anthracnose resistance and pod coiling traits from *Medicago arborea* to *Medicago sativa* by sexual reproduction. *Theor. Appl. Genet.* **2008**, *117*, 149–156. [CrossRef] [PubMed]
28. Bingham, E. *Medicago arborea* Project at the University of Wisconsin. *Medicago* Genetic Reports 5, 1–7 and Several Reports in Volums 7, 9, 10 and 11 and Subsequent Years. Available online: www.medicago-reports.org (accessed on 1 July 2023).
29. Julier, B.; Fajoulot, S.; Barre, P.; Cardinet, G.; Santoni, S.; Huguet, T.; Huyghe, C. Construction of two genetic linkage maps in cultivated tetraploid alfalfa (*Medicago sativa*) using microsatellite and AFLP markers. *BMC Plant Biol.* **2003**, *3*, 9. [CrossRef] [PubMed]
30. Mun, J.; Kim, D.; Choi, H.; Gish, J.; Debelle, F.; Mudge, J.; Denny, R.; Endre, G.; Saurat, O.; Dudez, A.-M.; et al. Distribution of microsatellites in the genome of *Medicago truncatula*: A resource of genetic markers that integrate genetic and physical maps. *Genetics* **2006**, *172*, 2541–2555. [CrossRef]
31. Mackie, J.; Musial, J.; Armour, D.; Phan, H.; Elwood, S.; Aitken, K.; Irwin, J. Identification of QTL for reaction to three races of *Colletotrichum trifolii* and further analysis of inheritance of resistance in autotetraploid lucerne. *Theor. Appl. Genet.* **2007**, *114*, 1417–1426. [CrossRef]
32. Irwin, J.; Armour, D.; Pepper, P.; Lowe, K. Heterosis in lucerne testcrosses with *Medicago arborea* introgressions and Omani landraces and their performance in synthetics. *Crop Pasture Sci.* **2010**, *61*, 450–463. [CrossRef]
33. De Wet, J.; Newell, C.; Brink, D. Counterfeit hybrids between *Tripsacum* and *Zea* (Gramineae). *Am. J. Bot.* **1984**, *71*, 245–251. [CrossRef]
34. Bingham, E.; Irwin, J. Evidence that 2n eggs explain partial hybrids between *Medicago sativa* and *Medicago arborea*. *Plants* **2022**, *11*, 1380. [CrossRef]

35. Allard, R. *Principles of Plant Breeding*; John Wiley and Sons: New York, NY, USA, 1960; p. 485.
36. Doebly, J. The genetics of maize evolution. *Annu. Rev. Genet.* **2004**, *38*, 37–59. [CrossRef]
37. Li, Z.; Heneen, W. Production and cytogenetics of intergeneric hybrids between the three cultivated *Brassica diploids* and *Orychophragmus violaceus*. *Theor. Appl. Genet.* **1999**, *99*, 694–704. [CrossRef] [PubMed]
38. Tu, Y.; Sun, J.; Ge, X.; Li, Z. Production and genetic analysis of partial hybrids from intertribal sexual crosses between Brassica napus and *Isatis indigotica* and progenies. *Genome* **2010**, *53*, 146–156. [CrossRef] [PubMed]
39. Pfeiffer, T.; Bingham, E. Abnormal meiosis in alfalfa, *Medicago sativa*; Cytology of 2N egg and 4N pollen formation. *Can. J. Genet. Cytol.* **1983**, *25*, 107–112. [CrossRef]
40. Brownfield, L.; Kohler, C. Unreduced gamete formation in plants: Mechanisms and prospects. *J. Exp. Bot.* **2010**, *62*, 1659–1668. [CrossRef] [PubMed]
41. Cheng, B.; Seguin-Swartz, G.; Somers, D. Cytogenetic and molecular characterization of intergeneric hybrids between *Brassica napus* and *Orychophragmus violaceus*. *Genome* **2002**, *45*, 110–115. [CrossRef] [PubMed]
42. Baraccia, G.; Albertini, D.; Rosellini, S.; Tavoletti, S.; Veronesi, F. Inheritance and mapping of 2n-egg production in diploid alfalfa. *Genome* **2000**, *43*, 528–537. [CrossRef] [PubMed]
43. Camadro, E.; Carputo, D.; Peloquin, S. Substitutes for genome differentiation in tuber-bearing *Solanum*; interspecific pollen-pistil incompatibility, nuclear-cytoplasmic male sterility, and endosperm. *Theor. Appl. Genet.* **2004**, *109*, 1369–1376. [CrossRef] [PubMed]
44. Boyes, J.; Thompson, W. The development of the endosperm and embryo in reciprocal crosses in cereals. *J. Genet. Breed.* **1937**, *34*, 203–227. [CrossRef]
45. Valentine, D. Studies in British primulas. V. The inheritance of seed compatibility. *New Phytol.* **1956**, *55*, 305–318.
46. Carputo, D.; Monti, L.; Werner, J.; Frusciante, L. Uses and usefulness of the endosperm balance number. *Theor. Appl. Genet.* **1999**, *98*, 478–484. [CrossRef]
47. Barnes, D.; Bingham, E.; Murphy, R.; Hunt, O.; Beard, D.; Skrdla, W.; Teuber, L. *Alfalfa Germplasm in the United States: Genetic Vulnerability, Use, Improvement and Maintenance*; Department of Agriculture, Agricultural Research Service: Washington, DC, USA, 1977.
48. Kidwell, K.; Austin, D.; Osborn, T. RFLP evaluation of nine *Medicago* accessions representing the original germplasm sources for North American alfalfa cultivars. *Crop Sci.* **1994**, *34*, 230–236. [CrossRef]
49. Shen, C.; Du, H.; Chen, Z.; Lu, H.; Zhu, F.; Chen, H.; Meng, X.; Liu, Q.; Liu, P.; Zheng, L.; et al. The chromosome- level genome sequence of the autotetraploid alfalfa and resequencing of core germplasms provide genomic resources for alfalfa research. *Mol. Plant* **2020**, *13*, 1250–1261. [CrossRef]
50. Elgin, J.; Ostazeski, S. Evaluation of selected alfalfa cultivars and related *Medicago* species for resistance to race-1 and race-2 anthracnose. *Crop Sci.* **1982**, *22*, 39–42. [CrossRef]
51. Tani, E.; Sarri, E.; Goufa, M.; Asimakopoulou, G.; Psychogiou, M.; Bingham, E.; Skaricis, G.; Abraham, E. Sedling growth and transcriptional responses to salt shock and stress in *Medicago sativa* L., *Medicago arborea* L., and their hybrid (Alborea). *Agronomy* **2018**, *8*, 231. [CrossRef]
52. Bingham, E.; Armour, D.; Irwin, J.; Jayaraman, D.; Ane, J. Weakening the Hybridization Barrier between Herbaceous *Medicago sativa* and Woody *M. arborea* by Genetic Selection and Bridge Crossing. Available online: http://www.medicago-reports.org/volumes09.php (accessed on 6 June 2023).
53. Annicchiarico, P.; Barrett, B.; Brummer, E.; Julier, B.; Marshall, A. Achievements and challenges in improving temperate perennial forage legumes. *Front. Plant Sci.* **2015**, *34*, 327–380. [CrossRef]
54. Brummer, E. Capturing heterosis in forage crop cultivar development. *Crop Sci.* **1999**, *39*, 943–954. [CrossRef]
55. Annicchiarico, P.; Pecetti, L. Exploiting heterosis of semi-hybrids and heterogeneity of cultivar mixtures to enhance alfalfa crop performance. *Field Crop Res.* **2022**, *283*, 108522. [CrossRef]
56. Kang, Y.; Seminario, A.; Udvardi, M.; Annicchiarico, P. Physiological and biochemical adaptive traits support the specific breeding of alfalfa (*Medicago sativa*) for severely drought-stressed or moisture-stressed environments. *J. Agron. Crop Sci.* **2022**, *209*, 132–143. [CrossRef]
57. Del Pozo, A.; Espinoza, S.; Barahona, V.; Inostroza, L.; Gerding, M.; Humphries, A.; Lobos, G.; Cares, J.; Ovalle, C. Ariel and ground-based phenotyping of an alfalfa diversity panel to assess adaptation to a prolonged drought period in a Mediterranean environment of central Chile. *Eur. J. Agron.* **2023**, *144*, 126751. [CrossRef]

applied biosciences

MDPI

Article

Sequencing, Fast and Slow: Profiling Microbiomes in Human Samples with Nanopore Sequencing

Yunseol Park [1,†], Jeesu Lee [1,†] and Hyunjin Shim [1,2,*]

1 Center for Biosystems and Biotech Data Science, Ghent University Global Campus, Incheon 21985, Republic of Korea; yunseol.park@ugent.be (Y.P.); jeesu.lee@ghent.ac.kr (J.L.)
2 Department of Biology, California State University, Fresno, 5241 N Maple Ave, Fresno, CA 93740, USA
* Correspondence: shim@mail.fresnostate.edu
† These authors contributed equally to this work.

Abstract: Rapid and accurate pathogen identification is crucial in effectively combating infectious diseases. However, the current diagnostic tools for bacterial infections predominantly rely on century-old culture-based methods. Furthermore, recent research highlights the significance of host–microbe interactions within the host microbiota in influencing the outcome of infection episodes. As our understanding of science and medicine advances, there is a pressing need for innovative diagnostic methods that can identify pathogens and also rapidly and accurately profile the microbiome landscape in human samples. In clinical settings, such diagnostic tools will become a powerful predictive instrument in directing the diagnosis and prognosis of infectious diseases by providing comprehensive insights into the patient's microbiota. Here, we explore the potential of long-read sequencing in profiling the microbiome landscape from various human samples in terms of speed and accuracy. Using nanopore sequencers, we generate native DNA sequences from saliva and stool samples rapidly, from which each long-read is basecalled in real-time to provide downstream analyses such as taxonomic classification and antimicrobial resistance through the built-in software (<12 h). Subsequently, we utilize the nanopore sequence data for in-depth analysis of each microbial species in terms of host–microbe interaction types and deep learning-based classification of unidentified reads. We find that the nanopore sequence data encompass complex information regarding the microbiome composition of the host and its microbial communities, and also shed light on the unexplored human mobilome including bacteriophages. In this study, we use two different systems of long-read sequencing to give insights into human microbiome samples in the 'slow' and 'fast' modes, which raises additional inquiries regarding the precision of this novel technology and the feasibility of extracting native DNA sequences from other human microbiomes.

Keywords: diagnostic for bacterial infection; long-read sequencing; human microbiome; nanopore Flongle; native DNA; antimicrobial resistance; taxonomic classification

Citation: Park, Y.; Lee, J.; Shim, H. Sequencing, Fast and Slow: Profiling Microbiomes in Human Samples with Nanopore Sequencing. *Appl. Biosci.* **2023**, *2*, 437–458. https://doi.org/10.3390/applbiosci2030028

Academic Editor: Robert Henry

Received: 7 June 2023
Revised: 17 July 2023
Accepted: 9 August 2023
Published: 17 August 2023

1. Background

Rapid and accurate identification of pathogens is crucial for effectively treating and managing infectious diseases. Traditional diagnostic methods for bacterial infections, such as culture-based techniques, have been largely unchanged in clinical practice for several decades [1,2]. However, these methods often take several days for identification and susceptibility testing of bacterial pathogens and are prone to false-negative results during antimicrobial therapy. For example, a recent study showed that the median time to pathogen identification for bloodstream infections using traditional culture-based methods takes around three days [3]. This delay in diagnostic procedures can result in inappropriate antibiotic therapy, which can negatively affect patient outcomes and lead to antibiotic resistance development [1]. Furthermore, culture-based techniques may not detect all bacterial infections, particularly if the patient is undergoing antimicrobial therapy.

Recent evidence suggests that it is important to have a comprehensive view of the microbial communities in which the pathogen resides to predict the progress of infection in clinical settings [4]. The human microbiota, consisting of a vast array of microorganisms, including bacteria, viruses, fungi, and protozoa, colonizes many different niches within the human body. Host–microbe interactions within the host microbiota play a vital role in determining the growth and establishment of pathogenic microbes. These interactions can range from beneficial to commensal to pathogenic, and a subtle shift in the balance of these interactions can have profound effects on the host's health, including inflammatory bowel disease (IBD) [5] and neurological disorders [6]. Recent research has highlighted the role of the host microbiota in shaping the host's immune response to pathogenic microorganisms, both through direct interactions with the immune system and through modulation of the host's innate and adaptive immune responses [7]. Studies have demonstrated that the host microbiota can provide colonization resistance against invading pathogens, limiting their growth and preventing their establishment within the host [8]. Furthermore, the host microbiota can also impact the virulence of pathogenic microorganisms through a variety of mechanisms, including competition for nutrients, secretion of antimicrobial compounds, and modulation of the expression of virulence factors [9]. Alterations in the composition of the host microbiota, such as those caused by antibiotics or changes in diet, can disrupt these finely balanced host–microbe interactions, leading to increased susceptibility to infections [10].

Understanding the intricate interactions between the host microbiota and pathogenic microorganisms is critical for the development of effective treatments for infectious diseases. This knowledge can inform the development of novel therapeutics that target specific bacterial species or modulate the host's immune response to promote the restoration of healthy microbiota and reduce the risk of disease [11]. Alterations in the composition of the microbiota can disrupt these interactions, leading to increased susceptibility to infections. Currently, we are in need of novel diagnostic methods that can both identify pathogens and profile the microbiome landscape in human samples rapidly and accurately. Given the expanding knowledge of the role of the microbiome in human health, this feature is a substantial advancement to traditional diagnostic methods that focus primarily on pathogen identification, such as culture-based diagnosis and MALDI-TOF mass spectrometry fingerprinting [1].

Molecular-based diagnostic methods, such as polymerase chain reaction (PCR) and next-generation sequencing (NGS), can detect a wider range of bacterial pathogens with greater sensitivity and specificity [12]. In recent years, the development of such technologies has enabled the rapid and accurate identification of pathogens. Particularly, NGS can generate vast amounts of sequence data in a short amount of time, providing rapid and accurate results for pathogen identification that enables clinicians to make informed decisions about antimicrobial treatments [13]. Several studies have shown the potential of NGS in clinical settings for the rapid identification of bacterial infections, such as the rapid identification of a bacterial outbreak in a neonatal intensive care unit [14] and in patients with sepsis [15]. Despite the potential of NGS, the technology is still not widely available in clinical settings, and there are several challenges related to quality control measures that need to be overcome. These challenges include the need for standardized protocols for sample preparation, sequencing, and data analysis, particularly since these methods produce short-read DNA sequences in metagenomic samples that require elaborate bioinformatic reconstructions [16].

Most recently, long-read sequencing technologies are revolutionizing genomics research by producing high-throughput sequencing of DNA reads longer than those obtained by traditional short-read sequencing methods. Nanopore sequencing is one of the long-read sequencing technologies, in which a DNA molecule is passed through a nanopore, and the electrical signal generated by the movement of the nucleotides through the pore is used to determine the sequence of the DNA molecule [17]. The development of long-read sequencing has had a significant impact on genomics research and has enabled the study of complex genomes, structural variation, and epigenetic modifications with unprecedented accuracy and resolution [18,19]. Long-read sequencing has the potential to further enhance the field of metagenomics by enabling the identification and characterization of unculturable microbes and the study of host–microbe interactions in complex microbial communities [20]. A recent study demonstrated the potential of nanopore sequencing as the clinical diagnosis of bacterial lower respiratory infections by directly sequencing sputum samples, providing comparable results to culture-based methods, but with significantly faster turnaround times [21,22].

In this study, we investigate the potential of long-read sequencing as a futuristic diagnostic tool to rapidly profile the microbiome landscape of diverse human samples. We use two different modes to sequence long-read native DNA sequences from diverse human microbiomes, including saliva and stool samples. The first part consists of a 'fast' mode, which aims to generate a biological interpretation of nanopore sequencing within 12 h from sampling to data analysis (Figure 1). This fast mode is aimed at providing ultrarapid analysis of nanopore sequence data such as pathogen identification and antimicrobial resistance (AMR) under a clinical scenario of tight time constraints where the streamlined pipeline of a diagnostic tool is essential for effective treatment. This mode is fast and automatic, and it enables clinicians to make quick identifications and decisions on antibiotic therapy based on the pathogen and its related microbes without much investment of time and effort. A 'slow' mode is aimed at providing deeper insight into the microbiome landscape of a patient for prognostic purposes, during which the microbial communities are analyzed more rigorously. This mode is slow and deliberate, and it engages the researchers to predict the long-term trajectory of an infection outcome using complex information such as host–microbe interactions and deep learning-based classification of unknown organisms. Overall, we aim to provide a comprehensive and insightful view of long-read sequencing as an innovative diagnostic tool for bacterial infections by rapidly profiling the microbiome landscape from various human samples.

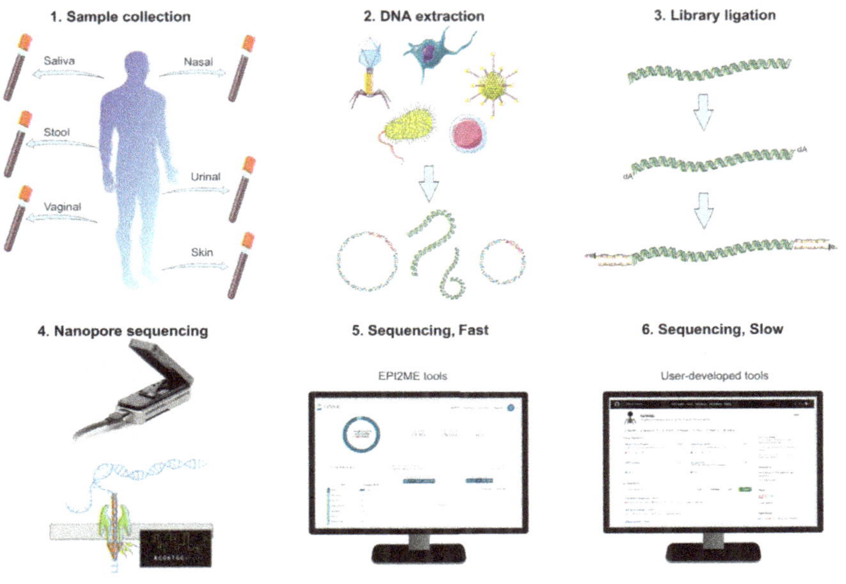

Figure 1. Experimental setup and process of nanopore sequencing. 1. The sample collection of human microbiome is collected with a sterile specimen swab. 2. Native DNA of the metagenomic sample is

extracted with a commercial kit with minimum shearing to obtain high molecular weight DNA. 3. The extracted DNA is ligated with a library kit provided by Oxford Nanopore that has been optimized for sequencing native and long-read DNAs in Flongle flow cells. 4. Flongle flow cells are fitted with an adaptor to a nanopore sequencer for rapid and cost-effective tests, running for at least 12 h to maximize yield per sample. 5. Basecalling is conducted in real-time for a fast sequencing mode, during which results can be obtained using cloud-based software tools for taxonomic classification and AMR analysis. 6. For a slow sequencing mode, user-developed tools can be used to conduct exploratory analysis on the same nanopore data, including host–microbe interaction assessment and deep learning-based classification of unidentified reads.

2. Results

2.1. Native DNA Could Be Extracted and Sequenced from the Saliva and Stool Samples

We used the QIAamp DNA Microbiome Kit to extract the native DNA of various human samples from healthy volunteers (Figure 1). The minimum DNA quantity needed for native DNA direct sequencing using Flongle flow cells is 500 ng according to the manufacturer's protocol, and the lowest DNA concentration achieved for the saliva and stool samples was 20.5 ng/μL, yielding sufficient quantities for nanopore sequencing (Table S1). However, the same kit was found to be ineffective for extracting native DNA from urine, nasal, and vaginal samples sourced from healthy volunteers, failing to provide the minimum DNA quantities required for nanopore sequencing (Table S1). These findings suggest that the QIAamp DNA Microbiome Kit may have limited utility for extracting native DNA from certain human microbiome sources, and alternative DNA extraction methods may need to be explored for these sample types.

Another disadvantage of using the QIAamp DNA Microbiome Kit to extract native DNA from various microbiome samples comes from the fragmentation of DNAs during the extraction step (Figure 1). Fragmentation of DNA during extraction can be caused by a variety of factors, including mechanical and enzymatic shearing. The QIAamp DNA Microbiome Kit uses a bead-beating step to lyse cells, which can result in excessive mechanical shearing of DNA. Due to the DNA fragmentation, the shortest and the longest estimated N50 values are 378 bases and 1090 bases, respectively (Table S2, Figure S1). N50 is a statistical measure commonly used in DNA sequencing to describe the quality of an assembly, and in the context of long-read sequencing, it is defined as the length of the shortest read within the set of the longest reads that constitute at least 50% of the sample [23]. Previously, it was shown that nanopore sequencers can produce long reads of around 10–30 kilobases (kb) reads in a typical sequencing experiment, while ultralong reads were shown to be around 3 megabases (Mb) with the N50 value of more than 100 kb [24].

2.2. Fast Sequencing Shows Oral and Gut Microbiomes Have Diverse Microbial Species

In this study, we ran one Flongle flow cell with each sample replicate for at least 12 h, to exhaust the capacity of nanopores to obtain as many DNA reads in each run as possible (Figure S2). Depending on the sample, nanopore sequencing using Flongle flow cells was saturated as early as 3 h (Figure S2; Saliva3_R2) and as late as 20 h (Figure S2; Stool1_R2). The recommended hours of sequencing for nanopore Flongle can vary depending on the desired experimental output. For example, a recent study reported using Flongle flow cells with a sequencing time of 24 h to generate high-quality, near-complete bacterial genomes of *Mycoplasma bovis* [25]. Similarly, another study utilized Flongle flow cells in a similar timeline to achieve high-quality, near-complete SARS-CoV-2 genome assemblies [26].

In this study, we used the high-accuracy basecalling program integrated into the MinKNOW software (v.4.5.4; 2021; Oxford Nanopore Technologies, Oxford, UK) in real-time, with a minimum Q-score of 9. The Phred score and Quality score (Q-score) are both measures of the quality of sequencing data, where the Phred score is a logarithmic measure of the error probability originated to identify fluorescently labeled DNA bases by comparing observed and expected chromatogram peak shapes and resolution [27], widely used in Sanger sequencing and Illumina sequencing. For nanopore sequencing, per-nucleotide quality scores are based on the outputs of the neural networks used to produce the basecall. Q-scores are per-read quality scores, calculated by averaging the per-nucleotide quality scores and by expressing on the Phred scale [28]. Importantly, Q-scores consider that the error rate in nanopore sequencing is not constant across the read and can vary depending on factors such as the sequence context and the quality of the signal.

After a complete sequencing run, we used the basecalled output for the quantitative and real-time identification of microbiome species from these metagenomic samples using the cloud-based data analysis platform (Table S3). This data analysis platform leverages long-read sequencing to enable the comparison of each read against databases containing reference genomes of bacteria, archaea, viruses, and fungi, achieved by constructing an indexing scheme that facilitates efficient searches of sequenced reads [29]. It rapidly classified and identified diverse species in each microbiome sample, even to the resolution of different strains of bacterial species (Table S3). The data analysis platform also rapidly determined the most reliable placement of these organisms in the taxonomy tree, assigning a score to each taxonomic placement (Figure S3).

The gut microbiomes contained the greatest number of species, while the oral microbiome contained varying amounts of microbial species (Table S3). For example, the Stool1 and Stool2 samples had more than 1000 and 500 known microbial species present, respectively. The most abundant species consists of *Lactobacillus ruminis* in Stool1 and *Megamonas funiformis* in Stool2. Among these abundant species, it was notable that the gut microbiome from Stool1 contained most bacteria species widely known to be beneficial [30], whereas that from Stool2 had most bacterial species recently found to be commensal [31]. For the saliva samples, both the diversity and number of microbial species were lower and the role of each species in the host–microbe interaction was less obvious (Table S3). The most abundant bacterial species include *Haemophilus parainfluenzae* in Saliva1 and Saliva2, whereas *Rothia mucilaginosa* in Saliva3. It was notable that the saliva microbiome from Saliva1 and Saliva2 contained most bacteria species widely known to be beneficial [32], whereas that from Saliva3 had most bacterial species recently found to be harmful [33]. Another notable observation includes the presence of viruses in these microbiome samples despite the use of an extraction kit that was not optimized for viral DNA extraction. The DNA reads that were classified as bacteriophage were of particular interest as the role and impact of these biological entities are just starting to get noticed in microbiome studies [34,35]. The most abundant virus species include Faecalibacterium phage in Stool1, crAss-like phage [36] in Stool2, Streptococcus phage in Saliva1 and Saliva2, and Shigella phage in Saliva3 (Table S3). These first analyses show the diversity and abundance of microbial communities in human samples could be rapidly profiled—however, we conducted more in-depth analyzes of the host–microbe interaction types subsequently (see below).

In the saliva and stool samples, varying amounts of human DNAs were present despite the host DNA depletion step of the QIAamp DNA Microbiome Kit. In the stool samples, the microbiome DNA was enriched compared to the human DNA, with almost 100% of reads classified as bacterial species in Stool1 (Table 1). However, the saliva samples tend to have a lower percentage of the microbiome DNA in the sequencing output, with almost 90% of reads classified as eukaryotic species in Saliva1 and Saliva2 (Table 1). The third saliva sample diluted with 1000 μL of PBS solution had a better percentage content of the microbiome DNA, which shows that the quantity of a sample does not always correlate with the quality of reads in long-read sequencing. We studied these human DNA reads to assess if they could provide valuable information about the host, such as some genetic markers that could give alternative insight into the host–microbe interactions, but the yield output of a Flongle flow cell with a maximum 2.8 Gb was not enough to generate any significant coverage. However, more high-throughput flow cells such as MinION and PromethION with a maximum output of 50 Gb and 290 Gb per flow cell, respectively, may be utilized to generate genomic data of the host and microbiome simultaneously, which may provide the most comprehensive view of the host–microbe interactions, given the recent findings of the interdependence of microbiome genomes and human genomes [37,38].

Table 1. The real-time analysis of nanopore reads with the cloud-based platform (EPI2ME) and its integrated software for sequence similarity-based taxonomic classification (WIMP). For each microbiome replicate, the data retrieved (Yield Data) in megabases, the average quality score from EPI2ME, the average sequence length, the total number of reads, and the number of reads classified into any operational taxonomic unit (OTU) are reported. The Superkingdom column gives the percentage of reads that have been classified into the kingdoms - Eukaryota, Bacteria, and Viruses.

Sample_Replicate	Yield Data (Mbases)	Average Quality Score	Average Sequence Length	Total Reads	Reads Classified	Superkingdom
Saliva1_R1	33.1	11.56	734	45,079	43,820	Eukaryota: 89% Bacteria: 11% Viruses: <1%
Saliva1_R2	25.0	11.12	847	29,462	29,059	Eukaryota: 95% Bacteria: 5% Viruses: <1%
Saliva2_R1	223.2	12.40	729	306,111	293,219	Eukaryota: 81% Bacteria: 19% Viruses: <1% Archaea: <1%
Saliva2_R2	66.1	10.92	672	98,330	94,121	Eukaryota: 86% Bacteria: 14% Viruses: <1%
Saliva3_R1	22.8	11.29	497	45,827	33,609	Bacteria: 89% Eukaryota: <11% Viruses: <1% Archaea: <1%
Saliva3_R2	10.7	11.47	445	24,099	17,798	Bacteria: 90% Eukaryota: <10% Viruses: <1%
Stool1_R1	37.3	11.70	428	87,146	50,254	Bacteria: 100% Eukaryota: <1% Viruses: <1% Archaea: <1%
Stool1_R2	58.3	11.63	451	129,091	74,810	Bacteria: 100% Eukaryota: <1% Viruses: <1% Archaea: <1%
Stool2_R1	15.9	11.21	431	36,832	22,744	Bacteria: 97% Viruses: 2% Archaea: <1% Eukaryota: <1%
Stool2_R2	21.1	11.04	566	37,172	22,425	Bacteria: 98% Viruses: <1% Archaea: <1% Eukaryota: <1%

2.3. *Slow Sequencing Shows Complex Host–Microbe Interaction Types*

We investigated the microbial species from these microbiomes further by assigning each microbial species or strain as a harmful, beneficial, or commensal organism in the oral or gut microbiome (Tables S4 and S5). This assessment of the host–microbe interaction was initially conducted by matching the name of each organism to the list generated by the previous studies to have a positive, negative, or neutral impact on the human host [32,39–45]. The ten most abundant species in each microbiome sample are shown with the host–microbe interaction type as beneficial, harmful, or commensal in Table

S4. This curated list shows that the ten most abundant species are consistently present in most replicates of the saliva samples. For example, the most abundant species of beneficial bacteria are found to be *Haemophilus parainfluenzae* in all the saliva replicates. In contrast, the most abundant species of harmful bacteria are found to be *Neisseria subflava* in Saliva1_R1, whereas *Prevotella melaninogenica* in Saliva1_R2 and Saliva2. In Saliva3, the most abundant harmful bacteria are found to be *Rothia mucilaginosa* in both replicates. Despite the difference in order, the ten most abundant species mostly match between two replicates of the microbiome sample. However, there was much more variation in the ten most abundant species in the stool samples. For example, the most abundant species of beneficial bacteria is *Lactobacillus ruminis* in both replicates of Stool1, whereas *Akkermansia muciniphila* in Stool2_R1. *Bifidobacterium adolescentis* is found in all stool samples as one of the most abundant beneficial bacteria. In both replicates of Stool1, the most abundant harmful bacteria is found to be *Acidaminococcus intestini*, which has been isolated from different clinical samples [46]. In both replicates of Stool2, the most abundant harmful bacteria is found to be *Desulfovibrio piger*, which are sulfate-reducing bacteria that may contribute to gastrointestinal diseases such as IBDs due to the production of hydrogen sulfide that is toxic to the gut epithelium [47].

Due to the extensive list of microbial species in the nanopore dataset, there were many microbes that were missing from the initial list of host–microbe interaction types, particularly in the gut microbiome which contained thousands of species. Thus, we further searched the most recent scientific literature to assess the impact of each microbial organism in these microbiomes (Table S5). In cases when there is contradicting evidence, we flagged the organisms as inconclusive. Furthermore, if the assessment level was higher than that of the genus, it was immediately assessed as inconclusive (as there is too much diversity) unless there was overwhelming evidence otherwise (Table 2).

Table 2. Assessment of host–microbe interaction types for each microbe species per microbiome replicate. Each read in the microbiome replicate was classified in the taxonomical level of genus, species, and strain for prokaryotes and virus for mobile genetic elements. These microbiomes were categorized as beneficial, harmful, commensal, and inconclusive depending on the type of host–microbe interaction defined in the previous literature. The number sign, #, is used to abbreviate the word "number".

Sample_Replicate	# Genus	# Species	# Strain	# Virus	Reads	Length	Beneficial	Harmful	Commensal	Inconclusive
Saliva1_R1	125	164	105	5	135	1597	31	101	52	146
Saliva1_R2	57	105	56	1	154	2767	16	75	35	62
Saliva2_R1	285	424	250	10	403	717	57	226	103	370
Saliva2_R2	134	198	127	3	249	790	36	135	56	164
Saliva3_R1	327	467	270	8	56	532	61	231	122	404
Saliva3_R2	227	285	184	3	43	470	47	174	87	250
Stool1_R1	514	686	426	9	76	485	82	115	246	705
Stool1_R2	702	921	603	8	82	526	92	144	275	1057
Stool2_R1	283	338	207	2	64	456	49	71	161	287
Stool2_R2	359	416	266	6	50	662	66	92	184	392

The comprehensive assessment of the host–microbe interaction types in the microbiome community is summarized in the bar chart of relative diversity (Figure 2). It shows that the oral microbiome tends to contain more diverse organisms that are known to be harmful than the gut microbiome. Moreover, a significant number of microbes exhibit inconclusive roles within the gut and oral microbiomes, underscoring the imperative to explore the impact of these microbes on microbiome communities in order to comprehensively map the landscape of the human microbiome. A bacterial species that have been isolated from a human gut may be beneficial or pathogenic depending on the individual or the health condition of the individual, leading to conflicting or inconclusive information about the host–microbe interaction type. Furthermore, one species may have many strains with completely different characteristics. In our dataset, there are some bacterial species such as *Escherichia coli* with dozens of strains, with a huge diversity in their genomic and functional characteristics. Therefore, even if one sequenced species was considered as one interaction type, there is no guarantee that the actual strain that was sequenced possesses the same interaction type.

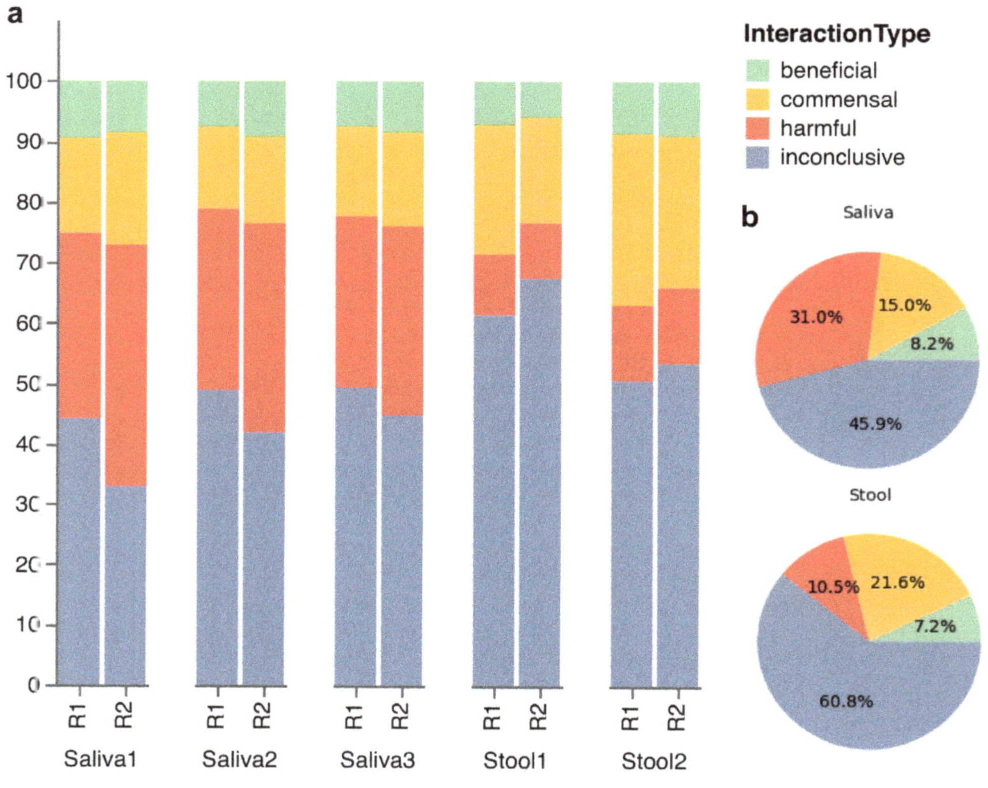

Figure 2. Stacked bar and pie charts of host–microbe interaction types in each microbiome replicate. (**a**) The stacked bars visually depict the species diversity within each replicate, with classification into beneficial, harmful, or commensal categories in the host microbiome based on existing literature. Any microbial species that have missing or conflicting information is categorized as inconclusive. (**b**) The pie charts depict the same classification of host–microbe interaction, with mean percentages for each interaction type within the saliva and stool samples. Components may not sum to totals because of rounding.

We found some microbes whose presence in the gut and oral microbiomes was particularly intriguing (Tables S4 and S5) [48]. Cellulolytic bacteria (in Caldicellulosiruptor) were sequenced in the microbiomes of Stool1 and Stool2_R2. No evidence was found for its host–microbe interaction type, but cellulolytic bacteria are important for mammals including humans, as they allow the digestion of plant materials and gain nutrients from plants. A previous study even shows the potential for these microbes to have antibacterial properties against pathogenic bacteria [49], which makes it difficult to assess their host–microbe interaction type as commensal or beneficial (list of cellulolytic bacteria includes *Caldicellulosiruptor bescii* DSM 6725, *Caldicellulosiruptor changbaiensis*, *Caldicellulosiruptor obsidiansis* OB47, and *Caldicellulosiruptor saccharolyticus* DSM 8903). We also found some plant bacteria, both beneficial and pathogenic to plants, whose role in human health has not been investigated. However, a plant pathogen that is also pathogenic to humans was found in Stool2_R1, such as *Pantoea ananatis*, whose presence is uncommon in the human microbiome [50]. There are some zoonotic bacteria found in the sample, including *Pasteurella multocida* [51]. Some bacteria have a natural affinity towards antimicrobial resistance, including *Clostridium boltae*, which is a commensal in the human gastrointestinal tract but also acts as a reservoir for antimicrobial resistance [52].

During the assessment, we noticed that defining the host–microbe interaction types as harmful, beneficial, or commensal is only a vague indicator of the microbial characteristics and should not be considered as an absolute measure. For example, *Corynebacterium matruchotii*, an oral microbe that is crucial in biofilm structure and may aid in the prevention of caries, has also been hypothesized to cause supragingival calculus formation if present in the oral microbiome [53]. The formation could lead to periodontal diseases, highlighting the dual nature of the host–microbe interaction as both beneficial and harmful. Many of the microbes that are commensal can also be harmful to immunocompromised patients [54], and the microbial pathogenicity or virulence can undergo changes due to the changes in the microbial DNA, the antimicrobial resistance, the surrounding environment, or the susceptibility of humans to particular diseases [55–57]. For example, *Acinetobacter baumannii* was pathogenic since the 1990s but its pathogenicity level has now increased to a critical level [55]. Furthermore, the composition of the microbiota is as important as the type, since the interplay between different microbes also changes the extent of beneficial or harmful effects [58].

2.4. Oral and Gut Microbiomes Have Numerous AMR Genes

We found numerous and diverse antibiotic resistance genes in all the microbiome samples, summarized in Table S6 and shown as a heatmap in Figure S4. There are several genes that are attributed to the antimicrobial resistance to a wide range of antibiotics, including beta-lactam, aminoglycoside, tetracycline, macrolide, and fluoroquinolone (Table S7). Bacteria can develop resistance against these antibiotics through multiple mechanisms. These antimicrobial resistance genes can be categorized into four Comprehensive Antibiotic Resistance Database (CARD) models depending on the type of resistance mechanisms: protein variant model, protein homolog model, protein wild type model, and rRNA mutation model (Figure S5).

One of the antibiotics, named aminoglycoside, is widely used to fight against bacteria, especially aerobic Gram-negative bacteria. Aminoglycoside inhibits peptide elongation at 30S ribosomal subunit, resulting in inaccurate mRNA translation which can halt protein synthesis or alter amino acid compositions at certain points [59]. However, when some mutations occur in the 30S ribosomal subunit, aminoglycosides no longer interact with the target [60]. In all the microbiome samples, the AMR genes conferring resistance to aminoglycoside were the most prevalent (Figure 3a). For instance, at least 60% of the AMR genes are related to the resistance against aminoglycoside in Saliva3_R2 (Table S6). Particularly, we found *Mycobacterium tuberculosis* in one of the saliva samples (Saliva3_R2), which is known to cause tuberculosis and it had 16S rRNA variant genes that confer multidrug resistance to streptomycin and amikacin, which belong to the family of aminoglycoside, posing a potential threat as these antibiotics are commonly used to treat tuberculosis [61]. In one of the stool samples (Stool1_R1), *Campylobacter jejuni*, known to cause gastroenteritis was found. It had *ant(6)-Ib* genes, which encode a family of aminoglycoside nucleotidyltransferase named ANT(6)-Ib. The expression of *ant(6)-Ib* can exacerbate the antimicrobial resistance in *Campylobacter jejuni*, as aminoglycosides and macrolides are the effective way to treat this disease [62].

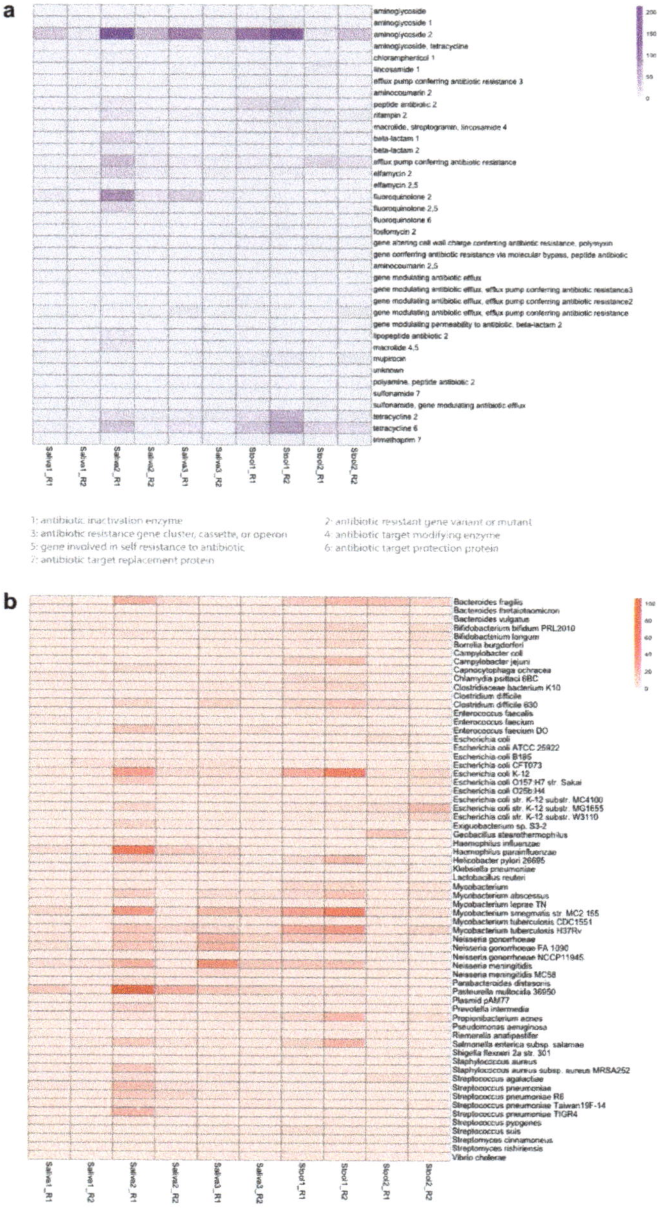

Figure 3. The Comprehensive Antibiotic Resistance Database (CARD) resistance ontology in each microbiome replicate based on (**a**) antibiotic category and (**b**) taxon conferring resistance to antibiotics. Antibiotics are classified based on their mechanism of action, spectrum of activity, or chemical structure. The antibiotic category shows all resistance pathways linking the gene to antibiotic molecules. The heatmap scale shows the number of alignments to the antibiotic category or the taxon conferring resistance to antibiotics.

As shown in our microbiome samples, many conventional antibiotics as well as some newer antibiotics are no longer effective in certain types of bacteria due to the spread of antimicrobial resistance. Recently, the World Health Organization (WHO) has designated antimicrobial resistance as one of the top threats against public health and published a list of pathogens that are in urgent need of novel antibiotics [2]. The WHO list is divided into three levels of priority (critical, high, and medium) according to the severity of antimicrobial resistance and the urgency for novel antibiotics. We compared the WHO list with the microbial species present in each microbiome sample, and we found three bacterial species (*Neisseria gonorrhoeae, Shigella flexneri, Streptococcus pneumoniae*) that matched the list (Table 3). In all the saliva samples, we found *Neisseria gonorrhoeae* included in the high-priority category (Table 3), which are found to be resistant to cephalosporin or fluoroquinolone (Table S5). Fluoroquinolones are one of the most important antibiotics listed by the WHO, as they inhibit DNA replication by preventing the ligase activity of the bacterial DNA gyrase and topoisomerase IV [63]. In Gram-negative bacteria, plasmid-mediated resistance genes produce proteins that can bind to the bacterial DNA gyrase, protecting it from the action of quinolones [64].

Table 3. The AMR-conferring taxa and their characteristics in the oral and gut microbiome of the human samples. Multidrug therapy implies that this pathogen requires multiple antibiotics to treat the related disease. N/A is used to abbreviate the phrase "not applicable".

Gram-Positive						
Phylum	**Taxon**	**Colony**	**Spore**	**Respiration**	**Disease**	**Antimicrobial Therapy**
Actinomycetota	*Bifidobacterium bifidum*	Rod	No	Anaerobic	N/A	N/A
Actinomycetota	*Bifidobacterium longum*	Rod	No	Anaerobic	N/A	N/A
Actinomycetota	*Cutibacterium acnes*	Rod	No	Anaerobic	Skin infections	Benzoyl peroxide
Actinomycetota	*Mycobacterium leprae*	Rod	No	Aerobic	Hansen's disease	Multidrug therapy
Actinomycetota	*Mycobacterium smegmatis*	Rod	No	Aerobic	N/A	N/A
Actinomycetota	*Mycobacterium tuberculosis*	Rod	No	Aerobic	Tuberculosis	Multidrug therapy
Actinomycetota	*Mycobacteroides abscessus*	Rod	No	Aerobic	Lung disease	Macrolide
Actinomycetota	*Mycobacteroides chelonae*	Rod	No	Aerobic	Skin infections	Macrolide
Actinomycetota	*Streptomyces cinnamoneus*	Filamentous	Yes	Aerobic	N/A	N/A
Actinomycetota	*Streptomyces rishiriensis*	Filamentous	Yes	Aerobic	N/A	N/A
Bacillota	*Clostridioides difficile*	Rod	Yes	Anaerobic	Colon infections	Glycopeptide
Bacillota	*Enterococcus faecium*	Cocci	No	Facultative anaerobic	Urinary tract infections	Glycopeptide
Bacillota	*Enterococcus faecium*	Rod	No	Facultative anaerobic	N/A	N/A
Bacillota	*Lactobacillus reuteri*	Rod	No	Anaerobic	N/A	N/A
Bacillota	*Staphylococcus aureus*	Cocci	No	Facultative anaerobic	Skin infections	Oxazolidinone

Table 3. *Cont.*

Bacillota	*Streptococcus agalactiae*	Cocci	No	Facultative anaerobic	Group B Streptococcal (GBS) infections	Ampicillin
Bacillota	*Streptococcus pneumoniae*	Diplococci	No	Facultative anaerobic	Pneumonia	Multidrug therapy
Bacillota	*Streptococcus pyogenes*	Cocci	No	Facultative anaerobic	Group A Streptococcal (GAS) Infections	Amoxicillin
Bacillota	*Streptococcus suis*	Cocci	No	Facultative anaerobic	Zoonotic disease	Aminopenicillin
			Gram-Negative			
Phylum	**Taxon**	**Colony**	**Spore**	**Respiration**	**Disease**	**Antimicrobial therapy**
Bacteroidota	*Bacteroides fragilis*	Rod	No	Anaerobic	Inflammatory bowel disease	Nitroimidazole
Bacteroidota	*Bacteroides vulgatus*	Rod	No	Anaerobic	Inflammatory bowel disease	Nitroimidazole
Bacteroidota	*Capnocytophaga ochracea*	Rod	No	Facultative anaerobic	Capnocytophaga infection	Multidrug therapy
Bacteroidota	*Parabacteroides distasonis*	Rod	No	Anaerobic	N/A	N/A
Bacteroidota	*Prevotella intermedia*	Rod	No	Anaerobic	Periodontal infections	Nitroimidazole
Campylobacterota	*Campylobacter jejuni*	Rod	No	Microaerophilic	Gastroenteritis	Macrolide
Campylobacterota	*Helicobacter pylori*	Helical	No	Microaerophilic	Stomach ulcers	Multidrug therapy
Chlamydiota	*Chlamydia psittaci*	Cocci	No	Anaerobic	Psittacosis	Macrolide
Pseudomonadota	*Escherichia coli*	Rod	No	Facultative anaerobic	Escherichia coli infection	Tetracycline
Pseudomonadota	*Haemophilus parainfluenzae*	Cocci	No	Facultative anaerobic	Pneumonia	Cephalosporin
Pseudomonadota	*Klebsiella pneumoniae*	Rod	No	Facultative anaerobic	Klebsiella pneumoniae infection	Carbapenem
Pseudomonadota	*Neisseria gonorrhoeae*	Diplococci	No	Anaerobic	Gonorrhea	Cephalosporin
Pseudomonadota	*Neisseria meningitidis*	Diplococci	No	Anaerobic	Meningitis	Cephalosporin
Pseudomonadota	*Pasteurella multocida*	Cocci	No	Facultative anaerobic	Subcutaneous infection	Aminopenicillin
Pseudomonadota	*Salmonella enterica*	Rod	No	Facultative anaerobic	Salmonellosis	Fluoroquinolone
Pseudomonadota	*Shigella flexneri*	Rod	No	Facultative anaerobic	Shigellosis	Fluoroquinolone
Pseudomonadota	*Vibrio cholerae*	Rod	No	Facultative anaerobic	Cholera infection	Tetracycline
Spirochaetota	*Borrelia burgdorferi*	Helical	No	Anaerobic	Lyme disease	Tetracycline

In some microbiome samples, we found some bacteria of a medium priority category from the WHO list, including *Shigella* which are also resistant to fluoroquinolone (Table S5). One of the stool samples (Stool2_R1) contains the same genus of bacteria named *Shigella flexneri* (Table 3). The point mutations in the DNA gyrase (*gyrA*) give rise to fluoroquinolone resistance, and we found the *gyrA* genes that confer resistance to fluoroquinolone in these bacteria (Table S6). It is intriguing to observe that the saliva and stool microbiomes all had these genes because they are known to cause cross-resistance to fluoroquinolones. For instance, recent research shows the *Mycobacterium tuberculosis* strain with the gene variant *gyrA* exhibits cross-resistance to six different fluoroquinolones, whereas the strain which does not have mutations in *gyrA* shows resistance specifically to the particular fluoroquinolones [65]. Another bacteria that matched the medium priority category is *Streptococcus pneumoniae* (Table 3), which is no longer susceptible to penicillin (Table S5). Bacteria can acquire resistance by synthesizing an enzyme such as beta-lactamase that attacks the beta-lactam ring of penicillin molecules. There are also other ways to become penicillin-resistant through mechanisms that decrease the binding affinity of the antibiotics. In all the saliva microbiome samples, *Streptococcus pneumoniae* has mutated variants of PBP1a, PBP2b, and PBP2x (Table S4). These penicillin-binding proteins (PBPs) are targeted by beta-lactam antibiotics [66], thus these mutations in the PBPs can lead to resistance against penicillin.

In the AMR analysis, we noticed that a wide variety of nonpathogenic bacteria have numerous and diverse AMR-related genes (Figure 3b). For example, *Mycobacterium smegmatis* are nonpathogenic bacteria but they are one of the most abundant bacteria that confer resistance to antibiotics in both the oral and gut microbiomes. *Haemophilus parainfluenzae* and *Bacteroides fragilis* are other examples of nonpathogenic bacteria that are present across all the microbiomes. These nonpathogenic bacteria are potential reservoirs for AMR-related genes through horizontal gene transfer, which is the primary mechanism for the spread of antibiotic resistance in bacteria [67]. Nonpathogenic bacteria which are in the same genus as pathogenic bacteria are of particular concern as their horizontal gene transfer is facilitated. For example, both *Mycobacterium tuberculosis* and *Mycobacterium smegmatis* with AMR-related genes are present in both the oral and gut microbiomes with high abundance [68].

2.5. Deep Learning-Based Classification of Unidentified Microbes Predicts Mobilome

The fast-sequencing mode of the nanopore data involves the taxonomic classification of metagenomic sequences in real time. This fast mode is enabled by a cloud-based platform integrated into the sequencing software, and it utilizes the benefits of long reads to enable rapid species identification and quantification from metagenomic samples based on the sequence similarity algorithm [29]. However, this sequence similarity-based approach does not fully exploit the potential of nanopore sequencing to produce long-read DNAs that can be regarded as a long stretch of DNA from a microbe, or even an individual. During the fast sequencing analysis, we noticed that there were many 'unclassified' reads in the classification results based on the sequence similarity algorithm. On average, the oral microbiome had around 10,000 unclassified reads and the gut microbiome had around 40,000 unclassified reads. We assumed that these unclassified reads are unidentifiable as unexplored organisms in the human microbiome, and we had a hypothesis that many of these unidentified reads are from mobile genetic elements such as bacteriophages and lasmids.

To test this hypothesis that these unidentified reads derive from mobile genetic elements, we searched for a different type of taxonomic algorithm that can classify a sequence without the presence of similar sequences in the search database. We found that deep learning-based algorithms can place de novo sequences in taxonomic categories with high accuracy when trained with a huge diversity and quantity of genetic sequences, exploiting the fact that different species have their specific patterns and characteristics engraved in their genetic information [69]. For example, eukaryotic genomes tend to have more noncoding regions compared to prokaryotic genomes, whereas bacteriophages are recently found to adapt alternative genetic coding to increase fitness and evolvability [70,71].

In the slow sequencing mode, we analyzed each unidentified read using a deep learning-based approach to assign taxonomic classification at the superkingdom level (Figure 4). The heatmaps show the predicted phylum of all the samples for each superkingdom, revealing the stool samples have more diversity in the four superkingdoms than the saliva samples as expected (Figure S6). The heatmap of the virus superkingdom is of particular interest, which is labeled with the predicted host phylum of each read. According to the deep learning-based approach, the oral and gut microbiomes are expected to have diverse viruses against archaea, bacteria, and eukaryotes, including against Actinobacteria, Crenarchaeota, and Arthropoda. Another interesting observation is that many DNA reads are still unknown even after the deep learning-based classification that does not utilize any database for inference. This reveals that some de novo reads in these microbiomes are completely devoid of any known patterns and characteristics, which is an intriguing observation to be investigated further.

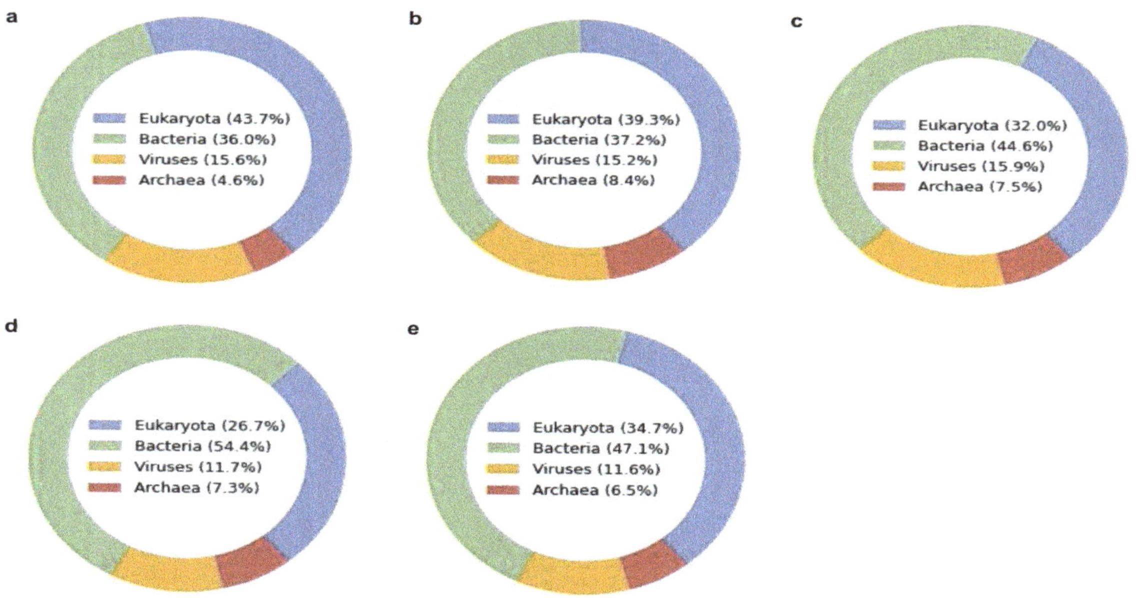

Figure 4. Superkingdom of unidentified reads predicted by the deep learning-based algorithm (BERTax) in each microbiome sample (**a**) Saliva1, (**b**) Saliva2, (**c**) Saliva3, (**d**) Stool1, and (**e**) Stool2. The two replicates per sample were combined for this exploratory analysis. Components may not sum to totals because of rounding.

Followingly, the predicted classification of unidentified reads from each microbiome sample is separated into four superkingdoms of archaea, bacteria, eukaryotes, and viruses, and summarized into the bar charts at the genus level (Figures S7–S12). The deep learning-based classification of this dataset at the genus level shows an intriguing range of diversity in the classification. Particularly, the diversity at the genus level of the eukaryotic organisms was rich in all the microbiome samples, but this may be due to the training dataset of the deep learning-based approach having a bias towards eukaryotic genomes [69]. The statistical summary of this analysis shows that many of the unidentified reads are classified into the virus category according to the deep learning-based algorithm (Table S8). This number is overrepresented as compared to the previous taxonomic classification of viruses based on sequence similarity (Table S3). The bar chart of the predicted virus at the genus level (Figure S12) is particularly interesting as they reveal the unexplored territory of mobilome in the human microbiome that is yet to be discovered for novel therapeutic tools and bacteriophage therapy [72–75]. We noticed some inconsistencies in the prediction at the different levels of superkingdom, phylum, and genus, thus this deep learning-based approach should be regarded more as an exploratory tool rather than a diagnostic tool.

3. Discussion

The development and implementation of rapid and accurate diagnostic tools for bacterial infections are essential in combating the current crisis of antimicrobial resistance (AMR) effectively. This requires a shift away from traditional culture-based techniques towards molecular-based diagnostic methods, which can provide faster and more accurate results, leading to better patient outcomes. Here, we focused on the ability of nanopore sequencing to generate long-read native DNAs from metagenomic samples of various human microbiomes. Nanopore sequencing enables direct analysis of DNA/RNA sequences by sensing changes in an electric current as they pass through a protein nanopore [76]. This new sequencing technology is revolutionizing genomics, as it can produce long-read DNA/RNA sequences allowing genomic analysis of microbes at individual levels. We explored the potential of nanopore sequencing as a futuristic diagnostic tool in clinical and laboratory setting [76,77], which could provide ultrarapid profiling of the human microbiome through real-time analysis such as species identification and antimicrobial resistance [2].

We further explored the potential of nanopore sequencing to be utilized in two different modes as a diagnostic tool: fast sequencing and slow sequencing. The fast mode enables real-time analysis of pathogen identification, metagenomic analysis of microbial communities, and antimicrobial resistance analysis. This mode is rapid and direct, requiring minimal inputs of human expertise and curation. In this fast analysis, we classified thousands of microbial species in the saliva and stool samples, with the most cost-effective but a lower-yield and single-use version of nanopore flow cells [78]. Furthermore, we rapidly identified the ten most abundant species that are known to be beneficial, harmful, or commensal in the oral and gut microbiome using the previously curated list. The slow mode enables in-depth analysis of host–microbe interactions and deep learning-based classification of unidentified reads. This mode is deliberate and exploratory, requiring the most advanced bioinformatic skills and expertise in microbiome research. A thorough exploration of host–microbe interaction types underscore the existing knowledge gaps regarding the impact of numerous microbes identified within the oral and gut microbiomes. Additionally, we evaluated a largely unexplored dataset of unclassified DNA reads from the sequence similarity-based analysis by utilizing a deep learning-based algorithm that does not require a match in the database to predict the superkingdom, phylum, and genus of these reads. The analysis further uncovers the potential existence of diverse organisms belonging to bacteria, archaea, and eukaryotes, with a significantly higher proportion of reads predicted to originate from virus genomes.

In this study, we aim to provide an exploratory application of nanopore sequencing as a future diagnostic tool for bacterial infection, which has resurfaced in the scientific community as an urgent global health issue due to the uncontrolled spread of antimicrobial resistance [79]. Nevertheless, it is important to acknowledge several caveats that were encountered during this exploratory application. Firstly, there are still debates about the accuracy of nanopore sequencers at simplex sequencing, which depends on the nanopore version, chemistry, and basecalling algorithms. According to the manufacturer, we used the flow cell version and chemistry (R9.4.1 and SQK-LSK110, respectively) with the expected raw-read accuracy of 98.3% modal. Regarding the accuracy of read classification, a recent study investigated that the taxonomic classification of long-read DNAs is satisfactory through controlled experiments using mock microbial communities [80]. This study further demonstrated that the expected microbial species corresponded at anticipated abundances, with the limit of detection observed at 4 reads and 5000 bp in length. However, we still had difficulties in determining the confidence level of very rare species, despite setting a high Q-score threshold while using the high-accuracy basecalling. Since the current state-of-the-art sequencing technologies cannot provide ground truth about the presence of these rare species in metagenomic samples, we utilized two replicates per sample to build confidence in the results of species identification.

Secondly, we used a specialized microbiome kit for extracting native DNA from different human microbiomes. While the kit was successful in extracting sufficient amounts of DNA from saliva and stool samples, it was not effective for extracting DNA from urine, nasal, and vaginal samples (Table S1). Alternative extraction methods that can extract small quantities of microbial DNA more efficiently may be necessary for these types of samples. Finally, the extracted DNA from some saliva samples using this kit still had a substantial fraction of human DNA despite having a host DNA depletion step. We suggest using other methods of human DNA depletion to enrich the microbiome DNA against the human DNA [81]. Adaptive sampling has emerged as a cutting-edge approach for selectively reducing host DNA content in human samples [82]. Adaptive sampling is a technique in nanopore sequencing that allows for selective sequencing of specific genomic regions of interest, optimizing the sequencing process by focusing on relevant regions and reducing sequencing time and cost [83]. It involves real-time analysis of the sequencing data and adjustment of the sequencing parameters to increase the coverage of targeted regions.

Lastly, our data exhibited some limitations, including instances where certain species were not consistently detected across samples or replicates, as well as the identification of species without established associations with the human microbiome. These discrepancies may be attributed to the current limitations in detection thresholds and error rates inherent in this particular version of the nanopore sequencing platform. It is anticipated that future advancements in long-read sequencing technology will enhance the detection threshold and accuracy of species identification. Additionally, incorporating validation through alternative sequencing methods such as next-generation sequencing (NGS) or polymerase chain reaction (PCR) can help mitigate the potential for false-negative results, particularly in identifying rare species. In future studies, it is important to account for other factors that contribute to variation in human microbiome compositions, including gender, age, medication usage, and dietary supplements. Considering these additional sources of variation will provide a more comprehensive understanding of the factors influencing the human microbiome and its relationship with health and disease [84].

4. Conclusions

In conclusion, rapid and accurate pathogen identification and microbial profiling are essential in combating infectious diseases effectively, and the development of new technologies, such as nanopore sequencing, offers great promise as innovative diagnostic tools. The main advantages of nanopore sequencing as a diagnostic tool include a cost-effective sequencer ($1000) and flow cell ($100 per sample) and flexible adaptation of downstream analysis as a fast mode (<12 h to pathogen identification) and a slow mode (several weeks) depending on the type of information needed. Nevertheless, addressing the existing challenges and ensuring the extensive utilization of these technologies in clinical settings necessitates further efforts and advancements.

5. Methods

5.1. Preparation of Non-Invasion Human Microbiome Sample

Human microbiomes were obtained from female healthy volunteers who provided written informed consent between March 2022 and July 2022. For the saliva samples (Saliva1_R1, Saliva1_R2, Saliva2_R1, Saliva2_R2, Saliva3_R1, and Saliva3_R2), saliva collected with sterile medical swabs were transferred to 1000 μL of PBS solution (P5119; Sigma-Aldrich, Darmstadt, Germany). For the stool samples (Stool1_R1, Stool1_R2, Stool2_R1, Stool2_R2), stool collected with sterile medical swabs was transferred to 1000 μL of PBS solution. For the urine, nasal, and vaginal samples (Urine1_R1, Urine1_R2, Nasal1_R1,

Nasal1_R2, Vaginal1_R1, and Vaginal1_R2), each sample was collected with sterile medical swabs and transferred to 1000 μL of PBS solution.

5.2. Microbiome DNA Extraction and Quality Control

Microbiome DNAs were extracted from the human samples on the same day of collection using the microbiome-specific kit (QIAamp DNA Microbiome Kit; Qiagen, Hilden, Germany). The DNA extraction was performed according to the manufacturer's protocol. The only modification to the protocol was to conduct all the centrifuge steps at the speed of 12,300× g instead of 20,000× g. The extracted DNA samples were quantified for the quantity (ng/μL) and quality (A260/A280 and A260/A230) using a spectrophotometer (NanoDrop 2000; Thermo Scientific, Waltham, MA, USA). The A260/A280 acceptable ratio was kept at 1.8–2.0, and the A260/A230 acceptable ratio was kept at 2.0–2.2 for the quality control for nanopore sequencing (Table S1). The quality-controlled DNA samples were kept at 4 °C until further treatment or analysis was performed.

5.3. Preparation of Sequencing Library Using Native DNA Ligation

The library ligation step was PCR-free without the need of primer choices, as we aimed to sequence native DNA from human microbiome samples to fully take advantage of long-read sequencing. The sequencing library was prepared from at least 500 ng of high molecular weight genomic DNA extracted from the human microbiome samples using the native DNA ligation kit (SQK-LSK110; Oxford Nanopore Technologies, Oxford, UK) according to the manufacturer's protocol. For Flongle flow cells, an expansion kit (EXP-FSE001; Oxford Nanopore Technologies, Oxford, UK) was additionally needed to prepare the sequencing mix. The NEBNext FFPE Repair Mix (M6630) and NEBNext Ultra II End repair/dA-tailing Module (E6056) reagents were prepared in accordance with the manufacturer's instructions. The sample purification was performed using magnetic beads (Agencourt AMPure XP; Beckman Coulter, Orange County, CA, USA) and a magnetic separator (DynaMag^{TM}-2 Magnet; Thermo Fischer Scientific, Waltham, MA, USA).

5.4. Nanopore Sequencing Using MinION and Flongle Adapter and Flow Cell

A Flongle flow cell (FLO-FLG001; Oxford Nanopore Technologies, Oxford, UK) was used for each sample, which was inserted into the Nanopore MinION sequencer (Mk1B MIN-101B; Oxford Nanopore Technologies, Oxford, UK) using the Flongle adapter (ADP-FLG001; Oxford Nanopore Technologies, Oxford, UK). Flongle flow cells were first checked for the minimum number of pores (at least 50 pores) before being primed with 119 μL of the priming mix prepared in accordance with the manufacturer's instructions. In the priming step, some liquid was left in the P200 pipette tip to ensure no air bubble was inserted. A total of 29 μL of the sequencing mix was loaded onto the Flongle flow cell immediately afterward, following the manufacturer's protocol. Finally, Flongle flow cells were sealed using the adhesive on the seal tab and the platform lid, and nanopore sequencing was performed for at least 12 h to obtain a maximum read output.

5.5. Real-Time High-Accuracy Basecalling and Cloud-Based EPI2ME Analysis

The MinKNOW software was used for raw data acquisition. The raw signal data in FAST5 files were basecalled real-time into the DNA reads in FASTQ files using the high-accuracy mode of the Guppy basecaller (v.5.1.13), integrated within the MinKNOW software.

For the rapid downstream analysis, a cloud-based analysis platform providing rapid analysis workflows called EPI2ME was used. Using the EPI2ME platform (v.3.5.7; Oxford Nanopore Technologies, Oxford, UK), species identification with the WIMP workflow (v.2021.11.26) such as fungi, bacteria, viruses, or archaea, was conducted in real-time based on the Centrifuge classification engine [29,84]. Next, antimicrobial resistance analysis was conducted in real-time with the ARMA workflow (v.2021.11.26.) to identify the genes responsible for antibiotic resistance in the DNA reads, based on the CARD database.

5.6. In-Depth Microbiome Analysis of Classified Reads

The WIMP workflow utilizes long reads from nanopore sequencing to rapidly identify and quantify microbial species from metagenomic samples. The WIMP results from each sample were downloaded as CSV files, which were processed into classified and unclassified categories. The classified reads from the WIMP workflow were saved separately, and the identified species were further categorized into four host–microbe interaction types (beneficial, commensal, harmful, and inconclusive). The initial list of host–microbe interaction types for several microbial species was curated by pooling a number of studies on the oral microbiome [32,39–41] and the gut microbiome [43,44]. However, many of the microbial species in the oral and gut microbiomes were missing from this curated list, which further required an extensive literature review on each microbial species to determine the host–microbe interaction type. When assessing these bacteria into different interaction types, the exact region in the human body was considered. For example, a commensal in the human gut may be assessed as a pathogen in the human skin.

5.7. In-Depth Microbiome Analysis of Unclassified Reads

The unclassified reads from the WIMP workflow based on the sequence similarity search were saved separately and analyzed with other methods. These latest algorithms for species identification include the BERTax taxonomic classification [85]. The BERTax taxonomic classification is a deep learning approach based on natural language processing [86] to classify the superkingdom and phylum of DNA sequences taxonomically. It achieves the assignment of unknown sequences to biological clades with shared ancestry in data-dependent training without the need for a genome similarity search of large genome databases. BERTax was shown to perform comparably to the state-of-the-art methods for sequences with close relatives in the database and superior for new species [69]. The unclassified reads from the human microbiome samples were run with the BERTax algorithm to assign the superkingdom, phylum, and genus given the patterns of DNA sequences.

Supplementary Materials: The following supporting information can be downloaded at: https://www.mdpi.com/article/10.3390/applbiosci2030028/s1.

Author Contributions: Experiments were primarily conducted by H.S. Analyses were primarily conducted by Y.P., J.L. and H.S. Specifically, EPI2ME analyses were performed by H.S., classified nanopore analyses were led by Y.P. and unclassified nanopore analyses were conducted by J.L. and H.S. The study was conceived by H.S. and all authors contributed to writing the manuscript. All authors have read and agreed to the published version of the manuscript.

Funding: The research and development activities described in this study were funded by Ghent University Global Campus (GUGC), Incheon, Republic of Korea.

Institutional Review Board Statement: The study was conducted in accordance with the Declaration of Helsinki, and approved by the Institutional Review Board of GUGC (IACUC 2022-008).

Informed Consent Statement: Informed consent was obtained from all subjects involved in the study.

Data Availability Statement: All codes related to this project are available under an open-source license at https://github.com/hshimlab/Nanopore_microbiome. For data analysis, Python v.3.6.4, NumPy v.1.17.5, SciPy v.1.1.0, seaborn v.0.9.0, Matplotlib v.3.3.4, pandas v.0.22.0 were used. For nanopore data acquisition, we used the MinKNOW v.21.11.8 and MinKNOW core v.4.5.4. For rapid nanopore data analysis, we used the EPI2ME platform v.3.5.7. For taxonomic analysis, we used the BERTax taxonomic classification.

Acknowledgments: We thank the members of the Center for Biotech Data Science at GUGC for their encouragement, support, and motivation. The research and development activities described in this study were funded by Ghent University Global Campus (GUGC), Incheon, Korea.

Conflicts of Interest: The authors declare no conflict of interest.

References

1. Maurer, F.P.; Christner, M.; Hentschke, M.; Rohde, H. Advances in Rapid Identification and Susceptibility Testing of Bacteria in the Clinical Microbiology Laboratory: Implications for Patient Care and Antimicrobial Stewardship Programs. *Infect. Dis. Rep.* **2017**, *9*, 6839.
2. Shim, H. Three Innovations of Next-Generation Antibiotics: Evolvability, Specificity, and Non-Immunogenicity. *Antibiotics* **2023**, *12*, 204. [CrossRef] [PubMed]
3. Tsalik, E.L.; Petzold, E.; Kreiswirth, B.N.; Bonomo, R.A.; Banerjee, R.; Lautenbach, E.; Evans, S.R.; Hanson, K.E.; Klausner, J.D.; Patel, R.; et al. Advancing Diagnostics to Address Antibacterial Resistance: The Diagnostics and Devices Committee of the Antibacterial Resistance Leadership Group. *Clin. Infect. Dis.* **2017**, *64*, S41–S47. [CrossRef] [PubMed]
4. Preidis, G.A.; Versalovic, J. Targeting the human microbiome with antibiotics, probiotics, and prebiotics: Gastroenterology enters the metagenomics era. *Gastroenterology* **2009**, *136*, 2015–2031. [PubMed]
5. Frank, D.N.; St Amand, A.L.; Feldman, R.A.; Boedeker, E.C.; Harpaz, N.; Pace, N.R. Molecular-phylogenetic characterization of microbial community imbalances in human inflammatory bowel diseases. *Proc. Natl. Acad. Sci. USA* **2007**, *104*, 13780–13785. [CrossRef]
6. Gonzalez, A.; Stombaugh, J.; Lozupone, C.; Turnbaugh, P.J.; Gordon, J.I.; Knight, R. The mind-body-microbial continuum. *Dialogues Clin. Neurosci.* **2011**, *13*, 55–62. [CrossRef]
7. Belkaid, Y.; Hand, T.W. Role of the microbiota in immunity and inflammation. *Cell* **2014**, *157*, 121–141. [CrossRef]
8. Buffie, C.G.; Pamer, E.G. Microbiota-mediated colonization resistance against intestinal pathogens. *Nat. Rev. Immunol.* **2013**, *13*, 790–801. [CrossRef]
9. Caballero, S.; Pamer, E.G. Microbiota-mediated inflammation and antimicrobial defense in the intestine. *Annu. Rev. Immunol.* **2015**, *33*, 227–256. [CrossRef]
10. Sommer, F.; Bäckhed, F. The gut microbiota--masters of host development and physiology. *Nat. Rev. Microbiol.* **2013**, *11*, 227–238. [CrossRef]
11. O'Toole, P.W.; Jeffery, I.B. Gut microbiota and aging. *Science* **2015**, *350*, 1214–1215. [CrossRef] [PubMed]
12. Barczak, A.K.; Gomez, J.E.; Kaufmann, B.B.; Hinson, E.R.; Cosimi, L.; Borowsky, M.L.; Onderdonk, A.B.; Stanley, S.A.; Kaur, D.; Bryant, K.F.; et al. RNA signatures allow rapid identification of pathogens and antibiotic susceptibilities. *Proc. Natl. Acad. Sci. USA* **2012**, *109*, 6217–6222. [CrossRef] [PubMed]
13. Caliendo, A.M.; Gilbert, D.N.; Ginocchio, C.C.; Hanson, K.E.; May, L.; Quinn, T.C.; Tenover, F.C.; Alland, D.; Blaschke, A.J.; Bonomo, R.A.; et al. Better tests, better care: Improved diagnostics for infectious diseases. *Clin. Infect. Dis.* **2013**, *57* (Suppl. 3), S139–S170. [CrossRef] [PubMed]
14. Salipante, S.J.; Sengupta, D.J.; Rosenthal, C.; Costa, G.; Spangler, J.; Sims, E.H.; Jacobs, M.A.; Miller, S.I.; Hoogestraat, D.R.; Cookson, B.T.; et al. Rapid 16S rRNA next-generation sequencing of polymicrobial clinical samples for diagnosis of complex bacterial infections. *PLoS ONE* **2013**, *8*, e65226.
15. Bradley, P.; Gordon, N.C.; Walker, T.M.; Dunn, L.; Heys, S.; Huang, B.; Pankhurst, L.J.; Anson, L.; de Cesare, M.; Piazza, P.; et al. Rapid antibiotic-resistance predictions from genome sequence data for Staphylococcus aureus and Mycobacterium tuberculosis. *Nat. Commun.* **2015**, *6*, 10063. [CrossRef] [PubMed]
16. Ayling, M.; Clark, M.D.; Leggett, R.M. New approaches for metagenome assembly with short reads. *Brief. Bioinform.* **2020**, *21*, 584–594. [CrossRef]
17. Branton, D.; Deamer, D.W.; Marziali, A.; Bayley, H.; Benner, S.A.; Butler, T.; Di Ventra, M.; Garaj, S.; Hibbs, A.; Huang, X.; et al. The potential and challenges of nanopore sequencing. *Nat. Biotechnol.* **2008**, *26*, 1146–1153. [CrossRef]
18. Quick, J.; Quinlan, A.R.; Loman, N.J. A reference bacterial genome dataset generated on the MinION^{TM} portable single-molecule nanopore sequencer. *GigaScience* **2014**, *3*, 2047-217X-3-22. [CrossRef]
19. Niedringhaus, T.P.; Milanova, D.; Kerby, M.B.; Snyder, M.P.; Barron, A.E. Landscape of next-generation sequencing technologies. *Anal. Chem.* **2011**, *83*, 4327–4341. [CrossRef]
20. Latorre-Pérez, A.; Villalba-Bermell, P.; Pascual, J.; Vilanova, C. Assembly methods for nanopore-based metagenomic sequencing: A comparative study. *Sci. Rep.* **2020**, *10*, 13588.
21. Charalampous, T.; Kay, G.L.; Richardson, H.; Aydin, A.; Baldan, R.; Jeanes, C.; Rae, D.; Grundy, S.; Turner, D.J.; Wain, J.; et al. Nanopore metagenomics enables rapid clinical diagnosis of bacterial lower respiratory infection. *Nat. Biotechnol.* **2019**, *37*, 783–792.
22. Petersen, L.M.; Martin, I.W.; Moschetti, W.E.; Kershaw, C.M.; Tsongalis, G.J. Third-Generation Sequencing in the Clinical Laboratory: Exploring the Advantages and Challenges of Nanopore Sequencing. *J. Clin. Microbiol.* **2019**, *58*, e01315-19. [CrossRef]
23. MacKenzie, M.; Argyropoulos, C. An Introduction to Nanopore Sequencing: Past, Present, and Future Considerations. *Micromachines* **2023**, *14*, 459. [CrossRef]
24. Jain, M.; Koren, S.; Miga, K.H.; Quick, J.; Rand, A.C.; Sasani, T.A.; Tyson, J.R.; Beggs, A.D.; Dilthey, A.T.; Fiddes, I.T.; et al. Nanopore sequencing and assembly of a human genome with ultra-long reads. *Nat. Biotechnol.* **2018**, *36*, 338–345.
25. Vereecke, N.; Bokma, J.; Haesebrouck, F.; Nauwynck, H.; Boyen, F.; Pardon, B.; Theuns, S. High quality genome assemblies of Mycoplasma bovis using a taxon-specific Bonito basecaller for MinION and Flongle long-read nanopore sequencing. *BMC Bioinform.* **2020**, *21*, 517. [CrossRef]

26. Napit, R.; Manandhar, P.; Chaudhary, A.; Shrestha, B.; Poudel, A.; Raut, R.; Pradhan, S.; Raut, S.; Rajbhandari, P.G.; Gurung, A.; et al. Rapid genomic surveillance of SARS-CoV-2 in a dense urban community of Kathmandu Valley using sewage samples. *PLoS ONE* **2023**, *18*, e0283664.
27. Ewing, B.; Hillier, L.; Wendl, M.C.; Green, P. Base-calling of automated sequencer traces using phred. I. Accuracy assessment. *Genome Res.* **1998**, *8*, 175–185.
28. Delahaye, C.; Nicolas, J. Sequencing DNA with nanopores: Troubles and biases. *PLoS ONE* **2021**, *16*, e0257521.
29. Kim, D.; Song, L.; Breitwieser, F.P.; Salzberg, S.L. Centrifuge: Rapid and sensitive classification of metagenomic sequences. *Genome Res.* **2016**, *26*, 1721.
30. O' Donnell, M.M.; Harris, H.M.B.; Lynch, D.B.; Ross, R.P.; O'Toole, P.W. Lactobacillus ruminis strains cluster according to their mammalian gut source. *BMC Microbiol.* **2015**, *15*, 80.
31. Sheng, S.; Yan, S.; Chen, J.; Zhang, Y.; Wang, Y.; Qin, Q.; Li, W.; Li, T.; Huang, M.; Ding, S.; et al. Gut microbiome is associated with metabolic syndrome accompanied by elevated gamma-glutamyl transpeptidase in men. *Front. Cell. Infect. Microbiol.* **2022**, *12*, 946757. [CrossRef]
32. Sedghi, L.; DiMassa, V.; Harrington, A.; Lynch, S.V.; Kapila, Y.L. The oral microbiome: Role of key organisms and complex networks in oral health and disease. *Periodontol. 2000* **2021**, *87*, 107–131.
33. Ahmed, U.; Chatterjee, T.; Kandula, M. Polyarteritis Nodosa: An unusual case of paraneoplastic process in renal cell carcinoma. *J. Community Hosp. Intern. Med. Perspect.* **2020**, *10*, 73.
34. De Sordi, L.; Khanna, V.; Debarbieux, L. The Gut Microbiota Facilitates Drifts in the Genetic Diversity and Infectivity of Bacterial Viruses. *Cell Host Microbe* **2017**, *22*, 801–808.e3.
35. Carr, V.R.; Shkoporov, A.; Hill, C.; Mullany, P.; Moyes, D.L. Probing the Mobilome: Discoveries in the Dynamic Microbiome. *Trends Microbiol.* **2021**, *29*, 158–170.
36. Dutilh, B.E.; Cassman, N.; McNair, K.; Sanchez, S.E.; Silva, G.G.Z.; Boling, L.; Barr, J.J.; Speth, D.R.; Seguritan, V.; Aziz, R.K.; et al. A highly abundant bacteriophage discovered in the unknown sequences of human faecal metagenomes. *Nat. Commun.* **2014**, *5*, 4498.
37. Kolde, R.; Franzosa, E.A.; Rahnavard, G.; Hall, A.B.; Vlamakis, H.; Stevens, C.; Daly, M.J.; Xavier, R.J. Host genetic variation and its microbiome interactions within the Human Microbiome Project. *Genome Med.* **2018**, *10*, 6.
38. Luca, F.; Kupfer, S.S.; Knights, D.; Khoruts, A.; Blekhman, R. Functional Genomics of Host-Microbiome Interactions in Humans. *Trends Genet.* **2018**, *34*, 30.
39. Zhang, Y.; Wang, X.; Li, H.; Ni, C.; Du, Z.; Yan, F. Human oral microbiota and its modulation for oral health. *Biomed. Pharmacother.* **2018**, *99*, 883–893.
40. Le Bars, P.; Matamoros, S.; Montassier, E.; Le Vacon, F.; Potel, G.; Soueidan, A.; Jordana, F.; de La Cochetière, M.-F. The oral cavity microbiota: Between health, oral disease, and cancers of the aerodigestive tract. *Can. J. Microbiol.* **2017**, *63*, 475–492.
41. Belstrøm, D. The salivary microbiota in health and disease. *J. Oral Microbiol.* **2020**, *12*, 1723975.
42. Willis, J.R.; Gabaldón, T. The Human Oral Microbiome in Health and Disease: From Sequences to Ecosystems. *Microorganisms* **2020**, *8*, 308. [CrossRef]
43. Matsue, M.; Mori, Y.; Nagase, S.; Sugiyama, Y.; Hirano, R.; Ogai, K.; Ogura, K.; Kurihara, S.; Okamoto, S. Measuring the Antimicrobial Activity of Lauric Acid against Various Bacteria in Human Gut Microbiota Using a New Method. *Cell Transplant.* **2019**, *28*, 1528–1541.
44. Vernocchi, P.; Del Chierico, F.; Putignani, L. Gut Microbiota Profiling: Metabolomics Based Approach to Unravel Compounds Affecting Human Health. *Front. Microbiol.* **2016**, *7*, 1144.
45. Yang, J.; Pu, J.; Lu, S.; Bai, X.; Wu, Y.; Jin, D.; Cheng, Y.; Zhang, G.; Zhu, W.; Luo, X.; et al. Species-Level Analysis of Human Gut Microbiota with Metataxonomics. *Front. Microbiol.* **2020**, *11*, 2029.
46. Jumas-Bilak, E.; Carlier, J.P.; Jean-Pierre, H.; Mory, F.; Teyssier, C.; Gay, B.; Campos, J.; Marchandin, H. Acidaminococcus intestini sp. nov., isolated from human clinical samples. *Int. J. Syst. Evol. Microbiol.* **2007**, *57*, 2314–2319. [CrossRef]
47. Loubinoux, J.; Bronowicki, J.-P.; Pereira, I.A.C.; Mougenel, J.-L.; Faou, A.E. Sulfate-reducing bacteria in human feces and their association with inflammatory bowel diseases. *FEMS Microbiol. Ecol.* **2002**, *40*, 107–112.
48. Lu, J.; Nogi, Y.; Takami, H. Oceanobacillus iheyensis gen. nov., sp. nov., a deep-sea extremely halotolerant and alkaliphilic species isolated from a depth of 1050 m on the Iheya Ridge. *FEMS Microbiol. Lett.* **2001**, *205*, 291–297.
49. Froidurot, A.; Julliand, V. Cellulolytic bacteria in the large intestine of mammals. *Gut Microbes* **2022**, *14*, 2031694.
50. Coutinho, T.A.; Venter, S.N. Pantoea ananatis: An unconventional plant pathogen. *Mol. Plant Pathol.* **2009**, *10*, 325–335.
51. Arashima, Y.; Kumasaka, K.; Okuyama, K.; Kawabata, M.; Tsuchiya, T.; Kawano, K.; Asano, R.; Hokari, S. Clinicobacteriological study of Pasteurella multocida as a zoonosis (1). Condition of dog and cat carriers of Pasteurella, and the influence for human carrier rate by kiss with the pets. *Kansenshogaku Zasshi. J. Jpn. Assoc. Infect. Dis.* **1992**, *66*, 221–224.
52. Dehoux, P.; Marvaud, J.C.; Abouelleil, A.; Earl, A.M.; Lambert, T.; Dauga, C. Comparative genomics of Clostridium bolteae and Clostridium clostridioforme reveals species-specific genomic properties and numerous putative antibiotic resistance determinants. *BMC Genom.* **2016**, *17*, 819.
53. Li, Q.; Zhou, F.; Su, Z.; Li, Y.; Li, J. A Confirmed Calcifying Bacterium With a Potentially Important Role in the Supragingival Plaque. *Front. Microbiol.* **2022**, *13*, 940643.

54. Ezeji, J.C.; Sarikonda, D.K.; Hopperton, A.; Erkkila, H.L.; Cohen, D.E.; Martinez, S.P.; Cominelli, F.; Kuwahara, T.; Dichosa, A.E.K.; Good, C.E.; et al. Parabacteroides distasonis: Intriguing aerotolerant gut anaerobe with emerging antimicrobial resistance and pathogenic and probiotic roles in human health. *Gut Microbes* **2021**, *13*, 1922241.
55. Sarshar, M.; Behzadi, P.; Scribano, D.; Palamara, A.T.; Ambrosi, C. Acinetobacter baumannii: An Ancient Commensal with Weapons of a Pathogen. *Pathogens* **2021**, *10*, 387. [CrossRef] [PubMed]
56. Cassir, N.; Benamar, S.; La Scola, B. Clostridium butyricum: From beneficial to a new emerging pathogen. *Clin. Microbiol. Infect.* **2016**, *22*, 37–45.
57. Krawczyk, B.; Wityk, P.; Gałęcka, M.; Michalik, M. The Many Faces of Enterococcus spp.—Commensal, Probiotic and Opportunistic Pathogen. *Microorganisms* **2021**, *9*, 1900. [CrossRef]
58. Parker, B.J.; Wearsch, P.A.; Veloo, A.C.M.; Rodriguez-Palacios, A. The Genus Alistipes: Gut Bacteria with Emerging Implications to Inflammation, Cancer, and Mental Health. *Front. Immunol.* **2020**, *11*, 906.
59. Mingeot-Leclercq, M.P.; Glupczynski, Y.; Tulkens, P.M. Aminoglycosides: Activity and resistance. *Antimicrob. Agents Chemother.* **1999**, *43*, 727–737.
60. Shakil, S.; Khan, R.; Zarrilli, R.; Khan, A.U. Aminoglycosides versus bacteria—A description of the action, resistance mechanism, and nosocomial battleground. *J. Biomed. Sci.* **2008**, *15*, 5–14.
61. Reeves, A.Z.; Campbell, P.J.; Sultana, R.; Malik, S.; Murray, M.; Plikaytis, B.B.; Shinnick, T.M.; Posey, J.E. Aminoglycoside cross-resistance in Mycobacterium tuberculosis due to mutations in the 5′ untranslated region of whiB7. *Antimicrob. Agents Chemother.* **2013**, *57*, 1857–1865. [PubMed]
62. Hormeño, L.; Ugarte-Ruiz, M.; Palomo, G.; Borge, C.; Florez-Cuadrado, D.; Vadillo, S.; Píriz, S.; Domínguez, L.; Campos, M.J.; Quesada, A. ant(6)-I Genes Encoding Aminoglycoside O-Nucleotidyltransferases Are Widely Spread Among Streptomycin Resistant Strains of Campylobacter jejuni and Campylobacter coli. *Front. Microbiol.* **2018**, *9*. 2515. [PubMed]
63. Hooper, D.C. Emerging mechanisms of fluoroquinolone resistance. *Emerg. Infect. Dis.* **2001**, *7*, 337–341.
64. Robicsek, A.; Jacoby, G.A.; Hooper, D.C. The worldwide emergence of plasmid-mediated quinolone resistance. *Lancet Infect. Dis.* **2006**, *6*, 629–640.
65. Cheng, A.F.B.; Yew, W.W.; Chan, E.W.C.; Chin, M.L.; Hui, M.M.M.; Chan, R.C.Y. Multiplex PCR amplimer conformation analysis for rapid detection of gyrA mutations in fluoroquinolone-resistant Mycobacterium tuberculosis clinical isolates. *Antimicrob. Agents Chemother.* **2004**, *48*, 596–601.
66. Schweizer, I.; Peters, K.; Stahlmann, C.; Hakenbeck, R.; Denapaite, D. Penicillin-binding protein 2x of Streptococcus pneumoniae: The mutation Ala707Asp within the C-terminal PASTA2 domain leads to destabilization. *Microb. Drug Resist.* **2014**, *20*, 250–257.
67. Arnold, B.J.; Huang, I.-T.; Hanage, W.P. Horizontal gene transfer and adaptive evolution in bacteria. *Nat. Rev. Microbiol.* **2021**, *20*, 206–218.
68. Sparks, I.L.; Derbyshire, K.M.; Jacobs, W.R., Jr.; Morita, Y.S. Mycobacterium smegmatis: The Vanguard of Mycobacterial Research. *J. Bacteriol.* **2023**, *205*, e0033722.
69. Mock, F.; Kretschmer, F.; Kriese, A.; Böcker, S.; Marz, M. Taxonomic classification of DNA sequences beyond sequence similarity using deep neural networks. *Proc. Natl. Acad. Sci. USA* **2022**, *119*, e2122636119.
70. Peters, S.L.; Borges, A.L.; Giannone, R.J.; Morowitz, M.J.; Banfield, J.F.; Hettich, R.L. Experimental validation that human microbiome phages use alternative genetic coding. *Nat. Commun.* **2022**, *13*, 5710.
71. Hammerling, M.J.; Ellefson, J.W.; Boutz, D.R.; Marcotte, E.M.; Ellington, A.D.; Barrick, J.E. Bacteriophages use an expanded genetic code on evolutionary paths to higher fitness. *Nat. Chem. Biol.* **2014**, *10*, 178–180.
72. Shim, H.; Shivram, H.; Lei, S.; Doudna, J.A.; Banfield, J.F. Diverse ATPase Proteins in Mobilomes Constitute a Large Potential Sink for Prokaryotic Host ATP. *Front. Microbiol.* **2021**, *12*, 691847.
73. Park, H.-M.; Park, Y.; Vankerschaver, J.; Van Messem, A.; De Neve, W.; Shim, H. Rethinking Protein Drug Design with Highly Accurate Structure Prediction of Anti-CRISPR Proteins. *Pharmaceuticals* **2022**, *15*, 310.
74. Shim, H. Investigating the genomic background of CRISPR-Cas genomes for CRISPR-based antimicrobials. *arXiv* **2022**, arXiv:2202.07171.
75. Park, H.-M.; Park, Y.; Berani, U.; Bang, E.; Vankerschaver, J.; Van Messem, A.; De Neve, W.; Shim, H. In silico optimization of RNA-protein interactions for CRISPR-Cas13-based antimicrobials. *Biol. Direct.* **2022**, *17*, 27.
76. Kasianowicz, J.J.; Brandin, E.; Branton, D.; Deamer, D.W. Characterization of individual polynucleotide molecules using a membrane channel. *Proc. Natl. Acad. Sci. USA* **1996**, *93*, 13770–13773.
77. Borges, A.S.G.; Basu, M.; Brinks, E.; Bang, C.; Cho, G.-S.; Baines, J.F.; Franke, A.; Franz, C.M.A.P. Fast identification method for screening bacteria from faecal samples using oxford nanopore technologies MinION sequencing. *Curr. Microbiol.* **2023**, *80*, 101.
78. Grädel, C.; Terrazos Miani, M.A.; Barbani, M.T.; Leib, S.L.; Suter-Riniker, F.; Ramette, A. Rapid and Cost-Efficient Enterovirus Genotyping from Clinical Samples Using Flongle Flow Cells. *Genes* **2019**, *10*, 659. [CrossRef]
79. Antimicrobial Resistance Collaborators. Global burden of bacterial antimicrobial resistance in 2019, a systematic analysis. *Lancet* **2022**, *399*, 629–655.
80. Nicholls, S.M.; Quick, J.C.; Tang, S.; Loman, N.J. Ultra-deep, long-read nanopore sequencing of mock microbial community standards. *GigaScience* **2019**, *8*, giz043.

81. Street, T.L.; Barker, L.; Sanderson, N.D.; Kavanagh, J.; Hoosdally, S.; Cole, K.; Newnham, R.; Selvaratnam, M.; Andersson, M.; Llewelyn, M.J.; et al. Optimizing DNA Extraction Methods for Nanopore Sequencing of Neisseria gonorrhoeae Directly from Urine Samples. *J. Clin. Microbiol.* **2020**, *58*, e01822-19. [CrossRef]
82. Marquet, M.; Zöllkau, J.; Pastuschek, J.; Viehweger, A.; Schleußner, E.; Makarewicz, O.; Pletz, M.W.; Ehricht, R.; Brandt, C. Evaluation of microbiome enrichment and host DNA depletion in human vaginal samples using Oxford Nanopore's adaptive sequencing. *Sci. Rep.* **2022**, *12*, 4000.
83. Martin, S.; Heavens, D.; Lan, Y.; Horsfield, S.; Clark, M.D.; Leggett, R.M. Nanopore adaptive sampling: A tool for enrichment of low abundance species in metagenomic samples. *Genome Biol.* **2022**, *23*, 11.
84. Tuddenham, S.; Koay, W.L.; Sears, C. HIV, sexual orientation, and gut microbiome interactions. *Dig. Dis. Sci.* **2020**, *65*, 800–817.
85. Hingamp, P.; Grimsley, N.; Acinas, S.G.; Clerissi, C.; Subirana, L.; Poulain, J.; Ferrera, I.; Sarmento, H.; Villar, E.; Lima-Mendez, G.; et al. Exploring nucleo-cytoplasmic large DNA viruses in Tara Oceans microbial metagenomes. *ISME J.* **2013**, *7*, 1678–1695.
86. Shim, H. Feature Learning of Virus Genome Evolution with the Nucleotide Skip-Gram Neural Network. *Evol. Bioinform.* **2019**, *15*, 1176934318821072.

applied biosciences

MDPI

Article

Determination of Target Crop Loads for Maximising Fruit Quality and Return Bloom in Several Apple Cultivars

Sally A. Bound

Tasmanian Institute of Agriculture, University of Tasmania, Private Bag 98, Hobart, TAS 7001, Australia; sally.bound@utas.edu.au; Tel.: +61-3-6226-2958

Abstract: In apple (*Malus domestica*), the level and timing of crop load have a major impact on the final fruit size and can also play a role in optimising internal fruit quality. Ideal crop loads vary with cultivar, but very few cultivars have recommended crop load targets that consider the effect of crop load on both return bloom and fruit quality. To address this issue, studies examining a range of crop loads and thinning times were undertaken on several apple cultivars. Return bloom and multiple fruit quality parameters were examined. The results of these studies demonstrate positive effects for early thinning, not only on fruit size but also on firmness and soluble solids content. Early-thinned fruit showed higher sugar levels than late-thinned fruit. Previously undemonstrated positive relationships between fruit sugar content and weight and between fruit firmness and weight in both 'Fuji' and 'Delicious', as well as between fruit sugar content and fruit firmness in 'Delicious', indicate that early thinning is a valuable tool in improving fruit quality. The current target crop load recommendations of 4–6 fruit cm^{-2} trunk cross-sectional area (TCSA) for 'Fuji' and 2–4 fruit cm^{-2} TCSA for 'Delicious' are confirmed by this study. New recommendations are proposed for the other cultivars in this study taking into account the impact of crop load on both fruit quality and return bloom. Both 'Pink Lady' and 'Gala' can support crop loads of up to eight fruit cm^{-2} TCSA without impacting return bloom, but fruit quality is compromised; hence, lower targets in the range of 4–6 fruit cm^{-2} TCSA are recommended. Large fruit size and good return bloom can be maintained in 'Jonagold' at crop loads of eight fruit cm^{-2} TCSA, while crop loads of four fruit cm^{-2} TCSA are suggested for 'Braeburn' to sustain regular bearing and good fruit size.

Keywords: thinning; fruit weight; total soluble solids; firmness; fruit shape; biennial bearing; hand thinning; crop load management

Citation: Bound, S.A. Determination of Target Crop Loads for Maximising Fruit Quality and Return Bloom in Several Apple Cultivars. *Appl. Biosci.* **2023**, 2, 586–606. https://doi.org/10.3390/applbiosci2040037

Academic Editor: Robert Henry

Received: 17 August 2023
Revised: 1 October 2023
Accepted: 19 October 2023
Published: 1 November 2023

1. Introduction

In commercial apple production, crop load is managed through the cultural practice of removing (thinning) excess flowers/fruitlets during the flowering and post-bloom periods. While mechanical thinning devices are being developed, chemicals are most often used for thinning; regardless of the thinning method employed, there is normally a need to follow up with hand thinning. This is usually undertaken after the second wave of natural fruit drop that occurs around 8–10 weeks after flowering. Crop load is commonly expressed as the number of fruit per square centimetre of trunk cross-sectional area (fruit cm^{-2} TCSA) [1]. The effect of crop load on fruit weight and size and on return bloom has been examined in multiple studies [2–6], but there is limited information available on the impact of crop load and time of thinning on other fruit quality attributes, such as fruit shape, skin colour, soluble solids content, and flesh firmness. Fruit soluble solids levels have been reported to be dependent on the leaf/fruit ratio [7]; hence, factors that result in an increase in leaf area and thus increased photosynthesis, such as lower crop loads, will aid in the accumulation of sugars in the fruit. Flesh firmness is determined by the number and size of cells within the cortex, with a large cell size resulting in softer fruit [6] and higher numbers of smaller cells producing firmer fruit [8]. Reducing crop load early in the season before the major period of cell division allows for a greater increase in cell number than later thinning [9].

Most studies relating crop load to fruit quality have involved the use of thinning chemicals as this is the most economical way of reducing crop load. Multiple reports demonstrate that thinning chemicals can also impact fruit quality [10–17], thus clouding the understanding of the impact of reducing crop load on fruit quality. As there are many situations where growers are loath to apply chemicals, particularly on younger trees or high-value cultivars, further examination of the impact of crop load on fruit quality is warranted. In commercial orchards, hand thinning may be undertaken either to complement inadequate chemical (or mechanical) thinning or in preference to the application of chemicals, but it is often not completed until 8–12 weeks after flowering. To investigate the effects of crop load on fruit quality independent of any possible direct influences of chemical thinners, the trials presented in this study examined the effects of the time and level of thinning performed without chemicals on fruit quality for several apple cultivars and, where available, different rootstocks.

The objective of this study was to evaluate the impact of different crop loads and thinning times on both return bloom and fruit quality in several apple cultivars to enable recommendations for ideal crop load targets for each cultivar, on the assumption that different cultivars would respond differently and, thus, the ideal crop load would vary between cultivars.

2. Materials and Methods

Eight trials were undertaken on six different cultivars over a four-year period. All trials were conducted in the Huon Valley, Tasmania (43°07′ S, 147°01′ E) on mature regular bearing trees. Details of cultivar, rootstock, tree age, height, and planting spacings are provided in Table 1.

Table 1. Details of apple cultivars used in each trial.

Trial	Cultivar	Rootstock	Height (m)	Age (Years)	Row Spacing (m)	Tree Spacing (m)
1	Naga-Fu No. 2 'Fuji'	MM106	2.5	9	4	3
2	Naga-Fu No. 2 'Fuji'	MM106	2.5	10	4	3
3	Oregon spur 'Delicious'	MM106	2.0	8	4	2.5
4	Oregon spur 'Delicious'	MM106	2.2	10	4	2.5
5	'Pink Lady'	M26 MM106	2.0 3.0	7	3	2
6	'Jonagold'	M26 MM106	2.0 3.0	7	3	2
7	'Braeburn'	M26 MM106	2.0 3.0	7	3	2
8	Royal 'Gala'	M26	2.0	6	3	1.5

Trees in all trials were trained to a central axis system. Apart from thinning, all trees were subjected to standard commercial orchard management practices.

In all trials, trees were selected in early spring based on uniformity of size and vigour, trunk girths were measured 10 cm above the graft union, and trunk cross-sectional areas (TCSA) calculated. Blossom clusters were counted on each tree and blossom density (number of blossom clusters cm^{-2} TCSA) was calculated. Trees were blocked according to blossom density and treatments were allocated at random to single-tree plots within each block.

To confirm that the results were not affected by seasonal conditions, the same design and treatments in trial 1 were repeated the following season in the same orchard block but on different trees (trial 2). Details of full bloom (FB) dates, number of replicates, time of thinning, and crop load level are provided in Table 2.

Table 2. Details of treatments (thinning time and crop load level), date of full bloom, and number of replicates in hand-thinning trials conducted over four seasons. AFB, after full bloom; TCSA, trunk cross-sectional area.

Trial	Season	Cultivar	Rootstock	Thinning Time (Weeks AFB)	Crop Load Levels (Fruit cm^{-2} TCSA)	Full Bloom Date	Replicate Number
1	1	'Fuji'	MM106	6	2, 4, 6, 8, 10	17 Oct	5
2	2	'Fuji'	MM106	6	2, 4, 6, 8, 10	14 Oct	5
3	1	'Delicious'	MM106	6	2, 4, 6, 8, 10	21 Oct	5
4	2	'Delicious'	MM106	1, 2, 4, 8, 12, 16	3, 6	18 Oct	3
5	3	'Pink Lady'	M26, MM106	2, 6, 10, 14	4, 6, 8	6 Oct	4
6	3	'Jonagold'	M26, MM106	2, 6, 10, 14	4, 6, 8	12 Oct	4
7	3	'Braeburn'	M26, MM106	2, 6, 10	2, 4, 6, 8	8 Oct	4
8	4	'Gala'	M26	2, 6, 10, 14	3, 6, 9	11 Oct	4

Crop loads were set by hand thinning and the retention of larger fruit was preferred over small and/or damaged fruit. Where possible, clusters were thinned to a single fruit, but if there was insufficient clusters on the tree to allow this, clusters were thinned to two or three fruit.

2.1. Assessments

2.1.1. Fruit Set Counts

Fruit set counts were completed in each trial in December after natural fruit drop and used to calculate the crop load variable, number of fruit cm^{-2} TCSA [18].

2.1.2. Fruit Weight and Size

Fruit was harvested by hand at normal commercial harvest time for each cultivar, based on measurements of the maturity indices total soluble solids (TSS) content, starch levels and skin background colour. Fruit from each tree were counted and weighed, and the mean fruit weight was calculated. Fruit was graded on a commercial size grader into increments of 5 mm in diameter ranging from 50 to 95 mm, and the percentage of fruit $\geq$ 75 mm in diameter was determined for trials 1–5 and 7. As 'Jonagold' produce large fruit, the percentage of fruit $\geq$ 85 mm in diameter was determined for trial 6, while the percentage of fruit $\geq$ 65 mm in diameter was used for 'Gala' in trial 8 as 'Gala' are a genetically small apple and the fruit size for this season was small. For laboratory quality assessments, samples of 28 fruit per replicate were collected from the grader by taking seven fruit at random from each of the 60, 65, 70, and 75 mm fruit sizes. These fruit were placed into labelled plastic bags and put into a cold storage room at 0 °C and 90–95% humidity for quality assessments in the laboratory the following day.

2.1.3. Fruit Quality

Fruit was assessed for length (L), diameter (D), TSS, and flesh firmness in all trials. Starch levels and fruit background skin colour were also assessed for 'Gala' in trial 8. Fruit shape was determined by measuring the length and diameter of the fruit using a Vernier calliper and calculating L/D ratios. Flesh firmness was measured on pared flesh with a Mecmesin AFG250 force gauge fitted with an Effegi 11 mm penetrometer probe connected to a Mecmesin 2500E motorised stand operating at a speed of 0.65 cm/second.

Juice expressed from the apples during the firmness measurements was collected and TSS concentration (°Brix) was assessed with an Atago PR-1 digital refractometer. For the determination of the starch pattern index (SPI), the cut surface of the calyx half of each fruit was dipped in iodine solution (10 g/L iodine and 40 g/L potassium iodide). The area of blue/black colouration was assessed according to the six-point index for the starch-staining pattern as described by Little [19]; the higher the starch index, the lower the percentage of starch present. Fruit background skin colour was measured visually using the scale presented by Frappell and O'Loughlin [20] (Supplementary Table S1).

2.1.4. Return Bloom

Return bloom was determined for all cultivars, except for 'Fuji' in trials 1 and 2, by counting blossom clusters on each tree during the spring following treatment and calculating blossom density (number of blossom clusters cm^{-2} TCSA).

2.2. Data Analysis

Data were subjected to analysis of variance using Genstat release 17.1 (VSN International Ltd., Hemel Hempstead, Hertfordshire, UK). Tests were performed within Genstat to check all data for normality and homogeneity of variance—all data were found to be normally distributed. Linear regressions were undertaken using the Simple Linear Regression option in Genstat.

Data are presented as mean values for each treatment combination. Results described as significant were at a probability level (p) of $\leq$0.05 and Fisher's least significant difference (LSD) ($p = 0.05$), calculated after Steel and Torrie [21], was used for comparison of treatment means.

Regressions were plotted where appropriate to illustrate linear responses to crop load or relationships between measured variables in trials 1, 2, and 3. In all cases, regressions shown are for treatment means and error bars are standard errors of the mean. Graphs were plotted using SigmaPlot 13.0 (Systat Software Inc., Palo Alto, CA, USA).

3. Results

The crop loads obtained in trials 1–3 were relatively close to the target crop loads (Supplementary Table S2). In trial 4, the final crop loads were higher than the target in all but one treatment. The mean crop loads achieved in trials 5–8 were within 0.8 fruit cm^{-2} TCSA of the target, with the exception of the 6 wAFB treatment in trial 8 where crop loads were higher.

3.1. Trial 1: 'Fuji'

A significant linear regression ($R^2 = 0.76$) was observed between crop load and mean fruit weight (Figure 1a), with a reduction of 15.25 g for every unit increase in crop load. The regressions between fruit size, represented as percentage of fruit $\geq$ 75 mm in diameter, and crop load (Figure 1b), as well as between fruit TSS and crop load (Figure 1c), were also significant ($R^2 = 0.75$ and 0.86, respectively).

Fruit with a significantly higher L/D ratio were produced at a crop load of two fruit cm^{-2} TCSA compared to other treatments, but there was no significant difference in the L/D ratio between higher crop loads (Table 3).

There was a significant regression between mean fruit weight and fruit sugar content (Figure 2a) and between mean fruit weight and fruit shape, represented by the fruit L/D ratio (Figure 2b) ($R^2 = 0.87$ and 0.90, respectively).

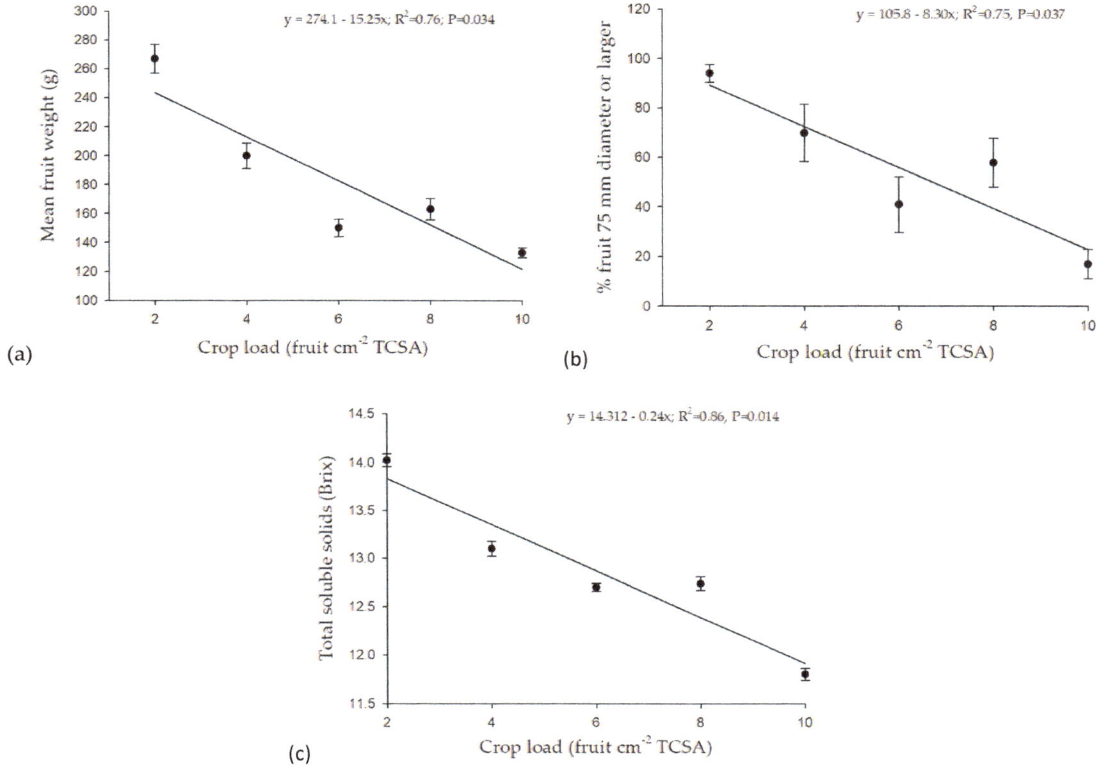

Figure 1. The effect of crop load on (**a**) mean fruit weight, (**b**) fruit size and (**c**) fruit soluble solids content of 'Fuji' apple (trial 1). Error bars represent the standard error of the mean (n = 700).

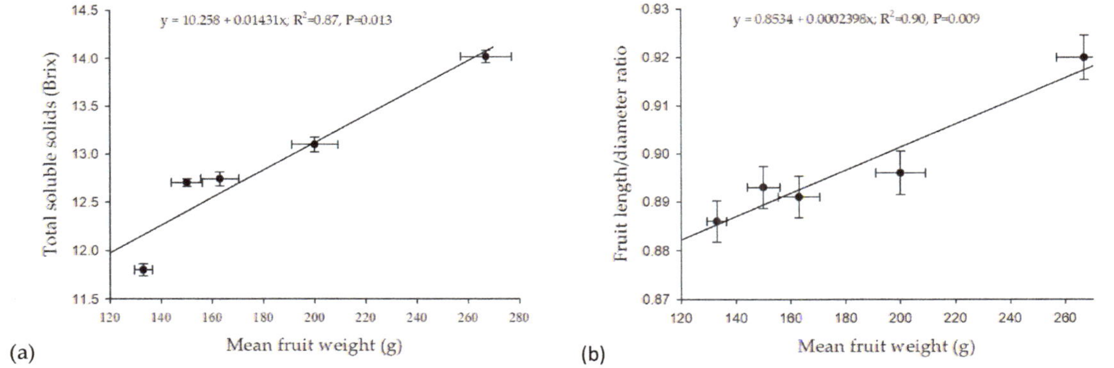

Figure 2. The effect of fruit weight on (**a**) fruit sugar content and (**b**) fruit shape (length/diameter ratio) of 'Fuji' apple (trial 1). Error bars represent the standard error of the mean (n = 700).

Table 3. The effect of crop load on fruit shape (length/diameter ratio) and flesh firmness of 'Fuji' apples hand-thinned 6 weeks after full bloom (Trial 1). TCSA, trunk cross-sectional area.

Crop Load	Length/Diameter Ratio	Flesh Firmness (kg)
Two fruit cm^{-2} TCSA	0.920 b	12.11 bc
Four fruit cm^{-2} TCSA	0.896 a	12.28 c
Six fruit cm^{-2} TCSA	0.893 a	11.74 a
Eight fruit cm^{-2} TCSA	0.891 a	11.95 ab
Ten fruit cm^{-2} TCSA	0.886 a	12.36 c

Means within each column with the same letter are not significantly different at the 5% level.

3.2. Trial 2: 'Fuji'

As with trial 1, in trial 2, there was a significant linear regression between crop load and mean fruit weight (Figure 3a), with a reduction of 11 g for every unit increase in crop load (R^2 = 0.90). There was also an inverse correlation between crop load and percentage of fruit ≥ 75 mm in diameter (Figure 3b) and a significant linear regression between crop load and fruit TSS (Figure 3c) (R^2 = 0.83 and 0.85 respectively).

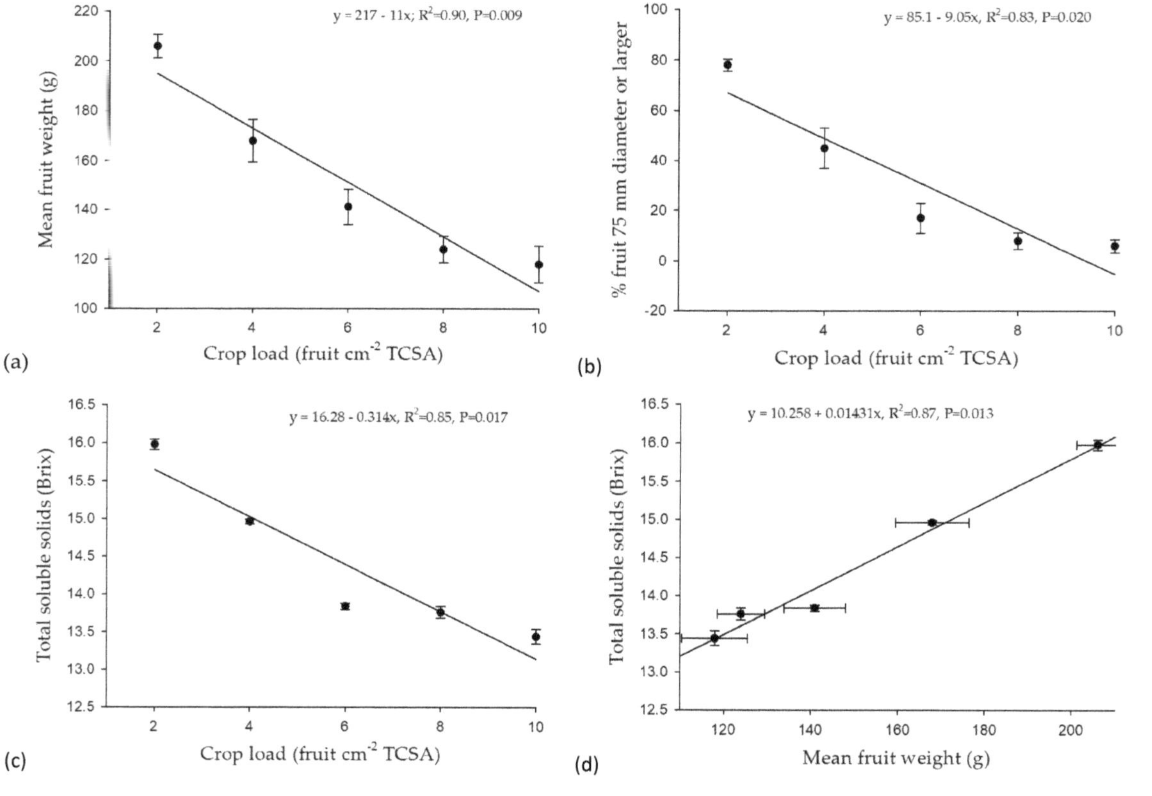

Figure 3. The effect of crop load on (**a**) mean fruit weight, (**b**) fruit size and (**c**) fruit soluble solids content, and (**d**) the relationship between fruit weight and sugar content of 'Fuji' apple (trial 2). Error bars represent the standard error of the mean (n = 700).

There was a significant linear regression between fruit weight and fruit sugar content (Figure 3d), with an increase of 0.014 °Brix for every gram increase in fruit weight (R^2 = 0.87).

The fruit L/D ratio was highest at the two lower crop loads of two and four fruit cm^{-2} TCSA (Table 4).

Table 4. The effect of crop load on fruit shape (length/diameter ratio) and flesh firmness of 'Fuji' apples hand-thinned 6 weeks after full bloom (Trial 2). TCSA, trunk cross-sectional area.

Crop Load	Length/Diameter Ratio	Flesh Firmness (kg)
Two fruit cm^{-2} TCSA	0.854 bc	8.25 c
Four fruit cm^{-2} TCSA	0.859 c	7.60 ab
Six fruit cm^{-2} TCSA	0.839 a	7.78 b
Eight fruit cm^{-2} TCSA	0.843 ab	7.82 b
Ten fruit cm^{-2} TCSA	0.841 a	7.51 a

Means within each column with the same letter are not significantly different at the 5% level.

Trees with a crop load of two fruit cm^{-2} TCSA produced significantly firmer fruit compared to heavier crop loads. Fruit was significantly softer in trees with a crop load of 10 fruit cm^{-2} TCSA than in trees with crop loads of 6 or 8 fruit cm^{-2} TCSA.

3.3. Trial 3: 'Delicious'

As for 'Fuji', there was a significant negative linear regression between crop load and fruit weight (R^2 = 0.85), with a reduction of 10.45 g for every unit increase in crop load (Figure 4a), and between crop load and fruit size (Figure 4b) (R^2 = 0.97).

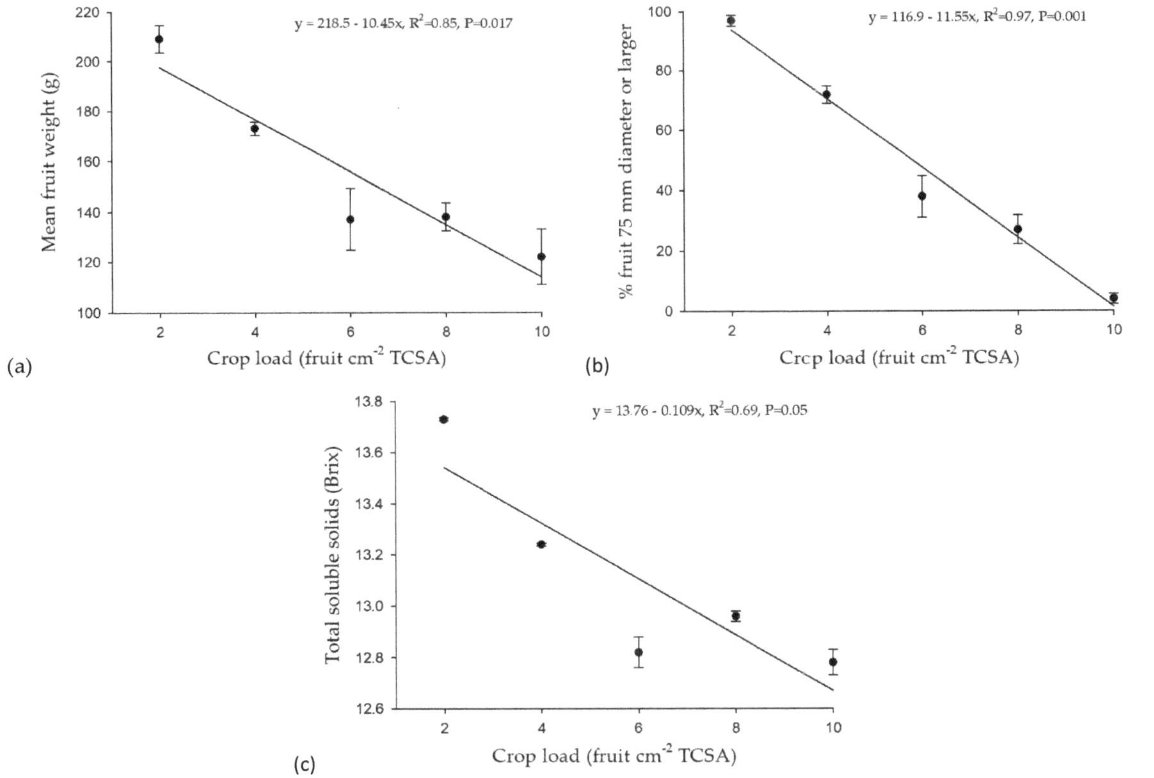

Figure 4. The effect of crop load on (**a**) mean fruit weight, (**b**) fruit size and (**c**) fruit soluble solids content of 'Delicious' apple (trial 3). Error bars represent the standard error of the mean (n = 700).

A significant negative regression was observed between fruit TSS and crop load (Figure 4c), with a reduction of 0.109 °Brix for every unit increase in crop load ($R^2 = 0.69$).

Fruit flesh firmness decreased with increasing crop load from two to six fruit cm^{-2} TCSA (Table 5). Increasing crop load had no significant effect on the fruit L/D ratio.

Table 5. The effect of crop load on fruit shape (length/diameter ratio) and flesh firmness of 'Delicious' apples hand-thinned 6 weeks after full bloom (trial 3). TCSA, trunk cross-sectional area.

Crop Load	Length/Diameter Ratio	Flesh Firmness (kg)
Two fruit cm^{-2} TCSA	0.984	11.18 c
Four fruit cm^{-2} TCSA	0.983	10.64 b
Six fruit cm^{-2} TCSA	0.969	10.25 a
Eight fruit cm^{-2} TCSA	0.973	10.47 ab
Ten fruit cm^{-2} TCSA	0.977	10.28 a

Means within each column with the same letter are not significantly different at the 5% level.

There was a significant positive linear regression between fruit weight and fruit sugar content (Figure 5a), between fruit weight and flesh firmness (Figure 5b), and between fruit sugar content and firmness (Figure 6) (R^2 = 0.97, 0.90 and 0.98, respectively).

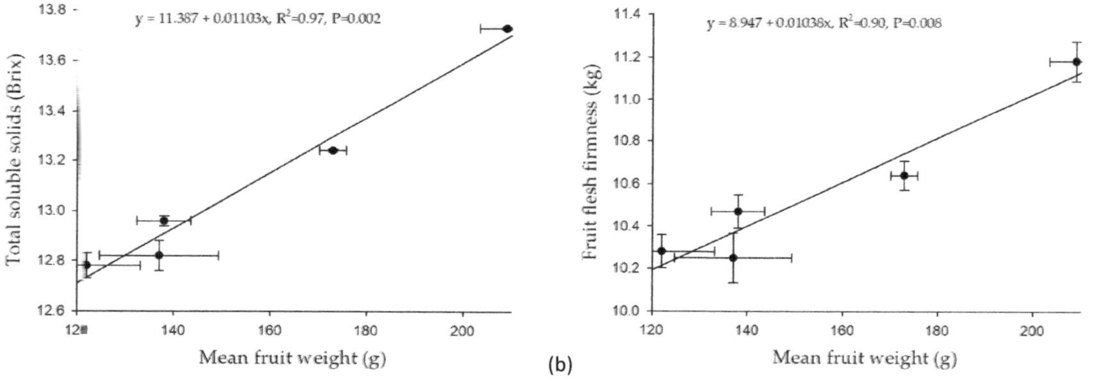

Figure 5. The effect of fruit weight on (**a**) fruit soluble solids content and (**b**) fruit firmness of 'Delicious' apple (trial 3). Error bars represent the standard error of the mean (n = 700).

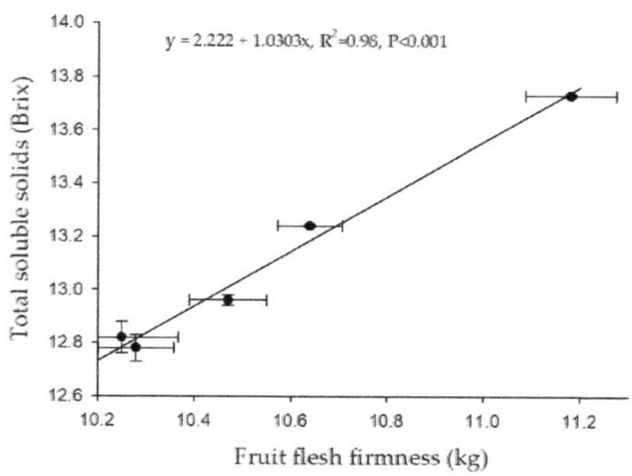

Figure 6. The relationship between fruit firmness and sugar content of 'Delicious' apple. Error bars represent the standard error of the mean (n = 700).

3.4. Trial 4: 'Delicious'

There were no significant interactions between crop load and time of thinning for mean fruit weight and return bloom (results not presented), but there were significant interactions for other parameters (Table 6).

Table 6. The interaction between crop load and time of thinning on fruit size (% fruit ≥ 75 mm in diameter), shape (length/diameter ratio), soluble solids content and flesh firmness of 'Delicious' apples. TCSA, trunk cross-sectional area; wAFB, weeks after full bloom; TSS, total soluble solids.

Crop Load (Fruit cm^{-2} TCSA)	Thinning Time (wAFB)	% Fruit ≥ 75 mm Diameter	Length/Diameter Ratio	TSS Content (°Brix)	Flesh Firmness (kg)
3	1	67 f	0.971 ab	13.43 f	8.58 f
6	1	63 f	0.994 de	13.23 de	8.01 cd
3	2	67 f	0.997 de	13.17 cd	7.88 bc
6	2	25 c	1.003 e	13.07 bc	7.57 a
3	4	59 ef	0.989 cde	13.90 h	8.32 ef
6	4	23 bc	0.984 abcd	12.70 a	7.79 abc
3	8	43 de	0.987 bcde	13.33 ef	8.44 ef
6	8	8 ab	0.969 a	13.03 b	8.25 de
3	12	43 de	0.998 de	14.13 i	8.59 f
6	12	17 abc	0.974 abc	13.73 g	7.72 ab
3	16	33 cd	0.999 de	13.77 g	9.47 h
6	16	6 a	0.984 abcd	13.00 b	8.90 g

Means within each column with the same letter are not significantly different at the 5% level.

Fruit size (% fruit ≥ 75 mm in diameter) was significantly higher at a crop load of three fruit cm^{-2} TCSA than at six fruit cm^{-2} TCSA at all thinning times, with the exception of 1 wAFB (Table 7). The treatments that produced the highest number of fruit ≥ 75 mm in diameter were the two 1 wAFB treatments and three fruit cm^{-2} TCSA thinned 2 or 4 wAFB.

Table 7. The effect of crop load and time of thinning on mean fruit weight and return bloom of 'Delicious' apples (trial 4). TCSA, trunk cross-sectional area; wAFB, weeks after full bloom.

	Mean Fruit Weight (g)	Return Bloom (Buds cm^{-2} TCSA)
(i) Crop load		
Three fruit cm^{-2} TCSA	170 b	14.1
Six fruit cm^{-2} TCSA	144 a	13.3
(ii) Time of thinning		
1 wAFB	172 b	16.7 b
2 wAFB	161 ab	15.4 b
4 wAFB	164 ab	17.9 b
8 wAFB	147 a	10.5 a
12 wAFB	153 a	10.6 a
16 wAFB	149 a	11.3 a

Means within each column with the same letter are not significantly different at the 5% level.

The fruit L/D ratio was significantly lower in the three fruit cm^{-2} TCSA treatment at 1 wAFB than the higher crop load (Table 7). At 8 and 12 wAFB, the fruit L/D ratio was significantly lower at the higher crop load than the lower crop load.

Fruit TSS was significantly lower at the higher crop load compared with the lower crop load at all thinning times, except for 2 wAFB. At all thinning times, except for 8 wAFB, fruit firmness was significantly higher at three fruit cm^{-2} TCSA than at six fruit cm^{-2} TCSA. Trees thinned 16 wAFB produced significantly firmer fruit than all other treatments.

Mean fruit weight was significantly higher at three fruit cm^{-2} TCSA than at six fruit cm^{-2} TCSA (Table 7). The time of thinning also influenced mean fruit weight, with the later thinning times of 8, 12, or 16 wAFB producing significantly smaller fruit than the trees thinned 1 wAFB (Table 7). Crop load had no significant effect on return bloom (Table 7). The time of thinning showed a significant effect, with thinning at or later than 8 wAFB resulting in a lower return bloom than earlier thinning.

3.5. Trial 5: 'Pink Lady'

No interactions were observed between crop load, time of thinning, and rootstock for mean fruit weight, size, or return bloom (results not presented). All three factors had a significant effect on both mean fruit weight and percentage of fruit ≥ 75 mm in diameter, while rootstock and time of thinning, but not crop load, had a significant effect on return bloom (Table 8).

Table 8. The effect of rootstock, crop load, and time of thinning on mean fruit weight, size (% fruit ≥ 75 mm in diameter), and return bloom of 'Pink Lady' apples. TCSA, trunk cross-sectional area; wAFB, weeks after full bloom.

	Mean Fruit Weight (g)	% Fruit ≥ 75 mm in Diameter	Return Bloom (Buds cm^{-2} TCSA)
(i) Rootstock			
M26	166 b	30 b	13.9 b
MM106	156 a	24 a	5.3 a
(ii) Crop load			
Four fruit cm^{-2} TCSA	167 b	32 b	9.7
Six fruit cm^{-2} TCSA	164 b	28 b	10.3
Eight fruit cm^{-2} TCSA	151 a	20 a	8.8
(iii) Time of thinning			
2 wAFB	169 b	36 b	10.7 b
6 wAFB	165 b	33 b	11.4 b
10 wAFB	156 a	18 a	8.8 a
14 wAFB	153 a	20 a	7.4 a

Means within each column with the same letter are not significantly different at the 5% level.

Trees on M26 rootstock produced significantly heavier fruit than those on MM106 (Table 8). Fruit weight was significantly reduced at crop loads of eight fruit cm^{-2} TCSA compared with lower crop loads, while trees thinned at 10 and 14 wAFB produced significantly lighter fruit than earlier-thinned trees (Table 8). Similar patterns were observed in the percentage of fruit ≥ 75 mm in diameter for all three factors. Return bloom was significantly higher in trees on M26 rootstock than on MM106. Crop load had no significant effect on return bloom, but earlier thinning (2 or 6 wAFB) resulted in a higher return bloom than later thinning times.

Significant interactions were observed between the thinning treatments for fruit L/D ratio, TSS and firmness (Table 9). Although there were significant differences between treatments in the L/D ratio, the results showed no clear pattern with no consistent effects of rootstock, crop load or time of thinning.

Fruit TSS decreased with increasing crop load on M26 rootstocks on trees thinned 2 wAFB, and on MM106 rootstocks thinned at 6 wAFB. TSS levels were significantly higher on M26 rootstocks than in the corresponding MM106 treatments.

Fruit firmness was significantly higher in the four and six fruit cm^{-2} TCSA 6 wAFB treatments on M26 than all other treatments. Increasing crop load resulted in a decrease in firmness at all thinning times on M26 rootstocks, but there were no distinct trends for MM106 stocks. Firmness was significantly higher on M26 rootstocks than in the corresponding MM106 rootstocks, except for the eight fruit cm^{-2} TCSA 2 and 6 wAFB treatments.

Table 9. The effect of rootstock, crop load and time of thinning on fruit shape (length/diameter ratio), sugar content and flesh firmness of 'Pink Lady' apples. TCSA, trunk cross-sectional area; wAFB, weeks after full bloom; TSS, total soluble solids.

Rootstock	Crop Load (Fruit cm^{-2} TCSA)	Thinning Time (wAFB)	Length/Diameter Ratio	TSS (°Brix)	Flesh Firmness (kg)
M26	4	2	0.936 efghi	15.96 lm	8.96 hi
M26	6	2	0.906 ab	15.45 j	8.98 i
M26	8	2	0.941 ghi	15.18 gh	7.99 abcd
M26	4	6	0.946 i	15.88 kl	9.60 j
M26	6	6	0.934 defghi	16.35 o	9.53 j
M26	8	6	0.942 hi	15.53 j	8.64 gh
M26	4	10	0.964 j	16.07 mn	9.10 i
M26	6	10	0.930 cdefghi	16.30 o	8.98 i
M26	8	10	0.934 defghi	16.18 no	8.56 fg
M26	4	14	0.916 bc	15.40 ij	8.92 def
M26	6	14	0.917 bcd	15.23 hi	8.55 fg
M26	8	14	0.923 bcdef	15.45 j	8.24 def
MM106	4	2	0.924 cdefg	15.16 gh	8.18 de
MM106	6	2	0.916 bc	14.05 c	8.06 bcde
MM106	8	2	0.897 a	14.35 de	8.04 bcd
MM106	4	6	0.923 bcdef	15.23 hi	7.98 abcd
MM106	6	6	0.921 bcde	15.00 g	8.15 cde
MM106	8	6	0.917 bcd	14.58 f	8.38 efg
MM106	4	10	0.920 bcde	14.50 ef	8.30 def
MM106	6	10	0.938 fghi	13.70 a	7.83 abc
MM106	8	10	0.929 cdefghi	14.38 de	7.75 ab
MM106	4	14	0.914 abc	13.99 b	8.00 abcd
MM106	6	14	0.925 cdefgh	14.23 cd	7.81 ab
MM106	8	14	0.918 bcd	14.36 de	7.68 a

Means within each column with the same letter are not significantly different at the 5% level.

3.6. Trial 6: 'Jonagold'

There were no interactive effects between crop load, time of thinning and rootstock for mean fruit weight (p = 0.315), size (p = 0.269) or return bloom (p = 0.198) (results not presented). However, analysis of the main effects showed that mean fruit weight was influenced by crop load and time of thinning, percentage of fruit ≥ 85 mm in diameter was affected only by the crop load, and return bloom was affected by rootstock but not by crop load or time of thinning (Table 10).

Table 10. The effect of rootstock, crop load and time of thinning on fruit weight, size and return bloom of 'Jonagold' apples. TCSA, trunk cross-sectional area; wAFB, weeks after full bloom.

	Mean Fruit Weight (g)	% Fruit ≥ 85 mm Diameter	Return Bloom (Buds cm^{-2} TCSA)
(i) Main effects—rootstock			
M26	229	33	6.7 a
MM106	231	41	11.9 b
(ii) Main effects—crop load			
Four fruit cm^{-2} TCSA	259 c	49 b	10.6
Six fruit cm^{-2} TCSA	223 b	36 a	9.0
Eight fruit cm^{-2} TCSA	207 a	27 a	8.3
(iii) Main effects—time of thinning			
2 wAFB	242 b	42	11.4
6 wAFB	240 b	37	8.4
10 wAFB	226 ab	41	7.7
14 wAFB	211 a	29	9.6

Means within each column with the same letter are not significantly different at the 5% level.

Mean fruit weight decreased significantly with increasing crop load (Table 10). Thinning at 2 and 6 wAFB produced heavier fruit than later thinning. The percentage of fruit ≥ 85 mm in diameter was significantly higher at the lower crop load compared with the two higher crop loads.

Return bloom was influenced by rootstock, with trees on MM106 rootstocks having a significantly higher return bloom than trees on M26 rootstocks.

Although the interactions between thinning treatments were significant for the fruit L/D ratio, no distinct trends were discernible (Table 11). On the M26 rootstocks, TSS levels decreased significantly with increasing crop load on trees thinned 6 and 14 wAFB, and at 2 and 14 wAFB on MM106 rootstocks.

Table 11. The effect of rootstock, crop load and time of thinning on fruit shape and total soluble solids (TSS) content of 'Jonagold' apples. TCSA, trunk cross-sectional area; wAFB, weeks after full bloom; TSS, total soluble solids.

Rootstock	Crop Load (Fruit cm^{-2} TCSA)	Thinning Time (wAFB)	Length/Diameter Ratio	TSS (Brix)
M26	4	2	0.920 de	13.70 b
M26	6	2	0.911 bcde	15.10 ij
M26	8	2	0.917 cde	13.65 b
M26	4	6	0.924 e	15.83 k
M26	6	6	0.916 cde	14.25 ef
M26	8	6	0.899 ab	13.75 bc
M26	4	10	0.905 bcd	14.45 fg
M26	6	10	0.898 ab	14.20 ef
M26	8	10	0.899 ab	14.13 def
M26	4	14	0.912 bcde	15.70 k
M26	6	14	0.905 bcd	14.65 gh
M26	8	14	0.911 bcde	13.85 bcd
MM106	4	2	0.899 ab	16.33 l
MM106	6	2	0.911 bcde	14.68 gh
MM106	8	2	0.904 bc	13.52 b
MM106	4	6	0.910 bcde	15.20 j
MM106	6	6	0.909 bcde	14.25 ef
MM106	8	6	0.916 cde	14.80 hi
MM106	4	10	0.910 bcde	15.20 j
MM106	6	10	0.905 bcd	14.80 hi
MM106	8	10	0.909 bcde	14.94 hij
MM106	4	14	0.921 e	15.68 k
MM106	6	14	0.902 bc	14.05 cde
MM106	8	14	0.885 a	13.15 a

Means within each column with the same letter are not significantly different at the 5% level.

3.7. Trial 7: 'Braeburn'

There were no significant interactions between crop load, time of thinning and rootstock for mean fruit weight ($p = 0.732$), fruit size ($p = 0.237$), return bloom ($p = 0.568$) or fruit L/D ratio ($p = 0.076$) (results not presented). Analysis of the main effects showed that the mean fruit weight, percentage of fruit ≥ 75 mm in diameter, return bloom and fruit shape (L/D ratio) were all influenced by rootstock and crop load but not by time of thinning (Table 12).

Table 12. The effect of crop load and time of thinning on fruit weight, size, return bloom and fruit shape of 'Braeburn' apples. TCSA, trunk cross-sectional area.

	Mean Fruit Weight (g)	% Fruit ≥ 75 mm Diameter	Length/Diameter Ratio	Return Bloom (Buds cm^{-2} TCSA)
(i) Main effects—rootstock				
M26	220 b	70 b	0.907 a	20.2 b
MM106	196 a	45 a	0.920 b	7.4 a
(ii) Main effects—crop load				
Two fruit cm^{-2} TCSA	244 c	76 c	0.926 c	23.5 c
Four fruit cm^{-2} TCSA	218 b	67 c	0.916 b	15.9 b
Six fruit cm^{-2} TCSA	191 a	51 b	0.904 a	9.4 a
Eight fruit cm^{-2} TCSA	179 a	37 a	0.907 a	6.3 a

Means within each column with the same letter are not significantly different at the 5% level.

The fruit from M26 rootstocks was significantly heavier than those from MM106 rootstocks (Table 12), while increasing crop load from 2 to 6 fruit cm^{-2} TCSA resulted in lower fruit weight (Table 12(ii)). M26 rootstocks produced significantly more fruit ≥ 75 mm in diameter than MM106 rootstocks. There was no difference between two and four fruit cm^{-2} TCSA in the percentage of fruit ≥ 75 mm in diameter, but increasing the crop load from four to eight fruit cm^{-2} TCSA resulted in a decrease in fruit ≥ 75 mm in diameter.

Return bloom was significantly higher on M26 rootstocks compared to MM106 rootstocks. Return bloom was reduced significantly with increasing crop load from two to six fruit cm^{-2} TCSA. MM106 rootstocks produced fruit with a significantly higher L/D ratio than the M26 rootstocks. Increasing crop load from two to six fruit cm^{-2} TCSA resulted in a significant reduction in L/D ratio. Time of thinning had no effect on fruit weight, size, or shape, or on return bloom (results not presented).

The treatment interactions were significant for fruit TSS (Table 13). The level of TSS decreased with increasing crop load for the M26 rootstocks at all thinning times, but no distinct pattern emerged for the MM106 rootstocks.

Table 13. The effect of crop load and time of thinning on total soluble solids content (Brix) of 'Braeburn' apples on two different rootstocks, M26 and MM106. TCSA, trunk cross-sectional area; wAFB, weeks after full bloom; TSS, total soluble solids.

Crop Load (Fruit cm^{-2} TCSA)	Thinning Time (wAFB)	TSS (Brix)	
		M26	MM106
2	2	15.60 j	13.82 de
4	2	14.26 f	14.05 ef
6	2	13.95 e	13.65 d
8	2	13.06 c	12.43 b
2	6	15.25 i	12.15 a
4	6	14.87 g	12.53 b
6	6	12.50 b	13.00 c
8	6	12.65 b	13.20 c
2	10	15.20 i	15.13 hi
4	10	14.93 gh	15.15 hi
6	10	13.88 de	12.03 a
8	10	13.25 c	13.12 c

Means within each column with the same letter are not significantly different at the 5% level.

3.8. Trial 8: 'Gala'

There were significant interactions between treatments for all the parameters assessed in 'Gala' (Tables 14 and 15).

Table 14. The effect of crop load and time of thinning on mean fruit weight, size (% fruit ≥ 65 mm in diameter), shape (length/diameter ratio) and return bloom of 'Gala' apples. TCSA, trunk cross-sectional area; wAFB, weeks after full bloom.

Crop Load (Fruit cm^{-2} TCSA)	Thinning Time (wAFB)	Mean Fruit Weight (g)	% Fruit ≥ 65 mm Diameter	Length/Diameter Ratio	Return Bloom (Buds cm^{-2} TCSA)
3	2	148 e	37 d	0.928 efg	19.7 e
6	2	138 de	16 c	0.929 fg	9.1 abc
9	2	118 bc	2 ab	0.915 def	10.6 abc
3	6	140 de	11 bc	0.934 g	18.5 de
6	6	125 cd	4 ab	0.933 g	11.5 bc
9	6	95 a	0 a	0.908 cd	8.2 abc
3	10	133 cde	7 abc	0.942 g	12.1 cd
6	10	94 a	1 ab	0.904 bcd	5.0 ab
9	10	98 a	0 a	0.894 abc	7.9 abc
3	14	104 ab	0 a	0.913 de	5.4 abc
6	14	95 a	0 a	0.887 a	7.7 abc
9	14	88 a	0 a	0.892 ab	3.9 a

Means within each column with the same letter are not significantly different at the 5% level.

Table 15. The effect of crop load and time of thinning on fruit sugar content, flesh firmness, starch index and background skin colour of 'Gala' apples. TCSA, trunk cross-sectional area; wAFB, weeks after full bloom; TSS, total soluble solids; SPI, starch pattern index.

Crop Load (Fruit cm^{-2} TCSA)	Thinning Time (wAFB)	TSS (°Brix)	Flesh Firmness (kg)	SPI	Background Skin Colour
3	2	15.38 g	8.16 a	4.0 f	4.5 d
6	2	14.50 c	8.67 bc	3.4 e	4.5 d
9	2	14.50 cd	8.40 ab	3.6 e	4.1 c
3	6	15.15 f	9.26 e	2.9 d	4.2 cd
6	6	14.69 e	8.85 cd	3.4 e	4.5 d
9	6	14.18 b	9.60 fg	3.6 e	3.6 b
3	10	14.44 c	9.17 de	2.7 cd	4.5 d
6	10	13.90 a	9.28 ef	2.9 d	4.2 b
9	10	15.20 f	9.73 g	2.9 d	3.3 d
3	14	14.23 b	10.20 h	1.9 a	4.5 d
6	14	14.60 de	9.63 g	2.3 b	3.2 a
9	14	13.85 a	9.47 ef	2.4 bc	4.2 cd

Means within each column with the same letter are not significantly different at the 5% level.

At 2 and 6 wAFB, mean fruit weight was significantly higher with crop loads of three or six fruit cm^{-2} TCSA than with nine fruit cm^{-2} TCSA (Table 14). At 10 wAFB, a crop load of three fruit cm^{-2} TCSA produced significantly heavier fruit than crop loads of either six or nine fruit cm^{-2} TCSA. There was no significant difference in mean fruit weight between the different crop loads at 14 wAFB.

Fruit size (the percentage of fruit ≥ 65 mm in diameter) was significantly larger in the early-thinned lowest crop load trees than any other treatment. At 2 wAFB, there was a significant reduction in fruit size with increasing crop load.

Crop load had no significant effect on fruit shape (L/D ratio) at the earliest thinning time. However, the lightest crop load resulted in higher L/D ratios than the heaviest crop loads. Return bloom was significantly higher at the lowest crop load than the other treatments at 2, 6 and 10 wAFB.

Reducing crop load to three fruit cm^{-2} TCSA produced fruit with the highest TSS compared with other treatments for the two earlier times (Table 15). TSS levels decreased significantly from three to six fruit cm^{-2} TCSA at all thinning times except for 14 wAFB. Fruit firmness was significantly lower in fruit thinned 2 wAFB than later-thinned fruit.

The starch levels were lower in the early thinned trees with the lowest crop load, but this pattern was reversed at 6 and 14 wAFB. The general trend was for starch levels to be higher with later thinning.

There were significant differences between treatments in fruit background skin colour, with fruit from the highest crop load treatments at 2, 6 and 10 wAFB being greener than the two lighter crop loads at these times.

4. Discussion

The trials reported here demonstrate that both crop load and time of thinning play an important role in determining external and internal fruit quality parameters at harvest; however, between cultivars, there were differences in the optimum crop load and effect of thinning time.

Similar trends were observed in the two 'Fuji' trials conducted in consecutive years, but there were marked differences in the actual figures obtained for each parameter studied. This suggests that while crop load has a major influence, climatic differences between years (Supplementary Data, Figures S1–S5) can result in a shift in actual values obtained, most likely through variations in the date and spread of flowering, pollination, and early growth of fruit.

4.1. Fruit Weight and Size

The reduction in fruit weight with increasing crop load observed with all cultivars in this study is in line with the observation by Costa et al. [22] that there is an inverse relationship between the final fruit size and number of fruit per tree. For 'Fuji', weight reductions of 15.25 g and 11.00 g were observed for every unit increase in crop load in consecutive years, and for 'Delicious', the reduction was 10.45 g.

Weights of 200 g per fruit were achieved at crop loads of four fruit cm^{-2} TCSA in 'Fuji' in trial 1; however, in the following year, this fruit weight was only achieved at the lower crop load level of two fruit cm^{-2} TCSA. Jones et al. [6] suggested that weights of 200 g per fruit were readily achievable with crop loads of 4–6 fruit cm^{-2} TCSA; however, they also recommended thinning at blossom time rather than post-bloom. Setting target crop loads of 5–7 fruit cm^{-2} TCSA, Bound et al. [15] obtained fruit weights of around 200 g per apple with more than 40% of the fruit larger than 80 mm in diameter following chemical thinning with ethephon and BA within 3 weeks of FB. However, BA has been demonstrated to increase fruit size even in the absence of any thinning [23,24]. The lower weights achieved in this study at four or six fruit cm^{-2} TCSA are most likely the result of delaying thinning to 6 weeks after flowering, leading to a loss of fruit size through competition with fruit that was later removed. According to Jones et al. [6], delaying thinning can result in a loss of as much as 10 g per fruit for every week's delay in thinning. Koike et al. [25] concluded that the primary thinning of 'Fuji' should be performed within 28 dAFB to ensure a good fruit size. For 'Fuji', crop loads of 4–6 fruit cm^{-2} TCSA are considered appropriate to avoid biennial bearing [1], and this study showed that large fruit of 200 g or more can be produced at these crop load levels, but if thinning is delayed, the crop load should be reduced in order to maximise the fruit size.

In 'Delicious', weights of at least 150 g per fruit were achieved at crop loads of 2–4 fruit cm^{-2} TCSA. However, crop loads of 6–10 fruit cm^{-2} TCSA produced fruit weights in the order of 125–145 g. These results confirm the conclusions of Koen et al. [26] that 2–4 fruit cm^{-2} TCSA is an ideal target range crop load for 'Delicious'.

Recommendations for target crop loads for 'Pink Lady', 'Jonagold', 'Braeburn' and 'Gala' are lacking. Data from this study suggest that in both 'Pink Lady' and 'Gala', fruit weight and size start to decline with crop loads greater than six fruit cm^{-2} TCSA. However, in large-fruited cultivars such as 'Jonagold', crop loads of eight fruit cm^{-2} TCSA will still produce fruit of 200 g or heavier without affecting return bloom. A good fruit weight and size were achieved in 'Braeburn' at crop loads of up to eight fruit cm^{-2} TCSA; however, return bloom was reduced by increasing the crop load. Hence, the desire for large fruit should be carefully considered against the risk of pushing trees into biennial bearing.

Fruit weight and size were also heavily influenced by time of thinning, confirming the postulation by Link [27] that the supply of carbon available to the fruit may be limited by competition from other fruits; hence, a marked influence of time of thinning on fruit size would be expected.

The results of this study support the conclusion of Jones et al. [6] and McArtney et al. [28] that earlier thinning can result in considerable increase in fruit weight. Working with the cultivar 'Empire', Lakso et al. [29] concluded that effective hand thinning for size increases could be performed as late as 20 dAFB, but earlier application of the chemical thinning agents NAA, BA and carbaryl at 15 dAFB inhibited fruit growth too much to allow maximum response to crop reduction. This finding reinforces the negative impact that thinning chemicals can have on fruit quality, despite thinning relatively early in the season. In a comparison of hand thinning at 5 weeks AFB with artificial bud extinction (ABE), a thinning practice that reduces the number of floral buds prior to bud burst, at similar crop loads on the cultivar 'Scilate', Sidhu et al. [30] reported increased fruit weights of up to 70 g per fruit in ABE-managed trees, again demonstrating the importance of earlier removal of excess floral buds to reduce competition for, and wastage of, carbohydrate resources. This also agrees with the conclusions of Robinson et al. [31]: that leaving too many floral buds when pruning results in a lower crop value than pruning to the optimum bud load. High fruit weights can be achieved for most cultivars at relatively high crop loads if thinning is completed early in the season, preferably before or during flowering. If thinning is delayed, crop loads need to be reduced in order to achieve these weights, resulting in reduced yield, which is a function of the number and size of the fruit on the tree.

Rootstocks affect apple fruit quality by influencing both tree vigour and crop load. At similar crop loads, trees on M26 rootstocks in this study produced larger, heavier fruit than on the more vigorous MM106 rootstocks for both 'Pink Lady' and 'Braeburn', but there was no rootstock effect on the triploid cultivar 'Jonagold'. The increased fruit size on the weaker M26 rootstock conflicts with the findings of Fallahi and Simons [32] and Riesen and Husistein [33]. However, these authors were comparing a range of dwarfing rootstocks and did not include any semi-vigorous or vigorous rootstocks in their studies.

4.2. Fruit Shape

In this study, fruit shape was influenced by thinning in some cultivars but not others. In those cultivars where there was an effect, higher crop loads generally produced flatter fruit. This is in agreement with the conclusions of Link [27] that thinning normally favours fruit development. However, it appears from the present study that fruit shape may also be influenced by the time of thinning in some cultivars, particularly 'Delicious', where thinning close to bloom reversed this trend towards flatter fruit. From a marketing perspective, fruit shape and typiness are important attributes in 'Delicious' and management practices that flatten the fruit impact on marketability [34,35].

4.3. Total Soluble Solids

For most cultivars, fruit soluble solids content decreased with increasing crop load. This is in agreement with the findings of Koike et al. [25] who reported a 14% increase in sugar levels in 'Fuji' fruit from hand-thinned trees compared with unthinned trees. A similar effect was also observed for hand-thinned 'Cox's Orange Pippin' [36]. In the 'Fuji' and 'Delicious' trials that were thinned at 6 wAFB, the positive correlation between sugar content and fruit weight suggests that early thinning can maintain fruit sugar levels in larger fruit.

A rootstock effect was observed in 'Pink Lady', with lower soluble solids on the more vigorous MM106 rootstocks. Fallahi and Simons [32] also reported that soluble solids at harvest were lower in fruit from trees on M26 rootstocks compared with the more dwarfing M27 and M9 rootstocks. These trends suggest that the rootstock effect may be related to tree vigour, with higher soluble solids in less vigorous trees. This leads to the assumption that less assimilate is used for vegetative growth in the more dwarfing trees. It is important to note, however, that this trend was reversed in the triploid cultivar 'Jonagold', with TSS levels higher on MM106 rootstocks than on M26.

4.4. Firmness

While this study did not include fruit firmness results for 'Braeburn' and 'Jonagold' due to equipment breakdown, fruit firmness in the cultivars 'Fuji', 'Delicious' and 'Pink Lady' decreased with increasing crop load, supporting the results of Garriz et al. [37], who found that fruit flesh firmness was significantly lower in 'Braeburn' trees carrying high crop loads than in trees with moderate or low crop loads. Jones et al. [38] also reported increased firmness with reduced crop load following chemical thinning of 'Pink Lady' and 'Jonagold' with ethephon and BA. Link [27] suggested that the reduced firmness often observed in heavily cropped trees could be due to carbohydrate supply for cell wall synthesis becoming limited. In this study, 'Gala' showed no clear trends relating firmness to crop load, but there was an effect with time of thinning, with thinning close to bloom producing softer fruit than trees thinned from 6 weeks after bloom. A possible explanation for this result is that early thinning causes fruit to mature earlier than later thinning, as noted by Johnson [36]—the increased soluble solids observed in early-thinned fruit also lends support to this explanation.

An unanticipated finding from this work was the positive relationship in both 'Fuji' and 'Delicious' between fruit firmness and mean fruit weight, and between sugar content and firmness in early-thinned fruit. This study provides evidence that early thinning has a major role to play in fruit quality considerations. Previous correlations of fruit softness and high TSS in large fruit are based on concepts of the contrast between vigorously growing off-year trees compared with less vigorous on-year trees in a biennial bearing cycle. The leaf/fruit ratio in off-year trees is higher than in on-year trees, as off-year trees tend to be more vigorous [39]. Hence, more resources are available to each fruit in off-year trees, enabling a greater expansion of cells, regardless of cell number, potentially resulting in larger cell size with larger intercellular spaces, and consequently, softer fruit. In this study, early-thinned regular bearing trees produced large fruit that were firmer and with higher TSS than later-thinned fruit. Not only does this finding conflict with current thoughts on firmness, sugar content and fruit size, but it demonstrates additional advantages for early thinning beyond fruit size. However, caution may be needed with early thinning in areas prone to late spring frosts. These results also show that large fruit can be of better quality than small fruit, providing it is from regular bearing or on-year trees where the excess fruit was thinned early.

Rootstock influenced fruit firmness in 'Pink Lady'. While no relationship was observed between firmness and crop load on MM106 rootstocks, M26 rootstocks produced firmer fruit than did MM106. Differences in firmness for 'Arlet' and 'Fiesta' fruit from trees with different rootstocks were also observed by Riesen and Husistein [33]. These authors suggested that the softer fruit, which also had higher sugar levels, were the result of advanced fruit maturity on some rootstocks. While this is a logical conclusion, 'Pink Lady' in this study produced softer fruit with a lower sugar content on MM106 rootstocks. As these fruits were also smaller than fruit from M26 rootstocks, this result is difficult to explain, as the expectation would be that fruit from MM106 rootstocks should be firmer. If fruit from MM106 rootstocks contained fewer and larger cells that those from M26 rootstocks, this would explain the difference in fruit firmness between the two rootstocks.

4.5. Starch and Background Skin Colour

Starch levels were examined in only one cultivar, 'Gala'. The increase in starch hydrolysis with increasing crop load at the earliest thinning time of 2 wAFB in this study agrees with the findings of Sidhu et al. [30], who reported slower conversion of starch to sugar at higher crop loads in the cultivar 'Scilate'. This slower conversion of starch to sugar combined with a greener skin colour with increasing crop load observed in this study, and as also observed by Sidhu et al. [30], indicate a retardation of fruit maturity at higher crop loads. Serra et al. [40] also reported that crop load can affect fruit maturity, with advanced fruit ripeness in low-crop-load trees.

However, time of thinning did influence starch levels, with earlier thinning resulting in lower starch levels, indicating increased hydrolysis of starch to sugar. This is most likely associated with fruit maturity, particularly when examined in conjunction with fruit soluble solids content, as earlier-thinned fruit also had higher soluble solids than later-thinned fruit. Johnson [36] suggested that early thinning can advance fruit maturity by up to 16 days.

4.6. Return Bloom

The effect of crop load and time of thinning on return bloom varied between cultivars. In 'Delicious', 'Gala', 'Pink Lady' and 'Jonagold', return bloom reduced once a particular level of cropping was reached, while in 'Braeburn', return bloom decreased with increasing crop load. These results demonstrate that different cultivars have different crop load thresholds and suggest that, if regular bearing is to be maintained, 'Braeburn' should not be cropped at levels higher than four fruit cm^{-2} TCSA, while both 'Pink Lady' and 'Jonagold' can maintain crop loads of at least eight fruit cm^{-2} TCSA. However, other factors such as fruit size should also be borne in mind if trees are to be cropped at these levels.

Time of thinning had no influence on return bloom in 'Braeburn' or 'Jonagold', but was important in the three cultivars 'Delicious', 'Pink Lady' and 'Gala', with thinning later than 6 weeks after bloom reducing return bloom. Although return bloom was not assessed on 'Fuji' in this study, Jones et al. [6] reported a decline in return bloom at 8 weeks after bloom, and Koike et al. [25] demonstrated the importance of thinning before 4 wAFB to ensure the return bloom of 'Fuji'.

Williams and Edgerton [41] noted that the two factors of greatest influence on annual bearing, and thus return bloom, are the number of flowering spurs and the amount of initial fruit set. These authors suggest that, for thinning to be most effective, all fruit should be removed from about half of the fruiting spurs rather than reducing the fruit load to one fruit per spur. According to Costa et al. [22], fruit thinning performed after fruit set is normally ineffective in eliminating biennial bearing, but fruit thinning performed before fruit set may prevent or overcome biennial bearing. The importance of time of thinning on return bloom is reinforced by studies on the impact of artificial bud extinction (ABE). In a study of ABE on Fiero 'Fuji' and three strains of 'Gala', Bound [42] reported that ABE-managed trees showed no signs of biennial bearing, with sufficient return bloom to set a crop load of six fruit cm^{-2} limb cross-sectional area based on a single fruit per bud, unlike the conventional trees in which bud numbers varied between seasons. Breen et al. [43] found that as floral bud density was reduced, the proportion of buds failing to set fruit declined and the proportion setting multiple fruit increased, concluding that the early removal of competitive sinks through thinning improves the initiation and development of new floral buds, thus improving return bloom. This information, combined with the results of this study, shows the importance of reducing crop load early in the season.

In the three cultivars where rootstock effects were also examined, rootstock had an influence on return bloom. In both 'Pink Lady' and 'Braeburn' return bloom was tripled on trees with M26 rootstocks compared with the more vigorous MM106 rootstocks. However, the effect was reversed for 'Jonagold', with MM106 rootstocks producing twice as much return bloom as M26 rootstocks. While it is difficult to find an explanation for this differing effect of rootstock on 'Jonagold', its triploid genetic make-up may be one reason why this cultivar behaves differently to most other cultivars.

5. Conclusions

Optimum crop loads vary with cultivar, but large fruit can be obtained at higher crop loads by thinning during flowering or the early phase of fruit development, regardless of the method of thinning, whether by hand, chemical or mechanical. Seasonal weather patterns during early spring should be considered when determining final crop loads as climatic differences between years can also impact on fruit size and quality.

Early thinning also had a positive effect on fruit quality. Fruit sugar levels were higher in early-thinned fruit than in late-thinned fruit. The positive relationship demonstrated between fruit firmness and weight and between fruit firmness and sugar content with early thinning illustrate additional advantages for early thinning beyond those already established in relation to fruit size. Large fruit can be of better quality than small fruit, providing that it is from regular bearing early-thinned trees. While caution may be required in areas prone to late spring frosts, reducing fruit numbers at or soon after flowering has the effect of reducing competition for resources between fruit, allowing individual fruit to develop greater cell numbers, thus maintaining fruit firmness, even in larger fruit.

For 'Fuji', target crop loads of 4–6 fruit cm^{-2} TCSA are considered appropriate to avoid biennial bearing [1,6], and this study showed that large fruit of 200 g or more can be produced at these crop load levels, but if thinning is delayed, the crop load should be reduced to maximise fruit size. The current target crop load recommendation by Koen et al. [26] of 2–4 fruit cm^{-2} TCSA for 'Delicious' is confirmed by this study. While return bloom was adequate at crop loads of eight fruit cm^{-2} TCSA in both 'Pink Lady' and 'Gala', a decline in fruit size at crop loads above six fruit cm^{-2} TCSA suggests that the recommended target crop load for both these cultivars should be in the range of 4–6 fruit cm^{-2} TCSA. Large fruit size and good return bloom can be maintained in 'Jonagold' at crop loads of eight fruit cm^{-2} TCSA. However, crop loads of four fruit cm^{-2} TCSA are more realistic in 'Braeburn' to sustain regular bearing and good fruit size.

The positive relationships between fruit sugar content and weight and between fruit firmness and weight in both 'Fuji' and 'Delicious', and between fruit sugar content and fruit firmness in 'Delicious' have not been demonstrated previously and demonstrate that early thinning is a valuable tool in improving fruit quality. Early thinning also means that photosynthates produced by the tree are directed into the fruit that will remain on the tree, maximising resources during the cell division period in the first six weeks after bloom.

Supplementary Materials: The following supporting information can be downloaded at: https://www.mdpi.com/article/10.3390/applbiosci2040037/s1, Figure S1: Minimum monthly temperatures for Huon Valley trial sites; Figure S2: Maximum monthly temperatures for Huon Valley trial sites; Figure S3: Number of rain days for Huon Valley trial sites; Figure S4: Average monthly rainfall for Huon Valley trial sites; Figure S5: Cumulative rainfall over the growing season for trial sites; Table S1: Rating scales used for background fruit colour (Frappell and O'Loughlin 1962); Table S2: Mean crop loads (±standard deviation) obtained for each treatment in Trials 1–8.

Funding: This research received no external funding.

Institutional Review Board Statement: Not applicable.

Informed Consent Statement: Not applicable.

Data Availability Statement: Data is contained within the article and supplementary materials.

Acknowledgments: Thanks are due to Caron Summers and Shenan Daniels for technical assistance.

Conflicts of Interest: The author declares no conflict of interest.

References

1. Costa, G.; Botton, A.; Vizzotto, G. Fruit thinning: Advances and trends. *Hortic. Rev.* **2019**, *46*, 185–226.
2. Quinlan, J.D.; Preston, A.P. Effects of thinning blossom and fruitlets on growth and cropping of Sunset apple. *J. Hortic. Sci.* **1968**, *43*, 373–381. [CrossRef]
3. Looney, N.E. Chemical thinning of apples: Some new strategies and important refinements to old procedures. *Acta Hortic.* **1986**, *179*, 597–604. [CrossRef]
4. Johnson, D.S. The effect of flower and fruit thinning on the firmness of 'Cox's Orange Pippin' apples at harvest and after storage. *J. Hortic. Sci.* **1992**, *67*, 95–101. [CrossRef]
5. Johnson, D.S. Influence of time of flower and fruit thinning on the firmness of 'Cox's Orange Pippin' apples at harvest and after storage. *J. Hortic. Sci.* **1994**, *69*, 197–203. [CrossRef]

6. Jones, K.M.; Bound, S.A.; Koen, T.B.; Oakford, M.J. Effect of time of thinning on the cropping potential of red Fuji apple trees. *Aust. J. Exp. Agric.* **1992**, *32*, 417–420. [CrossRef]
7. Kupferman, E. Critical Aspects of Harvest and Quality Management. Washington State University—Tree Fruit Research and Extension Centre, Postharvest information Network. 2002. Available online: http://postharvest.tfrec.wsu.edu/EMK2002A.pdf (accessed on 1 July 2004).
8. Martin, D.; Lewis, T.L.; Cerny, J. Apple fruit cell numbers in relation to cropping alternation and certain treatments. *Aust. J. Agric. Res.* **1964**, *15*, 905–919. [CrossRef]
9. Goffinet, M.C.; Robinson, T.L.; Lakso, A.N. A comparison of 'Empire' apple fruit size and anatomy in unthinned and hand-thinned trees. *J. Hortic. Sci.* **1995**, *70*, 375–387. [CrossRef]
10. Link, H. Chemisches Ausdunnen mit Pantrin und Geramid Neu [Abstract]. *Mitt. Obstbau* **1967**, *11*, 96–101.
11. Wertheim, S.J. Chemical thinning of apple. In *Report for the Research Station for Fruit Growing, Wilhelminadorp for 1973*; Foundation Research Station for Fruit Growing: Wilhelminadorp, The Netherlands, 1974; pp. 17–18.
12. Flore, J.A. Chemical thinning of Paulared and Jerseymac [apple cultivars]. *Compact. Fruit Tree* **1978**, *11*, 115–119.
13. Williams, M.W.; Edgerton, L.J. Fruit thinning of apples and pears with chemicals. In *Agriculture Information Bulletin No. 289*; United States Department of Agriculture: Washington, DC, USA, 1981.
14. Jones, K.M.; Koen, T.B.; Longley, S.B.; Oakford, M.J. Thinning 'Golden Delicious' apples with naphthalene acetic acid in relation to spray concentration, volume and time of day. *J. Hortic. Sci.* **1988**, *63*, 27–30. [CrossRef]
15. Bound, S.A.; Jones, K.M.; Graham, B.; Oakford, M.J.; Tichon, M. Modelling the effects of timing and rates of application of benzyladenine as a secondary thinner of 'Fuji' apple after ethephon. *J. Hortic. Sci.* **1993**, *68*, 967–973. [CrossRef]
16. Greene, D.W. A review of the use of benzyladenine (BA) as a chemical thinner for apples. *Acta Hortic.* **1993**, *329*, 231–236. [CrossRef]
17. Byers, R.E. Effects of bloom-thinning chemicals on apple fruit set. *J. Tree Fruit Prod.* **1997**, *2*, 13–31. [CrossRef]
18. Koen, T.B.; Jones, K.M.; Oakford, M. Model building for prediction of ethephon thinning effects. *Acta Hortic.* **1986**, *179*, 645–652. [CrossRef]
19. Little, C. *Apple and Pear Maturity Manual*; Colin R Little: Sherbrooke, VIC, Australia, 1999; p. 118.
20. Frappell, B.D.; O'Loughlin, J.B. The harvest and storage of pears. *Tasmanian J. Agric.* **1962**, *33*, 70–77.
21. Steel, R.G.D.; Torrie, J.H. *Principles and Procedures of Statistics: A Biometrical Approach*, 2nd ed.; McGraw-Hill: New York, NY, USA, 1980.
22. Costa, G.; Blanke, M.M.; Widmer, A. Principles of thinning in fruit tree crops—Needs and novelties. *Acta Hortic.* **2013**, *998*, 17–26. [CrossRef]
23. Greene, D.W.; Autio, W.R.; Miller, P. Thinning activity of benzyladenine on several apple cultivars. *J. Am. Soc. Hortic. Sci.* **1990**, *115*, 394–400. [CrossRef]
24. Koike, H.; Tamai, H.; Ono, T.; Shigehara, I. Influence of time of thinning on yield, fruit quality and return flowering of 'Fuji' apple. *J. Am. Pomol.* **2003**, *57*, 169–173.
25. Koen, T.B.; Jones, K.M.; Longley, S.B. Spray thinning strategies for 'Red Delicious' apple using naphthalene acetic acid and ethephon. *J. Hortic. Sci.* **1988**, *63*, 31–35. [CrossRef]
26. Link, H. Significance of flower and fruit thinning on fruit quality. *Plant Growth Regul.* **2000**, *31*, 17–26. [CrossRef]
27. McArtney, S.; Palmer, J.W.; Adams, H.M. Crop loading studies with 'Royal Gala' and 'Braeburn' apples: Effect of time and level of hand thinning. *N. Z. J. Crop Hortic. Sci.* **1996**, *24*, 401–407. [CrossRef]
28. Lakso, A.N.; Robinson, T.L.; Goffinet, M.C.; White, M.D. Apple fruit growth responses to varying thinning methods and timing. *Acta Hortic.* **2001**, *557*, 407–412. [CrossRef]
29. McLaughlin, J.M.; Greene, D.W. Effects of BA, GA_{4+7}, and daminozide on fruit set, fruit quality, vegetative growth, flower initiation and flower quality of 'Golden Delicious' apples. *J. Am. Soc. Hortic. Sci.* **1984**, *109*, 34–39. [CrossRef]
30. Sidhu, R.S.; Bound, S.A.; Hunt, I. Crop Load and Thinning Methods Impact Yield, Nutrient Content, Fruit Quality, and Physiological Disorders in 'Scilate' Apples. *Agronomy* **2022**, *12*, 1989. [CrossRef]
31. Robinson, T.L.; Gonzalez, L.; Cheng, L.; Ziang, Y.; Peck, G.; Arnoldussen, B.; Gomez, M.; Guerra, M.; Sazo, M.M.; Kahlke, C.; et al. Studies in precision crop load management of apple. *Acta Hortic.* **2023**, *1366*, 219–225. [CrossRef]
32. Fallahi, E.; Simons, B.R. Rootstock, time, technique, and quantity of nitrogen effects on production, mineral nutrition, and postharvest quality of 'Fuji' apple. *HortScience* **1995**, *30*, 787. [CrossRef]
33. Riesen, W.; Husistein, A. Influence of rootstocks on apple fruit quality. *Acta Hortic.* **1998**, *466*, 161–166. [CrossRef]
34. Unrath, C.R. The commercial implication of gibberellin A4+7 plus benzyladenine for improving shape and yield of Delicious apples. *J. Am. Soc. Hortic. Sci.* **1974**, *99*, 381–384. [CrossRef]
35. Veinbrandts, N.; Miller, P. Promalin improves the shape of Delicious apples in Victoria. *Aust. J. Exp. Agric. Anim. Husb.* **1981**, *21*, 623–630. [CrossRef]
36. Johnson, D.S. Effect of flower and fruit thinning on the maturity of 'Cox's Orange Pippin' apples at harvest. *J. Hortic. Sci.* **1995**, *70*, 541–548. [CrossRef]
37. Garriz, P.I.; Colavita, G.M.; Alvarez, H.L. Influence of crop level on growth and quality of 'Braeburn' apple fruit. *HortScience* **2000**, *35*, 418–419. [CrossRef]

38. Jones, K.M.; Bound, S.A.; Summers, C.R.; Oakford, M.J. Preliminary examination of thinning strategies on young 'Jonagold' and 'Pink Lady' apples. *Aust. J. Exp. Agric.* **1997**, *37*, 377–382. [CrossRef]
39. Jones, K.; Bound, S.; Miller, P. *Crop Regulation of Pome Fruit in Australia*; Tasmanian Institute of Agricultural Research: Hobart, TAS, Australia, 1998; ISBN 1-86295-027-X.
40. Serra, S.; Leisso, R.; Giorani, L.; Kalcsits, L.; Musacchi, S. Crop load influences fruit quality, nutritional balance, and return bloom in 'Honeycrisp' apple. *HortScience* **2016**, *51*, 236–244. [CrossRef]
41. Williams, M.W.; Edgerton, L.J. Biennial bearing of apple trees. *Proc. 19th IHC Warszawa* **1974**, *3*, 343–352.
42. Bound, S.A. Precision crop load management of apple (Malus x domestica Borkh.) without chemicals. *Horticulturae* **2019**, *5*, 3. [CrossRef]
43. Breen, K.C.; van Hooijdonk, B.M.; Tustin, D.S.; Wilkie, J.D.; Bound, S.A.; Middleton, S.G.; Close, D.C. Changes in fruit set of 'Gala' apple in response to environment and artificial spur extinction. *Acta Hortic.* **2014**, *1058*, 77–83. [CrossRef]

applied biosciences

MDPI

Perspective

The Chaperone Hsp90, a Key Player in Salivary Gland Tumorigenesis

Charbel A. Basset [1], Inaya Hajj Hussein [2], Abdo R. Jurjus [3], Francesco Cappello [1,4], Everly Conway de Macario [4,5], Alberto J. L. Macario [4,5] and Angelo Leone [1,*]

1 Department of Biomedicine, Neuroscience and Advanced Diagnostics, Institute of Human Anatomy and Histology, University of Palermo, 90127 Palermo, Italy; cab19@mail.aub.edu (C.A.B.); francapp@hotmail.com (F.C.)
2 Department of Foundational Medical Studies, Oakland University William Beaumont School of Medicine, Rochester, MI 48309, USA; hajjhuss@oakland.edu
3 Department of Anatomy, Cell Biology and Physiological Sciences, Faculty of Medicine, American University of Beirut (AUB), Beirut 1107 2020, Lebanon; aj00@aub.edu.lb
4 Euro-Mediterranean Institute of Science and Technology (IEMEST), 90139 Palermo, Italy; econwaydemacario@som.umaryland.edu (E.C.d.M.); ajlmacario@som.umaryland.edu (A.J.L.M.)
5 Department of Microbiology and Immunology, School of Medicine, University of Maryland at Baltimore-Institute of Marine and Environmental Technology (IMET), Baltimore, MD 21202, USA
* Correspondence: angelo.leone@unipa.it

Abstract: The chaperone system (CS) is emerging as a key multistage participant in carcinogenesis. The CS chief components are the molecular chaperones (some of which are named heat shock proteins or Hsp), which are typically cytoprotective but if abnormal in structure, location, or quantity, can become etiopathogenic and cause diseases, known as chaperonopathies, including some cancers. For example, abnormal Hsp90 expression is associated with tumorigenesis and poor prognosis. Hsp90 is positioned at the center of several key oncogenic pathways by stabilizing and activating oncogenic kinases responsible for driving cell proliferation and survival. Consequently, inhibition of Hsp90 is being investigated as a possible anti-cancer strategy and some results are encouraging. However, the 5-year survival rate for patients suffering from salivary gland carcinomas is still unsatisfactory. Because of the rarity of these malignancies, they may have been overlooked and understudied and, thus, novel therapies (e.g., inhibition of CS components like Hsp90 and others) are urgently needed. In this review, we also summarize the histopathological quantitative patterns and the intra- and extra-cellular location characteristics of Hsp90 in tumors of salivary glands, pointing to their potential for differential diagnosis, prognostication, and patient monitoring.

Keywords: molecular chaperones; Hsp90; chaperone system; salivary gland tumors; carcinoma; diagnosis; prognosis; Hsp90 inhibition; salivary gland cancer treatment

Citation: Basset, C.A.; Hajj Hussein, I.; Jurjus, A.R.; Cappello, F.; Conway de Macario, E.; Macario, A.J.L.; Leone, A. The Chaperone Hsp90, a Key Player in Salivary Gland Tumorigenesis. *Appl. Biosci.* **2023**, *2*, 607–616. https://doi.org/10.3390/applbiosci2040038

Academic Editors: Robert Henry and Demetrios A. Arvanitis

Received: 21 July 2023
Revised: 27 September 2023
Accepted: 27 October 2023
Published: 7 November 2023

1. Introduction

Salivary glands are part of the body's exocrine system because they secrete their biological product, the saliva, through a system of ducts. The salivary glands are in the oral cavity and are classified into two groups: major and minor. The major salivary glands occur in pairs and are grouped based on their anatomical location in the oral cavity and on their size, with the parotid gland (PG) being the largest followed by the submandibular gland (SMG) and the sublingual being the smallest [1–3]. Despite being smaller than the parotid, the submandibular gland accounts for 70% of the saliva output. The minor glands consist of hundreds of small clusters of glands scattered throughout the mucosa of the aerodigestive tract being more concentrated along the buccal, labial, and lingual mucosae, soft and hard palates, and mouth floor [2,4]. The acini are the secretory units of the salivary glands, containing secretory granules that produce the saliva (Figure 1). Although saliva is formed in the acini, it undergoes several physiological and chemical processes along its passage through the ductal system to become the final product that is secreted in the buccal cavity.

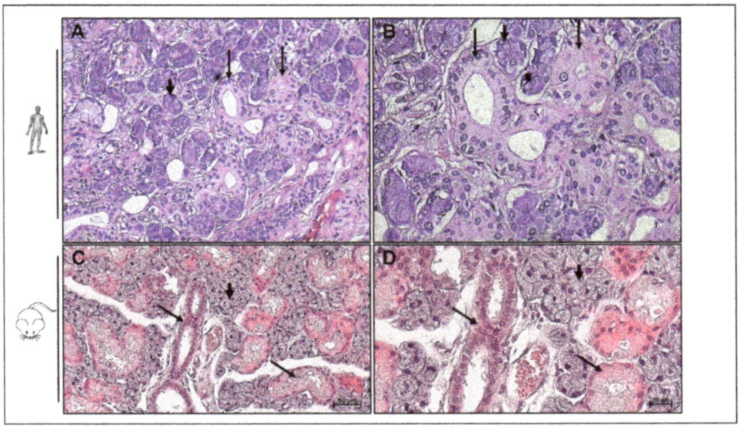

Figure 1. Histological structure of salivary glands. Hematoxilin and eosin staining of adult human (upper panel) and adult mouse (lower panel) submandibular gland. Acini (short arrow) and ducts (long arrow) are featured in both human and mouse submandibular glands. (**A**,**C**) Bar 50 µm. (**B**,**D**) Bar 20 µm.

Saliva is composed of 99.5% water and 0.5% glycoproteins, mucus, electrolytes, enzymes, immunoglobulins, and antibacterial composites [5]. Saliva composition is detailed in Table 1. As food enters the mouth, digestion is initiated by salivary α-amylase that catalyzes the breakdown of starch into glucose and maltose. Saliva has other physiological functions including: (1) mouth lubrication for facilitating speech and mastication, (2) protection of teeth from cavities, (3) antibacterial activity, (4) enhancing taste sensation, (5) hormonal activity, and (6) generation of the food bolus, enabling its swallowing for further processing [1–3,6,7].

Table 1. Saliva composition.

Water 99.5%
Solid constituents 0.5%
Organic solid constituents 0.3%: mucin, serum albumin, serum globulin, amino acids, amylase, lysozyme, IgA, IgG, glucose, citrate, lactate, ammonia, urea, uric acid, creatinine, cholesterol, cyclic nucleotides: cyclic adenosine monophosphate (cAMP) and cyclic guanosine monophosphate (cGMP)
Inorganic solid constituents 0.2%: NaCl; KCl; $NaHCO_3$; Na_2; HPO_4; $CaCO_3$; KSCN

Hsp90 (Heat shock protein 90) is recognized as a crucial player in various aspects of cellular function, including protein folding, stabilization, and degradation. Its role in cancer, including salivary gland tumorigenesis, has been an area of active research.

Hsp90 functions as a chaperone protein, assisting in the correct folding and maturation of other proteins that are involved in key signaling pathways regulating cell growth, survival, and proliferation [8]. In cancer, the overexpression and increased activity of Hsp90 have been observed in several tumor types, including salivary gland tumors [8]. This increased expression is thought to support the folding and stabilization of oncogenic client proteins, enabling the survival and proliferation of cancer cells.

Researchers in many laboratories (including ours) have explored Hsp90 as a potential therapeutic target for cancer treatment. By inhibiting Hsp90's activity, the stabilization of oncogenic proteins could be disrupted, leading to the degradation of these proteins and impairing tumor growth [8]. Several Hsp90 inhibitors have been investigated in preclinical studies and clinical trials for various cancers, including some types of salivary gland tumors [8].

2. Salivary Glands Cancer: Epidemiology and Pathogenesis

The epidemiology and pathogenesis of salivary gland cancer have been subjects of ongoing research. Here is an overview of the general understanding to date. Salivary gland cancer can occur at any age, but it is most commonly diagnosed in individuals between the ages of 50 and 70. Salivary gland tumors are very heterogenous and histologically diverse; more than 30 distinct pathological types of salivary gland tumors have been categorized [9]. Typically, salivary gland tumors are classified based on cytologic, histopathologic, and immunohistochemical characteristics. Although these established parameters for identifying and classifying salivary glands tumors are still considered the gold standard, lately genetic features have been added to their definition [10]. In their latest edition of the classification of tumors of head and neck published in 2022, the world health organization (WHO) has added six new entities [10]. Salivary gland cancers are a heterogeneous group of tumors that can arise from different cell types within the salivary glands. The exact pathogenesis can vary depending on the specific subtype of salivary gland cancer. Some common subtypes include: (1) mucoepidermoid carcinoma (MUC): this is the most common subtype and is believed to arise from ductal or glandular cells. Genetic alterations, such as gene fusions involving MAML2 and CRTC1 genes, have been implicated in the development of mucoepidermoid carcinoma. (2) Adenoid cystic carcinoma (ACC): this subtype is thought to arise from the myoepithelial cells of the salivary glands. Chromosomal translocations involving the MYB and NFIB genes have been identified in many cases of adenoid cystic carcinoma. (3) Acinic cell carcinoma: this subtype arises from the acinar cells of the salivary glands. Specific genetic alterations, including rearrangements of the genes ETV6 and NTRK3, have been associated with acinic cell carcinoma. (4) Polymorphous adenocarcinoma: the pathogenesis of this subtype is not fully understood, but it is believed to originate from the terminal duct cells of the salivary glands. Research into the molecular and genetic basis of salivary gland cancer has advanced our understanding of its pathogenesis, and ongoing studies aim to identify potential therapeutic targets for these tumors.

Despite being considered rare, a significant increase in salivary gland tumors incidence has occurred. Salivary gland cancers accounted for 6.3% of all head and neck carcinomas (HNC) in 1974–1976 and increased to 8.1% in 1998–1999 [11]. Statistical studies conducted in Europe in 2013 estimate that salivary gland malignancies represent 8.5% of all HNC [4]. Major salivary gland malignant neoplasia increased by 54% from 1973 to 2009 [12]. The etiology of salivary gland tumors remains largely unknown, although several risk factors have been established, with irradiation being the most dangerous. Advances in the medical field and the use of modern tools for diagnosis and for routine screening at medical and dental clinics have increased our daily exposure to ionizing radiation and may have played a role in the increase of salivary gland tumor occurrence [11]. Other risk factors have been associated with salivary gland cancer development, including: (1) gender: certain subtypes of salivary gland cancer, such as mucoepidermoid carcinoma and adenoid cystic carcinoma, tend to have a higher incidence in females. (2) Age: the risk of salivary gland cancer tends to increase with age. (3) Occupational exposures: occupational exposure to certain substances, such as asbestos and nickel refining, has been suggested as a potential risk factor for salivary gland cancer. (4) Prior benign tumors: some benign salivary gland tumors, such as pleomorphic adenomas, may have an increased risk of developing into malignant tumors.

3. Salivary Gland Cancer Treatment and Patient Management

Salivary gland cancer treatment and patient management depend on various factors, including the tumor type, stage, location, and the patient's overall health. The management often involves a multidisciplinary approach, with a team of specialists collaborating to create a personalized treatment plan. The primary treatment options for salivary gland cancer include surgery, radiation therapy, and in some cases, chemotherapy or targeted therapy. Here is a summary of the common treatment approaches and patient management strategies:

(1) Surgery: Surgery is the primary treatment for most salivary gland cancers. The extent of surgery depends on the tumor size, location, and whether it has spread to nearby lymph nodes or other tissues. The goal of surgery is to remove the tumor while preserving as much healthy tissue and salivary gland function as possible.
(2) Radiation therapy: Radiation therapy may be used as the primary treatment for small tumors, in combination with surgery, or after surgery to destroy any remaining cancer cells. External beam radiation delivers focused radiation to the tumor site while sparing surrounding healthy tissues.
(3) Chemotherapy: Chemotherapy is not commonly used for all types of salivary gland cancer but may be considered for certain aggressive or advanced cases. It involves the use of drugs to kill cancer cells or stop their growth.
(4) Targeted therapy: Targeted therapy is a newer approach that focuses on specific molecular targets within cancer cells. It is used in some salivary gland cancers that have specific genetic mutations or alterations.
(5) Immunotherapy: Immunotherapy is being explored in clinical trials for some salivary gland cancers. It involves boosting the patient's immune system to recognize and attack cancer cells more effectively.
(6) Adjuvant and neoadjuvant therapies: Adjuvant therapy refers to additional treatment after the primary treatment (surgery or radiation) to reduce the risk of cancer recurrence. Neoadjuvant therapy is given before the main treatment to shrink the tumor and improve the chances of successful surgery or radiation.
(7) Palliative care: For advanced or metastatic cases where a cure may not be possible, palliative care focuses on improving the patient's quality of life. It aims to manage symptoms, alleviate pain, and provide emotional and psychological support to both the patient and their family.
(8) Patient management also includes regular follow-up visits with the medical team to monitor treatment response, assess for any recurrence, and manage potential side effects. Supportive care, including speech and swallowing therapy, nutrition counseling, and psychological support, is vital for patients dealing with the effects of treatment on salivary gland function and overall well-being.

Overall, the treatment and management of salivary gland cancer require a comprehensive and individualized approach to optimize the chances of successful outcomes while minimizing potential side effects and maintaining the patient's quality of life. As research continues to advance, more targeted and personalized treatments are likely to emerge, improving the outlook for patients with this rare form of cancer.

The rarity of salivary gland tumors and their variety are probably the reason of the scarcity of pertinent animal models and cell lines, which in turn slows down progress in the understanding of carcinogenic mechanisms and, consequently, blocks development of efficacious treatments. Therefore, no protocol for salivary gland tumor treatment and management has yet been standardized [13,14]. Generally, salivary gland tumors are surgically excised, which may be accompanied by adjuvant radiotherapy (RT) or chemotherapy (CT) [13]. Because of its low efficiency, CT is usually considered the last resort, palliative, in locally recurrent or metastatic salivary gland cancers that are not amenable to further surgery and irradiation [13]. For diagnosis, ultrasound-guided fine needle aspiration cytology or core biopsy is applied. [15]. For benign tumors, excision of the SMG in a supracapsular plane is the recommended therapeutic approach, while for PG, a partial or radical excision with the facial nerve is usually recommended. RT is not recommended in most benign tumor cases, especially for younger patients, except for intra-operative tumor spillage, where it may be considered after a long-term follow up. For malignant tumors of SMG, wide excision is recommended with a 2 cm margin of healthy tissue, with neck dissection for patients with clinical metastasis in the neck. Elective neck dissection should be considered in high-grade tumors of SMG more than 4 cm in size in addition to a 6-week post-operative adjuvant RT. Primary radiotherapy should be considered for specific cases in which surgical treatments cannot be applied. For malignant tumors of PG, conservative or partial parotidectomy is recommended as the method of treatment, the latter for low-grade tumors (<1.5 cm). If the facial nerve is properly functioning pre-operatively, it is recommended that a direct microsurgical repair or primary nerve graft be performed within a year of the surgery. Neck dissection should be performed if clinical or radiological evidence of nodal diseases is found. In high-grade tumors, a selective/prophylactic neck dissection may be applied and, finally, radiotherapy may be applied to tumors that are >4 cm in size [15].

4. Hsp90 in Cancer

Recent advances in cancerology have encompassed various areas of research, including genetic inheritance, RNA varieties and roles (including diagnostic and therapeutic applications), drug design and delivery, extracellular vesicles, and chaperonology [16]. The latter pertains to the study of the chaperone system (CS), a physiological system akin to other systems, e.g., the immune system [17]. The CS is composed of molecular chaperones, chaperone co-factors, co-chaperones, chaperone interactors, and receptors. The molecular chaperones are the chief components of the CS and are typically cytoprotective, under physiologically normal conditions and under stress. However, when abnormal, molecular chaperones or any of the other CS components can become etiopathogenic and cause diseases, i.e., chaperonopathies [18,19]. These diseases have been classified like many other disorders into genetic and acquired, with the former caused by a change in a gene encoding a CS member, and the latter caused by post-transcriptional or post-translational modifications of the gene's product. Molecular chaperones are the chief members of the CS, but they typically do not act alone and form teams and networks. This is exemplified by the chaperone Hsp90, which forms teams and networks with various other components of the CS [8,18]. Hsp90 represents a group of molecular chaperones that have been named heat shock proteins (Hsp) accompanied by a number which is its molecular weight (e.g., 90 kDa for Hsp90). We have been studying the CS in the salivary glands, mapping some of its components in normal and pathological, including cancerous, tissues [8,20–23]. Heat shock proteins are molecular chaperones that maintain protein integrity and function to counteract the effects of stressors such as a heat shock, irradiation, inflammation, heavy metals, antibiotics, alcohols, oxidation, pH change, osmotic alteration, and mechanical trauma [8,22]. Under physiological conditions, molecular chaperones maintain protein homeostasis by promoting folding of newly synthesized proteins into their functional conformation [22]. Thus, the canonical role of the CS is maintaining protein homeostasis, which, for some functions requires interaction with the ubiquitin–proteasome system (UPS) and/or with the chaperone-mediated autophagy (CMA) machinery. Apart from their canonical role, molecular chaperones have other non-canonical functions and for this they may interact with the immune system [23–25]. For instance, some Hsp mediate immune reactions by binding to antigenic peptides and chaperone them towards MHC class I molecules at the cell surface for presentation to lymphocytes, a process known as cross-presentation [23]. As tumors develop and grow, protein synthesis is increased and thus the demand for molecular chaperones, including Hsp90, increases and neoplasms become "chaperone addicted" [8].

Hsp90 is the master regulator of various growth, inflammation, and survival pathways [8]. Pronounced expression of Hsp90 in several cancer types has been reported [26–30]. Hsp90 has been proposed as a potential biomarker because its increased expression in neoplastic tissue correlates with clinically advanced stages and poor prognosis [31–36]. As reported for other Hsp, Hsp90 pattern of expression can be tissue-specific as shown by a decrease in Hsp90 tissue levels in infiltrative lobular carcinoma [37].

Hsp90-targeted therapy has gained considerable attention as a non-invasive alternative/adjuvant intervention for cancer treatment [8]. Several natural and synthetic compounds have been assessed for their efficacy against cancer. For instance, Hsp90 inhibitors are considered promising candidates for use to abate cancer progression and some of them have entered stage III clinical trials [8].

5. Hsp90 in Salivary Gland Tumors

Studies have revealed that Hsp90 is often overexpressed in salivary gland tumors, promoting the correct folding, stability, and activity of oncogenic client proteins [38–40]. This activity allows cancer cells to survive, proliferate, and resist apoptosis (cell death), leading to tumor growth and progression. The overexpression of Hsp90 has been associated with more aggressive tumor behavior and poorer prognosis in some cases.

Researchers have investigated Hsp90 as a potential therapeutic target for salivary gland tumors. By inhibiting Hsp90's chaperone function, it is possible to disrupt the stabilization of oncogenic proteins, leading to their degradation and hindering tumor growth. Hsp90 inhibitors have been studied in cell models as a potential treatment strategy for salivary gland cancers as an initial step that precedes preclinical models and clinical trials with the hope of improving patient outcomes and developing more effective therapies.

Hsp90 histopathological expression and role in pathophysiology has been studied in a range of cancers but not as much in those of the salivary glands. In this section, we present findings retrieved from the literature about Hsp90 in salivary glands tumors. A study investigated the pattern of distribution and expression of a panel of Hsp (Hsp27, Hsp60, Hsp84, and Hsp86) in 81 specimens of salivary gland tumors [38]. Benign tumors encompassed adenolymphoma, mixed tumors and others while malignant tumors included ACC, MUC, and others. The spatial distribution of Hsp84 and Hsp86, members of the Hsp90 family, was mainly confined to the cytoplasmic portion of the epithelium of the tumors and partly expressed in the nucleus [38]. Semi-quantitative analysis showed that Hsp84 and Hsp86 levels were pronounced in both malignant and benign tumors while being higher in malignant tumors [38]. In benign tumors, Hsp84 and Hsp86 showed higher levels in mixed tumors than in adenolymphoma [38]. Hsp86 immunopositivity intensity was positively correlated with neural invasion, metastasis, malignant grading, tumor site, and proliferation index [38]. Two similar studies assessed the immunohistochemical levels of Hsp27 and Hsp60 as well as Hsp10 and Hsp90 in 20 and 66 cases of salivary gland tumors, respectively [20,21]. Illustrative images of chaperone Hsp90 in tumor tissues from salivary glands are shown in Figure 2. Warthin's tumor (WT) and pleomorphic adenoma (PA) represent the benign tumors while ACC, MUC and ex-pleomorphic adenoma (EX-PA) illustrate the malignant lesions. The panel of the aforementioned molecular chaperones was distributed in the cytoplasmic portion of the epithelium in all specimens [20]. The results reported corroborate the findings that were discussed in the previous study [38]. Hsp90 levels, unlike Hsp10, Hsp27, and Hsp60, were significantly increased in ACC, MUC, and EX-PA and significantly diminished in WT and PA when compared to healthy salivary glands [20]. In the previous study [38], a comparison of the Hsp90 levels in the tumors with those in healthy salivary glands was not reported, therefore, insights on the clinical applicability of Hsp90 measurements as a diagnostic marker to differentiate between healthy and neoplastic salivary glands and also to differentiate between benign and malignant tumors were not obtained. Another group examined Hsp90 molecular expression in benign and malignant tumors of the salivary glands and their findings are in accordance with the immunohistological data highlighted above as Hsp90 mRNA levels are prominent in both tissue types [39].

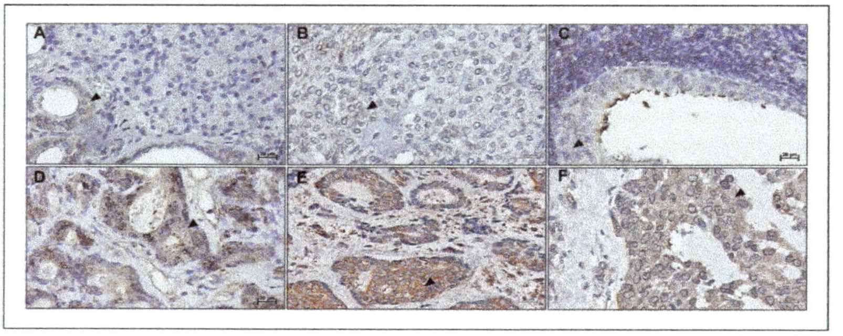

Figure 2. Hsp90 tissue levels help discern between different physiological and pathological states of salivary glands. (**A**) Healthy tissue from the periphery of a Warthin's tumor. Benign tumors pleomorphic adenoma (PA) (**B**) and Warthin's tumor (WT) (**C**). Hsp90 tissue levels assessed by immunohistochemistry show the highest signal intensity and cellular positivity (arrowhead) in the malignant epithelial cells of mucoepidermoid carcinoma (MUC) (**D**), adenoid cystic carcinoma (ACC) (**E**), and ex-pleomorphic adenoma (EX-PA) (**F**) when compared with A, B, and C. Hsp90 levels confined to the ducts of healthy salivary glands show a higher positivity and intensity than the neoplastic epithelium of both PA and WT. Bar 20 μm. Image (**B**) was reproduced from Ref. [20] under the terms and conditions of the Creative Commons Attribution License (CC BY 4.0).

Hsp90 plays a major role in promoting the activity of the PI3K-Akt proliferation pathway and the NF-kB inflammatory pathway, both pathways being upregulated in cancer, Figure 3 [8]. Hsp90 inhibition in ACC has been assessed in vitro for its potential anti-cancer properties. Hsp90 inhibition via ganetespib, the most potent second generation Hsp90 inhibitor, diminished cellular viability, proliferation, and migration while exacerbating apoptosis [20]. Hsp90-targeted inhibition further altered the activation of PI3K-Akt and NF-kB pathways by up-regulating the activity of Akt and down-regulating NF-kB protein levels [20]. Although ganetespib promoted cytotoxicity in ACC cells via apoptosis, cleaved caspase-3 levels were not altered, suggesting that in this instance apoptosis may not be mediated through the mitochondrial pathway. Oncogenic kinases are intrinsically unstable [40]. Hsp90 interaction with its co-chaperone, Cdc37, is essential for the stabilization of the targeted kinases and is required for their oncogenic activity. C-myc is an oncogene prevalent in many cancers [40]. C-myc was expressed in vivo in rodents, using a mouse mammary tumor virus-c-myc (MMTV-c-myc) transgenic mouse model. Mammary adenocarcinomas formed after ectopic expression of c-myc [40]. The introduction of double transgenic MMTV-Cdc37/c-myc mice increased the number of mammary adenocarcinomas and simultaneously induced salivary gland adenocarcinomas formation [40]. Despite their anti-cancer effect, no Hsp90 inhibitor has yet been FDA approved, because they trigger a regulatory feedback loop in the form of a "heat shock response" (HSR) promoting cellular survival [41,42]. This brings the need for higher doses to maintain the inhibitory effect of the drug which is accompanied by increased toxicity. Efforts are being re-directed into exploring isoform-specific Hsp90 inhibition. This is expected to circumvent the drawback of HSR and potentially offer less toxicity than conventional Hsp90 "pan" inhibitors [41,43,44]. One of those drugs, Pimitespib (TAS-116), has recently entered phase III clinical trials [44]. Although, as this is a rather new area of research, more time is needed for novel drugs to be synthesized and tested. Isoform-specific Hsp90 inhibitors may prove to be potent drugs for cancer treatment, especially for salivary glands as chemotherapy in this case has not been effective.

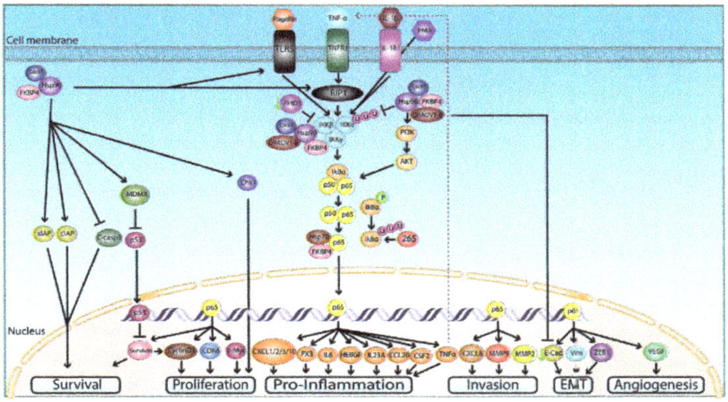

Figure 3. Hsp90 molecular mechanisms and regulation of the NF-kB signaling pathway in cancer. This diagram captures the various signal transduction cascades through which Hsp90 modulates NF-kB activity in tumorous cells and tissue to promote growth, progression, invasion, and metastasization. For the detailed figure caption, please refer to Ref. [8]. This figure is reproduced from Ref. [8] under the terms and conditions of the Creative Commons Attribution License (CC BY 4.0).

As research in this field continues, a deeper understanding of the specific mechanisms by which Hsp90 influences salivary gland tumorigenesis is likely to emerge. Targeting Hsp90 and its associated signaling pathways may offer promising avenues for novel and targeted therapies to combat these rare tumors.

6. Conclusions

Diagnosis, treatment, and patient follow up of benign and malignant tumors, including those of the salivary glands discussed here, can now benefit from the advances in our understanding of the chaperone system (CS) and its participation in carcinogenesis. Differential patterns of expression of Hsp90 (a major component of the CS) in salivary gland tissue may be used in clinical pathology to discern between healthy, benign, and malignant specimens. Hsp90 tissue levels could be used by pathologists as markers to enhance diagnosis and identification of salivary gland tumors. Interaction of Hsp90 with another member of the CS, Cdc37, is emerging as a driving factor behind salivary gland tissue transformation and progression via stabilizing and activating oncogenic kinases.

Author Contributions: Conceptualization, C.A.B.; methodology, C.A.B.; software, C.A.B.; validation, I.H.H.; formal analysis, C.A.B.; investigation, C.A.B.; resources, C.A.B.; data curation, A.J.L.M. and E.C.d.M.; writing—original draft preparation, C.A.B. and A.L.; writing—review and editing, A.J.L.M. and E.C.d.M.; visualization, A.L.; supervision, A.R.J.; project administration, F.C. All authors have read and agreed to the published version of the manuscript.

Funding: This research received no external funding.

Acknowledgments: A.J.L.M. and E.C.d.M. were partially supported by IEMEST and IMET.

Conflicts of Interest: The authors declare no conflict of interest.

References

1. Holmberg, K.V.; Hoffman, M.P. Anatomy, biogenesis and regeneration of salivary glands. *Monogr. Oral Sci.* **2014**, *24*, 1–13. [PubMed]
2. Kessler, A.T.; Bhatt, A.A. Review of the Major and Minor Salivary Glands, Part 1: Anatomy, Infectious, and Inflammatory Processes. *J. Clin. Imaging Sci.* **2018**, *8*, 47. [CrossRef] [PubMed]
3. Holsinger, F.C.; Bui, D.T. Anatomy, Function, and Evaluation of the Salivary Glands. In *Salivary Gland Disorders*; Myers, E.N., Ferris, R.L., Eds.; Springer: Berlin/Heidelberg, Germany, 2007; pp. 1–16.

4. Gatta, G.; Guzzo, M.; Locati, L.D.; McGurk, M.; Prott, F.J. Major and minor salivary gland tumours. *Crit. Rev. Oncol. Hematol.* **2020**, *152*, 102959. [CrossRef] [PubMed]
5. Radha, R.; John Blesswin, A.; Selva Mary, G. A Simple Innovative Approach DNA-Based Saliva Security System for User Authentication. *Indian J. Sci. Technol.* **2016**, *9*, 1–6. [CrossRef]
6. Zolotukhin, S. Metabolic hormones in saliva: Origins and functions. *Oral Dis.* **2013**, *19*, 219–229. [CrossRef]
7. Gröschl, M. The Physiological Role of Hormones in Saliva. *BioEssays* **2009**, *31*, 843–852. [CrossRef]
8. Basset, C.A.; Conway de Macario, E.; Leone, L.G.; Macario, A.J.L.; Leone, A. The chaperone system in cancer therapies: Hsp90. *J. Mol. Histol.* **2023**, *54*, 105–118. [CrossRef]
9. Lin, H.H.; Limesand, K.H.; Ann, D.K. Current State of Knowledge on Salivary Gland Cancers. *Crit. Rev. Oncog.* **2018**, *23*, 139–151. [CrossRef]
10. Skálová, A.; Hyrcza, M.D.; Leivo, I. Update from the 5th Edition of the World Health Organization Classification of Head and Neck Tumors: Salivary Glands. *Head Neck Pathol.* **2022**, *16*, 40–53. [CrossRef]
11. Guzzo, M.; Locati, L.D.; Prott, F.J.; Gatta, G.; McGurk, M.; Licitra, L. Major and minor salivary gland tumors. *Crit. Rev. Oncol. Hematol.* **2010**, *74*, 134–148. [CrossRef]
12. Del Signore, A.G.; Megwalu, U.C. The Rising incidence of major salivary gland cancer in the United States. *Ear Nose Throat J.* **2017**, *96*, E13–E16. [CrossRef] [PubMed]
13. Wang, X.; Luo, Y.; Li, M.; Yan, H.; Sun, M.; Fan, T. Management of salivary gland carcinomas—A review. *Oncotarget* **2017**, *8*, 3946–3956. [CrossRef] [PubMed]
14. Geiger, J.L.; Ismaila, N.; Beadle, B.; Caudell, J.J.; Chau, N.; Deschler, D.; Glastonbury, C.; Kaufman, M.; Lamarre, E.; Lau, H.Y.; et al. Management of Salivary Gland Malignancy: ASCO Guideline. *J. Clin. Oncol.* **2021**, *39*, 1909–1941. [CrossRef]
15. Sood, S.; McGurk, M.; Vaz, F. Management of Salivary Gland Tumours: United Kingdom National Multidisciplinary Guidelines. *J. Laryngol. Otol.* **2016**, *130*, S142–S149. [CrossRef]
16. Macario, A.J.L.; Conway de Macario, E. Chaperonins in Cancer: Expression, function, and migration in extracellular vesicles. *Semin. Cancer Biol.* **2022**, *86*, 26–35. [CrossRef] [PubMed]
17. Macario, A.J.L.; Conway de Macario, E. Chaperone Proteins and Chaperonopathies. In *Stress: Physiology, Biochemistry, and Pathology*; Fink, G., Ed.; Academic Press: Cambridge, MA, USA, 2019; Volume 3, pp. 135–152. [CrossRef]
18. Johnson, J.L. Mutations in Hsp90 Cochaperones Result in a Wide Variety of Human Disorders. *Front. Mol. Biosci.* **2021**, *8*, 787260. [CrossRef] [PubMed]
19. Macario, A.J.L.; Conway de Macario, E. Sick chaperones, cellular stress, and disease. *N. Engl. J. Med.* **2005**, *353*, 1489–1501. [CrossRef]
20. Basset, C.A.; Rappa, F.; Barone, R.; Florena, A.M.; Porcasi, R.; Conway de Macario, E.; Macario, A.J.L.; Leone, A. The Chaperone System in Salivary Glands: Hsp90 Prospects for Differential Diagnosis and Treatment of Malignant Tumors. *Int. J. Mol. Sci.* **2022**, *23*, 9317. [CrossRef]
21. Basset, C.A.; Rappa, F.; Lentini, V.L.; Barone, R.; Pitruzzella, A.; Unti, E.; Cappello, F.; Conway de Macario, E.; Macario, A.J.L.; Leone, A. Hsp27 and Hsp60 in human submandibular salivary gland: Quantitative patterns in healthy and cancerous tissues with potential implications for differential diagnosis and carcinogenesis. *Acta Histochem.* **2021**, *123*, 151771. [CrossRef]
22. Basset, C.A.; Cappello, F.; Rappa, F.; Lentini, V.L.; Jurjus, A.R.; Conway de Macario, E.; Macario, A.J.L.; Leone, A. Molecular Chaperones in Tumors of Salivary Glands. *J. Mol. Histol.* **2020**, *51*, 109–115. [CrossRef]
23. Basset, C.A.; Cappello, F.; Rappa, F.; Jurjus, A.R.; Conway de Macario, E.; Macario, A.J.L.; Leone, A. Chaperonin Hsp60 and Cancer Therapies. In *Heat Shock Proteins in Human Disease*; Asea, A.A.A., Kaur, P., Eds.; Springer: Cham, Switzerland, 2020; Volume 21, pp. 31–52.
24. Fouani, M.; Basset, C.A.; Mangano, G.D.; Leone, L.G.; Lawand, N.B.; Leone, A.; Barone, R. Heat Shock Proteins Alterations in Rheumatoid Arthritis. *Int. J. Mol. Sci.* **2022**, *23*, 2806. [CrossRef] [PubMed]
25. Macario, A.J.L.; Conway de Macario, E. The chaperone system in autoimmunity, inflammation, and virus-induced diseases: Role of chaperonins. In *Stress: Immunology and Inflammation*; Fink, G., Ed.; Elsevier/Academic Press: San Diego, CA, USA, 2023; Volume 5, pp. 119–128.
26. Caruso Bavisotto, C.; Cipolla, C.; Graceffa, G.; Barone, R.; Bucchieri, F.; Bulone, D.; Cabibi, D.; Campanella, C.; Marino Gammazza, A.; Pitruzzella, A.; et al. Immunomorphological Pattern of Molecular Chaperones in Normal and Pathological Thyroid Tissues and Circulating Exosomes: Potential Use in Clinics. *Int. J. Mol. Sci.* **2019**, *20*, 4496. [CrossRef]
27. Rappa, F.; Sciume, C.; Lo Bello, M.; Caruso Bavisotto, C.; Marino Gammazza, A.; Barone, R.; Campanella, C.; David, S.; Carini, F.; Zarcone, F.; et al. Comparative analysis of Hsp10 and Hsp90 expression in healthy mucosa and adenocarcinoma of the large bowel. *Anticancer Res.* **2014**, *34*, 4153–4160. [PubMed]
28. Barone, R.; Caruso Bavisotto, C.; Rappa, F.; Gargano, M.L.; Macaluso, F.; Paladino, L.; Vitale, A.M.; Alfano, S.; Campanella, C.; Gorska, M.; et al. JNK pathway and heat shock response mediate the survival of C26 colon carcinoma bearing mice fed with the mushroom Pleurotus eryngii Var. eryngii without affecting tumor growth or cachexia. *Food Funct.* **2021**, *12*, 3083–3095. [CrossRef]
29. Gorska-Ponikowska, M.; Kuban-Jankowska, A.; Marino Gammazza, A.; Daca, A.; Wierzbicka, J.M.; Zmijewski, M.A.; Luu, H.H.; Wozniak, M.; Cappello, F. The Major Heat Shock Proteins, Hsp70 and Hsp90, in 2-Methoxyestradiol-Mediated Osteosarcoma Cell Death Model. *Int. J. Mol. Sci.* **2020**, *21*, 616. [CrossRef] [PubMed]

30. Kamm, A.; Przychodzeń, P.; Kuban-Jankowska, A.; Marino Gammazza, A.; Cappello, F.; Daca, A.; Żmijewski, M.A.; Woźniak, M.; Górska-Ponikowska, M. 2-Methoxyestradiol and Its Combination with a Natural Compound, Ferulic Acid, Induces Melanoma Cell Death via Downregulation of Hsp60 and Hsp90. *J. Oncol.* **2019**, *2019*, 9293416. [CrossRef]
31. Ansa-Addo, E.A.; Thaxton, J.; Hong, F.; Wu, B.X.; Zhang, Y.; Fugle, C.W.; Metelli, A.; Riesenberg, B.; Williams, K.; Gewirth, D.T.; et al. Clients and Oncogenic Roles of Molecular Chaperone gp96/grp94. *Curr. Top. Med. Chem.* **2016**, *16*, 2765–2778. [CrossRef]
32. Chiu, C.C.; Lin, C.Y.; Lee, L.Y.; Chen, Y.J.; Lu, Y.C.; Wang, H.M.; Liao, C.T.; Chang, J.T.C.; Cheng, A.J. Molecular chaperones as a common set of proteins that regulate the invasion phenotype of head and neck cancer. *Clin. Cancer Res.* **2011**, *17*, 4629–4641. [CrossRef]
33. Chen, Y.; Chen, C.; Ma, C.; Sun, S.; Zhang, J.; Sun, Y. Expression of heat-shock protein gp96 in gallbladder cancer and its prognostic clinical significance. *Int. J. Clin. Exp. Pathol.* **2015**, *8*, 1946–1953.
34. Feng, J.; Xie, G.; Zhan, Y.; Lu, J.; Xu, L.; Fan, S.; Wang, W. Elevated HSP90 associates with expression of HIF-1α and p-AKT and is predictive of poor prognosis in nasopharyngeal carcinoma. *Histopathology* **2019**, *75*, 202–212. [CrossRef]
35. Maddalena, F.; Simeon, V.; Vita, G.; Bochicchio, A.; Possidente, L.; Sisinni, L.; Lettini, G.; Condelli, V.; Matassa, D.S.; Li Bergolis, V.; et al. TRAP1 protein signature predicts outcome in human metastatic colorectal carcinoma. *Oncotarget* **2017**, *8*, 21229–21240. [CrossRef] [PubMed]
36. Pick, E.; Kluger, Y.; Giltnane, J.M.; Moeder, C.; Camp, R.L.; Rimm, D.L.; Kluger, H.M. High HSP90 expression is associated with decreased survival in breast cancer. *Cancer Res.* **2007**, *67*, 2932–2937. [CrossRef] [PubMed]
37. Zagouri, F.; Sergentanis, T.; Nonni, A.; Papadimitriou, C.; Pazaiti, A.; Michalopoulos, N.V.; Safioleas, P.; Lazaris, A.; Theodoropoulos, G.; Patsouris, E.; et al. Decreased Hsp90 expression in infiltrative lobular carcinoma: An immunohistochemical study. *BMC Cancer* **2010**, *10*, 409. [CrossRef] [PubMed]
38. Wang, G.; Gu, X.; Chen, L.; Wang, Y.; Cao, B.; Qun, E. Comparison of the expression of 5 heat shock proteins in benign and malignant salivary gland tumor tissues. *Oncol. Lett.* **2013**, *5*, 1363–1369. [CrossRef]
39. Shigeishi, H.; Sugiyama, M.; Tahara, H.; Ono, S.; Bhawal, U.K.; Okura, M.; Kogo, M.; Shinohara, M.; Shindoh, M.; Shintani, S.; et al. Increased telomerase activity and hTERT expression in human salivary gland carcinomas. *Oncol. Lett.* **2011**, *2*, 845–850. [CrossRef]
40. Stepanova, L.; Finegold, M.; DeMayo, F.; Schmidt, E.V.; Harper, J.W. The oncoprotein kinase chaperone CDC37 functions as an oncogene in mice and collaborates with Both c-Myc and cyclin D1 in transformation of multiple tissues. *Mol. Cell Biol.* **2000**, *20*, 4462–4473. [CrossRef]
41. Sanchez, J.; Carter, T.R.; Cohen, M.S.; Blagg, B.S.J. Old and New Approaches to Target the Hsp90 Chaperone. *Curr. Cancer Drug Targets* **2019**, *20*, 253–270. [CrossRef]
42. Wang, L.; Zhang, L.; Li, L.; Jiang, J.; Zheng, Z.; Shang, J.; Wang, C.; Chen, W.; Bao, Q.; Xu, X.; et al. Small-molecule inhibitor targeting the Hsp90-Cdc37 protein-protein interaction in colorectal cancer. *Sci. Adv.* **2019**, *5*, eaax2277. [CrossRef]
43. Blair, L.J.; Genest, O.; Mollapour, M. The multiple facets of the Hsp90 machine. *Nat. Struct. Mol. Biol.* **2019**, *26*, 92–95. [CrossRef]
44. Honma, Y.; Kurokawa, Y.; Sawaki, A.; Naito, Y.; Iwagami, S.; Baba, H.; Komatsu, Y.; Nishida, T.; Doi, T. Randomized, double-blind, placebo (PL)-controlled, phase III trial of pimitespib (TAS-116), an oral inhibitor of heat shock protein 90 (HSP90), in patients (Pts) with advanced gastrointestinal stromal tumor (GIST) refractory to imatinib (IM), sunitinib (SU) and regorafenib (REG). *J. Clin. Oncol.* **2021**, *39*, 11524. [CrossRef]

applied biosciences

MDPI

Review

The Integration of Artificial Intelligence into Clinical Practice

Vangelis D. Karalis [1,2]

[1] Department of Pharmacy, School of Health Sciences, National and Kapodistrian University of Athens, 15784 Athens, Greece; vkaralis@pharm.uoa.gr; Tel.: +30-(210)-7274267

[2] Institute of Applied and Computational Mathematics, Foundation for Research and Technology Hellas (FORTH), 70013 Heraklion, Greece

Abstract: The purpose of this literature review is to provide a fundamental synopsis of current research pertaining to artificial intelligence (AI) within the domain of clinical practice. Artificial intelligence has revolutionized the field of medicine and healthcare by providing innovative solutions to complex problems. One of the most important benefits of AI in clinical practice is its ability to investigate extensive volumes of data with efficiency and precision. This has led to the development of various applications that have improved patient outcomes and reduced the workload of healthcare professionals. AI can support doctors in making more accurate diagnoses and developing personalized treatment plans. Successful examples of AI applications are outlined for a series of medical specialties like cardiology, surgery, gastroenterology, pneumology, nephrology, urology, dermatology, orthopedics, neurology, gynecology, ophthalmology, pediatrics, hematology, and critically ill patients, as well as diagnostic methods. Special reference is made to legal and ethical considerations like accuracy, informed consent, privacy issues, data security, regulatory framework, product liability, explainability, and transparency. Finally, this review closes by critically appraising AI use in clinical practice and its future perspectives. However, it is also important to approach its development and implementation cautiously to ensure ethical considerations are met.

Keywords: artificial intelligence; clinical practice; machine learning; neural networks; clinical decision; personalized medicine

Citation: Karalis, V.D. The Integration of Artificial Intelligence into Clinical Practice. *Appl. Biosci.* **2024**, *3*, 14–44. https://doi.org/10.3390/applbiosci3010002

Academic Editor: Robert Henry

Received: 29 September 2023
Revised: 7 December 2023
Accepted: 15 December 2023
Published: 1 January 2024

1. Introduction

Artificial intelligence (AI) refers to the emulation of human intelligence in machines designed to exhibit cognitive abilities and acquire knowledge akin to human beings [1,2]. The ancient Greeks attributed a distinctiveness to human beings by virtue of their possession of faculties of reasoning. The notion of the soul was introduced by various religious scholars, who posited it as an enduring and intrinsic essence bestowed upon humanity by a divine creator [3]. According to Plato, it is conceivable for an individual to possess intelligence while simultaneously lacking substantial knowledge about the external world or, more significantly, one's self. Aristotle, the student of Plato, pioneered the formulation of a distinct set of principles that govern the logical aspect of human cognition. In 1936, Alan Turing authored a scholarly article wherein he elucidated the concept of "*Entscheidungsproblem*" and put forth the notion of "effective calculability" as a means to address this quandary. The authors established the groundwork for computational models known as algorithms [4]. The initial development of an artificial neural network (ANN) composed of electrical circuits occurred in 1943, with the aim of simulating the interactions between neurons in the brain [5]. The inception of AI took place in 1956 at Dartmouth College. After a span of three years, the initial computer research using an ANN was successfully conducted, utilizing models referred to as "Adaline" and "Madaline" [6]. Computer-aided diagnosis was initially implemented in the examination of pulmonary nodules identified in chest radiographs in 1963 [7]. Researchers made a significant observation regarding AI's applicability in the bioscience field approximately fifteen years after its inception. This observation was particularly evident in the Dendral experiments [8]. Nevertheless, the utilization of AI in the field of medicine was constrained by technological limitations until 1998, when the United States Food and Drug Administration (FDA) granted approval for the first mammography computer-aided detection (CAD) system [9]. A schematic representation of some important milestones in the evolution of AI is depicted in Figure 1.

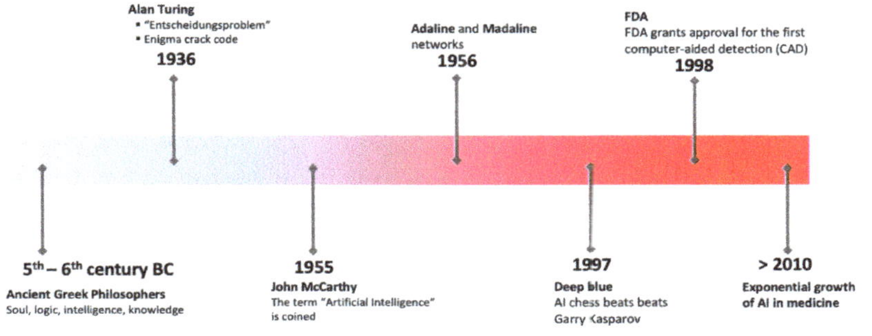

Figure 1. The progression of concepts in artificial intelligence and significant milestones.

Today, as stated by the expert group on AI within the European Commission's digital strategy, AI systems refer to software and potentially hardware systems. These systems are designed to operate in either physical or digital conditions, with the ability to perceive their surroundings through data acquisition. In recent years, there has been significant progress in AI, leading to its widespread adoption across various industries such as healthcare, finance, transportation, drug discovery, and quite recently in pharmacokinetics [10–22]. Over the past few years, notable progress has been made in the field of AI, characterized by the emergence of algorithms and computer programs that exhibit human-like cognitive abilities.

One area where AI has shown great promise is in clinical practice [23]. The incorporation of AI into clinical settings introduces a range of advantages and challenges, accompanied by notable implications for ethical and legal considerations [1]. AI holds the potential to enhance diagnostic precision, streamline administrative tasks, and personalize treatment plans. Through the analysis of extensive medical data, AI systems can discern patterns and correlations that may elude human observation, leading to more accurate and timely interventions [1,2]. Moreover, AI has the capacity to contribute to cost-effective healthcare solutions, ultimately improving overall patient outcomes. The integration of AI technology facilitates informed clinical decision-making processes, thereby promising advancements such as quicker and more accurate diagnoses, personalized treatment plans, and reduced healthcare costs. While the potential benefits of AI in clinical practice are substantial, ethical and legal complexities emerge. The utilization of AI in clinical decision-making raises concerns about transparency, accountability, and the potential bias within algorithms. Safeguarding patient privacy and ensuring data security becomes crucial, necessitating robust ethical guidelines and legal frameworks. Achieving a delicate equilibrium between fostering innovation and protecting patient rights requires thoughtful consideration of the ethical implications of AI in clinical practice, coupled with the development of adaptable legal frameworks capable of keeping pace with technological advancements in the healthcare sector. Addressing AI integration's ethical and legal challenges in clinical practice mandates a comprehensive approach encompassing legal frameworks and regulations, transparent and explainable AI, ethical guidelines and standards, regular audits and assessments, incentives for ethical practices, and international collaboration.

This literature review aims to provide a fundamental synopsis of current research on AI within the domain of clinical practice. Apart from the widespread role of AI in diagnostic methods, the applications of AI in several medical specialties like cardiology, anesthesiology, surgery, pneumology, neurology, urology, gynecology, hematology, and pediatrics are also discussed. It should be emphasized that the purpose of this review is not solely to provide a synopsis of a specific field (e.g., specialty) but rather to attempt to offer an overview of the current applications of AI in medicine.

2. Materials and Methods

The scope of this investigation was confined to articles written in English and subjected to peer review that fulfilled at least one of the following prerequisites: (a) being published within the timeframe of the last ten years and (b) being seminal papers in the field of AI that built what we know today as artificial intelligence.

A literature search was conducted utilizing the PubMed and Scopus databases from 14 July 2023 to 31 August 2023. Additionally, textbooks on AI were consulted. Two sets of keywords were utilized to recognize terms within the title, abstract, and keywords of the articles.

a. The initial set of keywords encompassed terms associated with artificial intelligence, such as "artificial intelligence", "machine learning", and "deep learning". Nevertheless, it is highly probable that research using these methodologies will incorporate terms such as "artificial intelligence" or "machine learning" in their abstracts or keywords;
b. The subsequent set of keywords encompassed concepts associated with the application in clinical practice and the legal status. In this case, composite searches were performed using the terms "Artificial intelligence" AND the medical specialty: "cardiology", "surgery", "anesthesiology", "gastroenterology and hepatology", "pneumonology", "nephrology", "urology", "dermatology", "orthopedics", "neurology", "gynecology, "ophthalmology", "pediatrics", "hematology", "intensive care unit", "diagnostic methods", "legal status", "liability", "regulatory framework".

Following the elimination of duplicate entries, a thorough assessment of the titles and abstracts of the identified articles was made in order to ascertain their suitability for inclusion:

The selection criteria for the evaluation of studies were systematically applied. After eliminating duplicate articles, the author assessed each study based on the following criteria: (i) journal, (ii) authorship, (iii) publication date, (iv) study design, (v) methods of analysis, (vi) results, and (vii) conclusions. The eligibility criteria encompassed articles written in English pertinent to the review objectives. An initial screening of abstracts was conducted, excluding studies that did not align with the eligibility criteria. To enhance data quality, all studies meeting the inclusion criteria underwent a comprehensive evaluation, focusing on aspects such as rationale, method design, results, discussion, and conclusions. Studies exhibiting any bias in methodology, results, or data interpretation that could impact the overall outcome were subsequently excluded.

The exclusion criteria encompassed the following: (a) studies that exclusively focused on the advancement and verification of clinical AI algorithms without any tangible implementation and (b) AI applications that predominantly provided automation functionalities, such as the automated delivery and monitoring of insulin, as opposed to offering decision support.

3. Results

3.1. General

Artificial intelligence has revolutionized the field of medicine and healthcare by providing innovative solutions to complex problems [1,5]. There are various types of AI, including deep learning (DL), machine learning (ML), and natural language (Figure 2). DL is a subset of artificial intelligence that focuses on training neural networks to learn and make decisions in a manner similar to the human brain (Figure 3). DL algorithms are designed to learn and improve from experience automatically, without the need for explicit programming [24–26]. This ability to analyze large amounts of data and extract meaningful patterns has made DL a powerful tool in fields such as image recognition and autonomous driving.

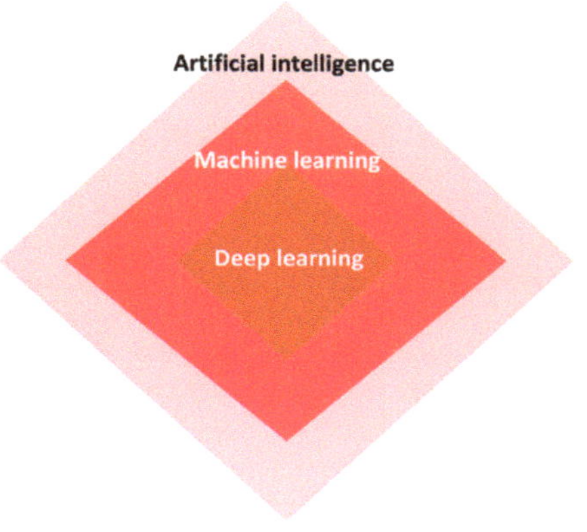

Figure 2. The interconnectedness among artificial intelligence, machine learning, and deep learning.

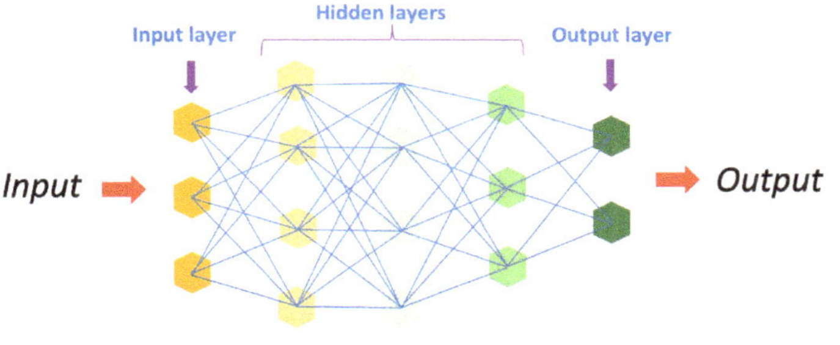

Figure 3. Schematic representation of an artificial neural network.

Machine learning primarily focuses on advancing algorithms and models that empower computers to acquire knowledge and generate predictions or decisions autonomously without the need for explicit programming [2]. ML can be broadly classified into several categories, such as supervised, unsupervised, and reinforcement learning (refer to Table 1). In supervised learning, an algorithm learns from labeled data to make predictions or decisions [2]. This approach trains the algorithm on a dataset comprising input variables and their corresponding output variables. The goal is to enable the algorithm to understand the relationship between the input and output variables, thereby facilitating precise predictions for novel and unobserved data instances. Various supervised learning algorithms are commonly used, including linear regression, logistic regression, support vector machines, and decision trees.

Table 1. A common classification of machine learning algorithms.

Supervised	Unsupervised	Reinforcement
Linear regression	Principal component analysis	Q-learning
Logistic regression	K-means clustering	SARSA
Linear discriminant analysis	KNN (k-nearest neighbors)	Policy iteration
Decision trees	Hierarchal clustering	Monte Carlo tree search
Naive Bayes	Anomaly detection	Bellman equations
Support vector machines	Neural networks	Markov decision process

Unsupervised learning represents a distinct subfield within machine learning, where the algorithm functions without the presence of labeled data [2,24,27,28]. Instead, its purpose is to autonomously identify patterns, structures, or relationships within the data. This learning type proves highly advantageous when a definitive target variable is absent or when the goal is to extract valuable insights from the data without predetermined predictions. Unsupervised learning algorithms include various methods, such as clustering algorithms like k-means and hierarchical clustering, as well as dimensionality reduction techniques like PCA and factor analysis.

The primary objective of reinforcement learning is to train autonomous agents to effectively make a series of decisions within a given environment, aiming to optimize the total cumulative reward obtained [29]. Unlike supervised learning, where the agent is provided with labeled data, or unsupervised learning, where the agent learns patterns and structures from unlabeled data, reinforcement learning operates on the principle of trial and error. Examples of reinforcement learning approaches include the value-based methods (e.g., Q-learning and SARSA), the policy-based methods (e.g., policy gradient and reinforce), and model-based methods (e.g., Monte Carlo tree search).

Natural language processing (NLP) is a field of study that centers on examining and understanding the interplay between computer systems and human language [2]. The field of study pertains to the advancement of algorithms and methodologies that facilitate machines in comprehending, interpreting, and producing human language in a manner that possesses significance and utility. NLP has become increasingly important in our digital age, as it allows computers to process and analyze vast amounts of text data, such as emails, social media posts, and news articles, to extract valuable insights and information.

One of the primary advantages of AI in clinical practice is its ability to rapidly and accurately analyze extensive volumes of data. This capability has given rise to a variety of applications that have not only improved patient outcomes but also lessened the workload on healthcare professionals [30]. In this section, we will explore some of the most promising applications of AI in clinical practice. The evolution of AI has undergone significant changes over the last few decades. The advent of machine learning (ML) and deep learning (DL) has expanded applications in the field of artificial intelligence in medicine, paving the way for personalized medicine rather than relying solely on algorithmic approaches. The use of predictive models holds promise for applications in disease diagnosis, forecasting therapeutic response, and potentially advancing the field of preventive medicine in the coming years. AI has the potential to enhance diagnostic precision, optimize the workflow of healthcare providers and clinical operations, facilitate more effective monitoring of diseases and therapies, improve the precision of medical procedures, and ultimately enhance patient outcomes.

3.2. Cardiology

The application of sophisticated computational algorithms and machine learning techniques in the field of cardiology is commonly referred to as AI. This approach aims to analyze and interpret cardiac data in a more advanced and efficient manner. It involves the development of intelligent systems that can learn from data, make predictions, and offer valuable insights to assist in diagnosing, treating, and managing cardiovascular diseases.

At present, two distinct positions for AI exist in the domain of cardiovascular imaging [31]. Automation refers to the process of replacing human involvement in various tasks, including but not limited to image segmentation and the assessment of structural and functional parameters. Another significant aspect is the identification of insights that hold clinical significance. The majority of documented applications were primarily centered around the implementation of task automation. Furthermore, there have been reports on developing algorithms capable of acquiring cardiac measurements.

AI has significantly impacted various facets of cardiovascular imaging, covering the entire spectrum from initial data acquisition to the final reporting phase [32,33]. Examples of this impact include the use of AI in advancing computed tomography and magnetic resonance imaging techniques for measuring lumen diameter, recognizing coronary calcium score, and identifying obstructive coronary disease. Furthermore, AI has been instrumental in automating processes such as acquisition, segmentation, and report generation [34,35]. In contrast to the methodologies mentioned earlier, a notable concern arises regarding the substantial observer variability observed in the interpretation of echocardiograms. AI holds the potential to address this issue by mitigating inter-observer variability and enhancing diagnostic precision within the field of echocardiography.

In recent years, numerous studies have been conducted to investigate cardiomyopathy screening, with a particular focus on the utilization of AI in conjunction with electrocardiography (ECG) for enhanced diagnostic capabilities [36,37]. The feasibility of the joint use of AI/ECG screening for amyloidosis, cardiomyopathy, and dilated cardiomyopathy remains intact, even in cases of mild left ventricular dysfunction [38,39]. The application of AI/ECG in regular clinical practice has increased the identification of left ventricular systolic dysfunction. In imaging, AI is utilized to automatically evaluate the thickness and properties of the myocardium to distinguish between different types of cardiomyopathies [39,40]. However, there is currently a lack of research investigating the prognostic potential of this AI technology. AI is also being utilized in cardiomyopathy genomics, particularly for predicting the pathogenicity of genetic variants and determining their clinical relevance [41–43].

3.3. Surgery

The application of AI and ML models holds significant potential in the field of surgery. These models demonstrate promising applications in both the preoperative phase, accurately diagnosing pancreatic conditions, and the postoperative phase, evaluating prognosis and predicting complications [44–46]. AI has also proven beneficial in assisting bariatric surgeries. The increasing integration of AI technologies in various healthcare subspecialties has led to promising developments in their application within bariatric surgery [47,48]. The management of patients who are candidates for bariatric surgery is a complex subject. The evaluation process requires the involvement of a multidisciplinary team comprising professionals from various fields, including internists, psychiatrists, general surgeons, and anesthesiologists. Physicians across various medical specialties engage in the comprehensive assessment of patients before, during, and after surgical procedures, a task that presents considerable difficulties due to the intricate nature of individuals afflicted with obesity [49].

Numerous potential applications of AI exist during the intraoperative period. It has the potential to be utilized in the management of pharmacotherapy, hemodynamic optimization, neuromuscular block monitoring, and anesthesia depth assessment [50]. One of the most notable reports pertains to predicting the early distribution kinetics of propofol. Indeed, the volume of drug distribution in individuals with obesity is subject to modification. Specifically, there is an increase in blood volume and cardiac output, alongside alterations in plasma transport proteins. A study utilized AI to handle the induction phase's kinetics effectively [51]. This was achieved through the utilization of a comprehensive pharmacokinetic dataset with high resolution. A comparative analysis was undertaken to evaluate the performance of a traditional four-compartment model, a recirculatory model, and a gated recurrent unit neural network. The study concluded that both a recirculatory model and a gated recurrent unit ANN demonstrated similar performance, surpassing a compartmental model in accurately representing high-resolution pharmacokinetic data of propofol [51].

In the same context, plastic surgeons frequently encounter clinical scenarios that lack definitive solutions. Achieving an ideal treatment approach necessitates utilizing a comprehensive decision model that effectively incorporates various influential factors, including clinical and demographic data. Before the advent of AI, the decision tree analysis technique was commonly used for constructing such models. The localization of significant anatomical landmarks in medical imaging plays a crucial role in preoperative planning and postoperative outcome evaluation [52]. Nevertheless, the current identification process is carried out either manually or by running the inserted auxiliaries, resulting in a time-consuming and imprecise procedure. In order to enhance the precision of landmark localization on the distal femur surface, scientists devised an algorithm that initially transformed three-dimensional images into three distinct sets of two-dimensional images [52]. Subsequently, the algorithm acquired the ability to recognize landmarks within these images and subsequently integrated these outcomes to accurately determine the spatial coordinates of the identified landmarks in three dimensions.

3.4. Anesthesiology

The application of AI has yielded remarkable outcomes in anesthesia and operating room management [53,54]. Throughout each phase of the perioperative process—specifically the preoperative [55–57], intraoperative [16–20], and postoperative phases [42,44]—distinct tasks can be executed using diverse techniques. The effectiveness of a neural network designed to identify esophageal intubation becomes unnecessary in the presence of continuous capnography [58,59]. In this case, a reliable clinical examination has revealed a previously concealed and highly detrimental complication. The use of video laryngoscopy requires the adjustment of an ML model designed to predict challenging intubation based on patient appearance. The expansion of airway management technology has resulted in an increased spectrum of acceptable outcomes in terms of laryngeal visualization.

Since the 1950s, the concept of an algorithm autonomously regulating the depth of anesthesia using EEG recordings has been a subject of ongoing research. Anesthesiologists have explored this possibility for a considerable period, but it continues to be an active area of investigation.

3.5. Gastroenterology and Hepatology

The field of gastroenterology and hepatology is witnessing significant growth in the potential implementation of AI and ML techniques. In recent years, there has been a burgeoning body of research focusing on examining AI applications in various medical contexts, particularly involving the utilization of computer-aided diagnosis (CAD). These applications encompass the use of CAD in diagnosing premalignant and malignant gastrointestinal lesions, predicting treatment response in patients with inflammatory bowel disease, conducting histopathological analysis of biopsy specimens, assessing the severity of liver fibrosis in individuals with chronic liver disease, developing models for liver transplant allocation, and exploring other related areas [60].

The domain of esophageal cancer prevention and early detection shows significant potential for advancements through the utilization of AI. Substantial research advancements have been made in this field, with a notable portion of esophageal cancer research in the United States dedicated to investigating technologies, including those involving AI, aimed at enhancing the early detection and treatment of Barrett's esophagus and esophageal adenocarcinoma [61,62].

AI possesses the capacity to assume a significant role in the decision-making process for the treatment of inflammatory bowel disease by accurately predicting treatment response at an earlier stage and providing guidance for personalized therapy selection. Within the field of inflammatory bowel disease, researchers have made advancements in the development of AI/ML computer vision tools. These tools have been specifically designed to assess the severity of diseases through endoscopic examination. The main goals of their study involve the differentiation of colitis from neoplasia and the distinction between sporadic adenomas and non-neoplastic lesions. AI algorithms have undergone training to forecast the response to treatment and assess the likelihood of disease recurrence [63,64]. There are numerous potential applications for AI and ML in the domain of hepatology. The objectives above encompass the assessment of hepatic fibrosis progression, the identification of non-alcoholic fatty liver disease, the recognition of individuals at risk for hepatocellular carcinoma development, and the enhancement of protocols for organ transplantation [65,66].

The prevention and control of colorectal cancer represent significant public health endeavors undertaken by gastroenterologists. The progress made in the field of ML has resulted in the utilization of computer vision techniques to assist in the detection of polyps during colonoscopy procedures. Empirical evidence has demonstrated the efficacy of CAD systems in enhancing the adenoma detection rate [67–70].

3.6. Pneumonology

AI, specifically the utilization of DL and ML algorithms for pattern recognition, holds significant promise for various applications within the field of pulmonary medicine. These applications encompass image analysis, decision-making processes, and the prediction of prognoses [5–7]. Lung cancer is a prevalent malignant neoplasm characterized by significant clinical morbidity and mortality rates [71]. Lung nodules are the prevailing imaging manifestations observed during the initial phase of lung cancer, posing challenges to manual film interpretation. AI recognition technology can conduct multi-parameter cluster analysis and streamline image processing, thereby assisting medical professionals in the early detection of lung cancer [72]. In recent years, reports have indicated that AI systems have demonstrated the capability to identify malignant pulmonary nodules by analyzing chest computed tomography (CT) images [73]. The model has been developed using DL technology, and AI is utilized for the analysis of CT films in order to support medical professionals in enhancing the accuracy of lung cancer screening. Another study constructed a predictive model by applying logistic regression analysis, integrating specific tumor markers into the model [74]. The study's results demonstrated that the developed predictive model showed significantly better performance when compared to the basic combined detection strategy involving tumor markers.

Research has demonstrated that AI can potentially enhance surgical risk prediction, thereby facilitating the selection of the most optimal surgical approach [75,76]. An example of a cognitive computing system, IBM Watson for Oncology, utilizes AI techniques for data analysis and image conversion. Its primary objective is to assist medical professionals in efficiently identifying crucial information within patients' medical records, presenting pertinent evidence, and facilitating the exploration of potential treatment options [77]. The application of deep neural networks in the identification of respiratory illnesses, specifically in chest radiographs and CT scans, has resulted in a noteworthy enhancement in diagnostic precision compared to subjective characteristics like tumor speculation, as well as objective characteristics such as shape and texture acquired through image analysis software [78].

3.7. Nephrology

The concept of progressive immunoglobulin refers to the gradual development and maturation of immunoglobulins and IgA nephropathy (IgAN) is an acknowledged etiology of renal failure. However, the ability of the nephrologist to anticipate the occurrence of kidney failure among patients at the time of diagnosis is challenging. Nevertheless, the capacity to discern these individuals would prove advantageous in terms of prognostication and treatment purposes. It has been postulated the existence of a function that establishes a relationship between clinical and biological parameters, such as age, sex, blood pressure, proteinuria, serum creatinine level, and anti-hypertensive treatments, at the time of IgAN diagnosis and the likelihood of developing progressive IgAN [79]. The researchers devised and executed the development of an ANN with the purpose of approximating the aforementioned function. The findings indicated that the ANN demonstrated superior accuracy in predicting the onset of progressive IgAN compared to experienced nephrologists [79]. Specifically, the ANN achieved correct predictions in 87% of cases, whereas the nephrologists achieved a lower accuracy rate of 69.4%. Furthermore, the ANN exhibited a higher sensitivity of 86.4% compared to the nephrologists' sensitivity of 72%, indicating its ability to correctly identify true positive cases. Similarly, the ANN displayed a higher specificity of 87.5% compared to the nephrologists' specificity of 66%, indicating its capacity to accurately identify true negative cases. These approaches can potentially be used in a wide range of progressive diseases, thereby aiding clinicians in the process of patient staging and management.

AI models have been applied for various purposes, including predicting the rate of decline in glomerular filtration rate in individuals with autosomal dominant polycystic kidney disease, enhancing anemia management in hemodialysis patients, estimating an appropriate duration for dialysis to achieve the desired level of urea removal, determining the optimal dry weight in patients undergoing hemodialysis, and identifying specific pathogens responsible for bacterial infections in patients with Parkinson's disease [80–83].

3.8. Urology

AI is predominantly used in the field of urology, particularly in the domain of genitourinary malignancies. In a study, AI was utilized to predict the outcomes of prostate biopsies, with a specific focus on prostate cancer. ML algorithms were applied to analyze recurrence-free probability and diagnostic evaluation for bladder cancer. There have been anecdotal reports concerning the staging and prediction of disease recurrence in cases of kidney and testis cancer. Recently, AI has found application in non-oncological diseases, specifically in areas such as stones and functional urology.

In recent decades, numerous scholarly investigations have examined the utilization of AI in the management of prostate cancer. These studies align with the contemporary paradigm of precision medicine and surgery [84]. Prostate cancer diagnosis encompasses a broad range of applications, which have experienced numerous advancements in recent years [85]. A seminal study was conducted in 1994 to determine the potential utility of ANN in predicting biopsy outcomes in males displaying abnormal prostate-specific antigen levels. Additionally, the study aimed to assess the effectiveness of ANN in predicting treatment outcomes following radical prostatectomy [85,86]. A study demonstrated the predictive accuracy of two distinct AI systems [87]. These systems were specifically designed using Vienna-based multicenter European referral database data. These AI systems aim to facilitate the early detection of prostate cancer in males. Another study found that a DL survival model exhibited the ability to predict the timeframe for urinary continence recovery after Robot-Assisted Radical Prostatectomy [88]. This prediction was achieved by incorporating Anatomical Pathology Markers (APMs) and patient-related factors. Furthermore, this particular model has successfully identified APMs of top surgeons that can effectively classify surgeons, surpassing the predictive ability of surgeon experience alone. The APMs were able to differentiate surgeons based on the quality of urinary continence recovery observed in their patients, distinguishing between those with superior and inferior outcomes.

In a seminal study twenty years ago, the authors conducted a comparative analysis of AI and Cox regression models to predict disease recurrence following surgery [89]. The results of the study demonstrated that Cox regression models exhibited superior performance in this regard. In conjunction with the increasing range of surgical indications for metastatic kidney cancer, a study was conducted to assess the predictive capacity of AI in determining the prognosis of patients with metastatic renal cell carcinoma who initiate systemic therapy [90]. The researchers provided their AI system with a dataset consisting of information from 175 patients who had undergone nephrectomy of the primary tumor prior to receiving systemic therapy. The objective of this study was to forecast the overall survival rate three years after initiating the initial treatment, utilizing parameters that are accessible at the commencement of first-line therapy. AI has demonstrated the potential to achieve a prediction accuracy of 95% in forecasting overall survival rates. This performance surpasses regression models, indicating the potential future application of AI as a risk stratification tool.

A urinary tract infection is a common bacterial infection that affects the urinary system, including the bladder and urethra. A notable study focused on urinary tract infections where an AI system was developed to assist in the diagnosis of such infections [91]. The study involved individuals diagnosed with either cystitis or nonspecific urethritis. Subjects underwent various procedures, including a medical history assessment, physical examination, analysis of urine samples, and the use of ultrasonography. The findings demonstrated the efficacy of AI in diagnosing urinary tract infections based solely on erythrocyte values in conjunction with symptoms such as suprapubic pain, pollakiuria, and urinalysis results. The AI model exhibited a remarkably high accuracy rate of 98.3%, suggesting it could serve as a cost-effective alternative to expensive laboratory and ultrasound tests.

Functional urology refers to the branch of urology that focuses on studying and managing the urinary tract. The exploration of AI potential applications has also extended to the domain of functional urology. A study compared an AI model and multiple linear regression in terms of their effectiveness in replacing preoperative urodynamic evaluation in women diagnosed with pelvic organ prolapse [92]. A total of 804 women diagnosed with pelvic organ prolapse were subjected to examination, revealing that both multivariate logistic regression and AI were determined to be less effective than urodynamic studies in evaluating urinary dysfunction. A kidney transplant is a surgical procedure in which a healthy kidney from a donor is transplanted into a recipient. Over the past few years, there has been a growing interest in utilizing AI predictive tools in kidney transplantation. Similarly, the potential application of AI in identifying risk factors and co-variates that contribute to the failure of renal transplantation has been explored [93]. The AI approach was compared with the traditional logistic regression model. The AI method demonstrated superior accuracy compared to logistic regression, as evidenced by data analysis from 378 patients.

3.9. Dermatology

Identifying skin diseases primarily relies on the apparent attributes exhibited by the lesions. However, dermatology encompasses a vast collection of over 2000 distinct types of dermatological diseases. Certain skin lesions associated with various diseases may exhibit similarities, posing challenges in accurate diagnosis and treatment [94,95]. Notably, there is a significant shortage of dermatologists, particularly in developing countries and remote regions, where increased medical resources, professional consultations, and clinical support are urgently needed [96,97].

The convergence of rapid iteration in big data, advancements in image recognition technology, and the global proliferation of smartphones present transformative potential for diagnosing and treating skin diseases [98,99]. AI, in particular, has the capacity to offer prompt diagnoses, facilitating a wider range of treatment options and enhancing accessibility, especially for marginalized regions and individuals with limited resources [100]. The integration of AI technology and algorithms has the potential to establish itself rapidly as a standard approach in the field of diagnosis and evaluation. The examination of the structure and form of a skin abnormality is a fundamental aspect of dermatological diagnosis. Advancements in AI have led to significant improvements in facial recognition and aesthetic analysis, rendering them more dependable [101].

The inception of AI in the field of dermatopathology can be traced back to 1987 when a text-based system known as TEGUMENT was utilized on a personal computer. The system was specifically developed with the purpose of classifying 986 histopathological diagnoses based on light microscopic images. It demonstrated a diagnostic accuracy of 91.8% compared to the assessments made by a qualified dermatopathologist [102]. During that particular time frame, the absence of necessary equipment and technologies for whole slide imaging led to the belief that the notion of human-independent image analysis was not viable. In recent years, the accurate classification of routine diagnoses by machine-based systems has become a tangible reality [103].

In a research study, 11 DL algorithms were developed to identify and classify whole slide images of dermal nevus, seborrheic keratoses, and nodular basal cell carcinoma [104]. The visual representations underwent a process of pixelation, resulting in the disintegration of the images, which were subsequently subjected to data analysis. A DL algorithm was developed for pathology that incorporates whole slide imaging. The algorithm effectively categorizes these images into four distinct diagnostic groups: basaloid, squamoid, melanocytic, and other. The implemented system utilizes a series of three consecutive convolutional neural networks to determine the diagnosis with the highest probability.

Distinguishing between malignant and benign lesions holds the highest importance for dermatopathologists due to the consequential divergence in therapeutic decisions. In this context, a study used a sample of 695 melanocytic neoplasms to distinguish between melanoma and nevus by means of classification [105]. The study included a comprehensive representation of all stages of melanoma, as well as various types of nevi. In the present investigation, it was observed that the convolutional neural network exhibited a statistically significant superiority over the pathologists in terms of accurately diagnosing nevi and melanoma through histopathological analysis. The observed discordance rate of 25–26% among dermatopathologists was found to be comparable to the aforementioned similarity.

In another research study, the objective was to assess the precision of a DL algorithm in diagnosing three dermatopathological conditions through the utilization of whole-slide imaging [106]. The study's findings indicated that the AI system demonstrated high accuracy, correctly classifying several types of carcinomas. In contrast to the straightforward binary classification involved in diagnosing melanoma and distinguishing it from pigmented nevi, the diagnosis of non-melanoma skin cancers presents a more challenging task. This challenge stems from the intricate categorization of these conditions and the inclusion of various benign and malignant diseases, along with inflammatory dermatoses, within the differential diagnoses. A study was conducted to assess the effectiveness of convolutional neural networks in precisely detecting and diagnosing non-pigmented lesions [107]. The findings were compared with the diagnoses rendered by a cohort of 95 clinicians, which included 62 dermatologists with appropriate qualifications. Convolutional neural networks did not exhibit superior accuracy in diagnosing medical conditions compared to human experts. However, they demonstrated greater accuracy in diagnosing prevalent skin cancers. Conversely, convolutional neural networks exhibited lower accuracy than clinicians in diagnosing uncommon non-pigmented malignancies, specifically amelanotic melanoma.

3.10. Orthopedics

Supervised ML can be applied to classify individuals into pain phenotypes based on brain MRI, considering the high prevalence of long-standing pain in the UK, estimated to be between 30% and 50% [108]. The absence of tissue pathology that corresponds to pain, as well as the dependence on self-reported measures for subgroup classification, pose a significant challenge in identifying the neural correlates of pain and provide a comprehensive overview of ML applications used in the context of chronic pain, which encompasses pain conditions beyond musculoskeletal disorders [109]. The authors specifically highlight using ML techniques to classify individuals into distinct pain phenotypes based on predictive models.

Another study established a correlation between frontal plane knee biomechanics and the ability to predict the risk of knee injuries [110]. In this study, inertial sensor data were used to categorize the performance of single-leg squats based on the extent of knee valgus [111]. The study sample consisted of 14 participants, and a total of 140 images were analyzed. Additionally, the researchers sought the opinions of three expert raters regarding the potential risk associated with the observed performances. Supervised learning was applied to perform classification among three distinct classes, namely "poor", "moderate", and "good". The study's findings indicate that the accuracy levels were observed to be significantly high when performing a 2-class classification task. However, when the complexity of the classification task was increased to a 3-class classification, the accuracy levels experienced a notable reduction of approximately 30%. There is a scarcity of instances where unsupervised learning techniques have been utilized within the domain of musculoskeletal research. According to a study, the chronic pain challenge assesses the likelihood of chronic pain based on assigned weights for various health behaviors [112]. The study included both supervised and unsupervised methods to demonstrate the precise prediction of pain levels, as measured by the visual analog scale and the Oswestry Disability Index. These predictions are made based on the corresponding scores for depression, nutrition, and physical activity. Nevertheless, although this emphasizes the potential of ML to categorize the risk of chronicity using patient-reported data, the effectiveness of unsupervised learning by itself has not been confirmed.

3.11. Neurology

Neuroimaging plays a pivotal role in clinical practice and scientific inquiry, facilitating the examination of the brain in various states of well-being and pathology. Similar to several other domains, neuroimaging is enhanced by the utilization of sophisticated analysis methodologies in order to harness imaging data effectively to investigate the brain and its functionality. In recent times, ML has made significant contributions. Additionally, it has played a crucial role in the prompt identification of acute conditions like stroke and in monitoring imaging changes over time. As our capacity to visualize and examine the brain progresses, so does our comprehension of its complex interconnections and their significance in making therapeutic decisions.

Despite being in the early stages of development, AI's utilization in neuro-oncology exhibits considerable potential. It is highly probable that AI algorithms will enhance our comprehension of brain tumors and play a pivotal role in fostering advancements in the field of neuro-oncology. The field of neuro-oncology has experienced a growing emphasis on the integration of molecular markers for the purpose of guiding therapeutic interventions [113]. AI algorithms have demonstrated notable efficacy in the noninvasive identification of significant molecular markers from diagnostic imaging, exhibiting remarkable accuracy. In various institutional datasets, AI algorithms have successfully determined the mutational status of several markers [114,115]. Moreover, it has been demonstrated that algorithms based on AI can effectively identify investigational molecular markers, even in smaller cohorts of patients [116].

The utilization of AI for the analysis of diagnostic imaging has proven to be beneficial in the clinical management of brain tumors. The utilization of AI to automate labor-intensive tasks holds great potential in the field of neuro-oncology. Multiple studies have demonstrated the efficacy of DL techniques in identifying brain metastases measuring in the millimeter range through MRI imaging. Furthermore, it has been observed that comparable DL models have demonstrated significant efficacy in the automated segmentation of tumors, thereby enhancing the efficiency of radiation therapy treatment planning [117–119]. AI has demonstrated potential in accurately differentiating various central nervous system malignancies without the need for invasive procedures, achieving comparable results to those of expert neuroradiologists [120,121]. The extensive application of these AI algorithms could prove to be highly beneficial in resource-constrained environments that lack access to specialized neuroradiologists.

3.12. Gynecology

Despite encountering various obstacles, the integration of AI in obstetrics and gynecology has exhibited remarkable progress. The utilization of AI in various domains has proven to be highly effective in addressing persistent issues related to diagnosis and treatment. According to a study, AI has the potential to enhance knowledge and provide assistance to medical professionals in the fields of gynecology and obstetrics [122]. The latest applications of AI models in gynecology involve identifying endometrial carcinoma, in vitro fertilization, uterine sarcoma, cervical intraepithelial neoplasia, and advancing anticancer medication [123,124].

The integration of AI technology into ultrasonography has the capacity to enhance the adoption of medical ultrasound in various clinical environments, facilitating its broader application by healthcare professionals. Therefore, the utilization of AI in the field of ultrasonography for prenatal care has the potential to assist medical professionals in efficiently prioritizing and accurately diagnosing the anatomical structures of pregnant individuals. In certain medical applications such as obstetric pelvic and echocardiography ultrasonography, where visual analysis and measurement play a crucial role, the utilization of video clips can provide a comprehensive set of structured data. This enables spatiotemporal analysis and enhances the advantages of ANNs [125]. A study investigated the efficacy of AI algorithms in ultrasonic diagnosis for pregnant patients with brain tumors. They specifically focused on evaluating the accuracy rate of this diagnostic approach [126]. The diagnostic accuracy achieved through the utilization of AI was recorded at 94.50%. Another research study was conducted, which involved a prospective and descriptive approach. The study focused on a sample of approximately 244 pregnant women in their first pregnancy trimester. The registered female participants were specifically queried regarding their utilization of iron, folic acid, potassium iodide, and multivitamin supplements throughout their pregnancies. The utilization of an ANN model that incorporates variables related to pregnancy checks, intake of iodized salt, iodized supplements, and iodine-rich foods can be used to predict iodine deficiency during the early stages of pregnancy. This predictive model can assist experts in making a more feasible diagnosis [127]. In their study, Sakai et al. utilized a newly developed DL-based explainable graph chart diagram representation to aid in fetal cardiac ultrasound screening. This screening process is known to have a relatively low rate of detecting congenital heart disease during the second-trimester stages, primarily due to the challenges associated with mastering the technique [128]. Consequently, the utilization of AI in the second and third trimesters of pregnancy for diagnostic purposes, specifically using diagram representation, enhances screening performance. The accuracy rate for experts increases from 96% to 97.50%, while non-experts improve from 82% to 89% [129].

3.13. Ophthalmology

The utilization of AI in diagnosing and managing ocular disease has become increasingly popular due to research findings emphasizing its potential to enhance personalized medicine and improve healthcare outcomes [130]. Numerous AI algorithms are currently under development for managing patients diagnosed with diabetes mellitus [131].

Due to advancements in the management of diabetes mellitus, there has been an enhancement in the monitoring of patients, resulting in a higher incidence of diabetic retinopathy and diabetic macular edema. The primary cause of significant visual impairment and blindness among individuals of working age is the presence of diabetic macular edema that has not been diagnosed or treated [132]. Hence, it is imperative to conduct extensive screening for diabetic retinopathy on a large scale in order to identify potentially detrimental alterations at an early phase, thereby facilitating effective management and treatment strategies.

Considering the prevailing patterns of population growth and the significant incidence of diabetic retinopathy and diabetic macular edema within the community, it is inevitable that automated screening and diagnosis will become increasingly prevalent in ophthalmic healthcare settings. Efforts have been made to explore automated retinal screening techniques for the diagnosis of diabetic retinopathy to enhance patient management and mitigate the societal impact. Various AI, ML, and DL methodologies have been used for the automated diagnosis and grading of diabetic retinopathy. The most efficacious automated systems have been developed based on comprehensive investigations conducted within the last three years. Recent research on diabetic retinopathy has shown that AI techniques have exhibited significant accuracy, sensitivity, and specificity in identifying and diagnosing diabetic retinopathy [133].

Automated application systems have the potential to enhance doctors' comprehension of diabetic retinopathy predictions and enhance the practicality of intelligent diagnostic models in real-world clinical settings [133]. Based on the aforementioned studies, it was observed that the automated analysis of retinal images exhibited a high level of accuracy, validity, sensitivity, and specificity in detecting diabetic retinopathy. Furthermore, the diagnostic performance of AI techniques was deemed clinically acceptable and demonstrated high reproducibility when applied to the validation data set.

Age-related macular degeneration is a chronic ocular condition that is recognized as a prominent contributor to visual impairment [134]. Prognostications AI algorithms exist to generate personalized predictions in age-related macular degeneration. These algorithms can make predictions regarding the presence of drusen beneath the retina in individuals with age-related macular degeneration. The AI algorithms offer automated detection capabilities for identifying drusen, fluid, and geographic atrophy in relation to age-related macular degeneration lesions. These algorithms leverage fundus images and spectral-domain optical coherence tomograph to enhance the diagnosis and treatment [135]. The utilization of AI in the automated detection of drusen holds promise for enhancing the diagnostic capabilities of ophthalmologists in the early and efficient assessment of fundus images [136]. The application of AI techniques in diagnosing and grading age-related macular degeneration has been extensively explored. Recent studies have shown that these automated approaches exhibit notable efficacy, demonstrating high accuracy, sensitivity, and specificity levels in detecting age-related macular degeneration [137].

Glaucoma, which ranks as the second most prevalent factor leading to visual impairment on a global scale, is distinguished by the gradual degeneration of retinal ganglion cells and the permanent depletion of axons within the optic nerve. The timely identification and management of glaucoma is of paramount significance in the prevention of preventable visual impairment. AI techniques have demonstrated exceptional efficacy in efficiently classifying glaucomatous and healthy eyes. Ophthalmologists have the ability to utilize these automated results as a reference point, enabling them to enhance their decision-making process within clinical practice. The utilization of automated AI applications has demonstrated significant efficacy and holds promise in addressing the imminent challenge of diabetic retinopathy, age-related macular degeneration, and glaucoma screenings in both developed and developing nations [138].

3.14. Pediatrics

Imaging techniques play a paramount role in the management of pediatric neurologic, neurosurgical, and neuro-oncological conditions [139]. Multi-parametric MRI techniques are gaining popularity, particularly when combined with radiogenomic analyses that establish connections between imaging features and molecular biomarkers associated with diseases. Nevertheless, incorporating this approach into regular clinical practice continues to be challenging. AI techniques can model extensive datasets related to childhood neurologic disease, including radiologic, biological, and clinical data. This capability enables the integration of such information into prognostic modeling systems at an early stage. Consequently, AI techniques offer a viable solution to address this issue [139].

In certain applications within the field of pediatric neuroradiology, ANNs have demonstrated notable efficacy in a focused manner. This concept is most effectively demonstrated through the utilization of ventricular size determination to categorize children into either a normal or hydrocephalic group. In a recent study, an analysis was performed on hydrocephalus and controls [140]. They achieved an accuracy score of 94.6% for hydrocephalus and 85.6% for controls using a training set of T2-weighted MRI images from around 399 children. Previous studies have reported comparable achievements in the field of pediatric hydrocephalus through the implementation of evolutionary modifications in ANN methodologies [141].

The application of a support vector machine for the categorization of children into normal or hypoxic-ischemic brain injury groups, based on the measurement of corpus callosum widths, yielded a classification accuracy of 95% [142]. Another study utilized a comparable methodology to examine a group of adolescents who had experienced traumatic brain injury. Specifically, they utilized edge-density imaging and support vector machines to classify the participants into two categories: normal and mild traumatic brain injury [143]. The aforementioned method, which achieved a precision rate of 94%, demonstrated superior performance compared to neurocognitive testing in this aspect [25]. Support vector machines have demonstrated successful classification of various magnetic resonance imaging abnormalities in the fetal brain. These abnormalities encompass functional connectivity, brain maturity, and severe fetal abnormalities. The classification accuracies achieved by support vector machines in these studies range from 79% to 84% [144].

Support vector machines (SVMs) have been utilized in magnetic resonance imaging texture analysis to examine brain tumors. This machine learning application aims to quantitatively analyze imaging data to generate an image texture that is generally imperceptible to human visual perception [145]. Texture analysis in clinical practice is advantageous for clinicians because it can incorporate comprehensive imaging data of the entire tumor. This approach takes into consideration the presence of intra-tumor heterogeneity, which may not be adequately represented by a single biopsy site or even multiple biopsy sites [145].

A study expanded on the application of texture analysis by integrating both linear discriminant analysis and a probabilistic neural network [146]. Their objective was to categorize posterior fossa tumors, specifically medulloblastoma, pilocytic astrocytoma, and ependymoma. The combined techniques achieved an accuracy ranging from 86% to 93% through a validation process. The utilization of AI in diagnosis offers a potential enhancement to the effectiveness of diagnoses.

Decision trees have also been used in another significant capacity within the field of ML in the context of pediatric neuroimaging. Specifically, they have been utilized for the purpose of data analysis in order to provide insights and information regarding neuroimaging in clinical trials. An instance can be observed in a study where a decision tree classifier was used within a randomized controlled trial conducted on children diagnosed with autism who were undergoing treatment with simvastatin [147]. The study utilized a random forest classifier to effectively categorize children from the control group who had undergone simvastatin treatment [147]. The classifier achieved a classification accuracy of 79%. This observation suggests the potential benefits that such applications may offer in the future.

3.15. Hematology

AI has been used to examine various types of medical data, including hematopathological, radiographic, laboratory, genomic, pharmacological, and chemical data. The purpose of using AI in these analyses is to enhance the accuracy and effectiveness of diagnosis, outcome prediction, and treatment planning and to expand our understanding of benign and malignant hematology.

Recent advancements in CNN-based models have shown the ability to effectively differentiate between various types of leukocytes on peripheral smears, indicating their potential for automating routine pathology practices [148]. Ongoing research is being conducted in the field of automated interpretation of bone marrow specimens [149]. CNNs have also exhibited their usefulness in characterizing qualitative and quantitative variations within specific cell lineages, such as the morphology of erythrocytes and the textural alterations observed in sickle cell disease [150]. The aforementioned achievements encompass the differential diagnosis of various diseases, as evidenced by the capacity of models to accurately diagnose acute myeloid leukemia, distinguish between different causes of bone marrow failure, and function as a screening tool for lymphoma in settings with limited resources [151].

AI has been applied in various domains to enhance diagnostic processes' dependability, convenience, and efficacy. Previous studies have shown that CNN methods have proven effective in diagnosing multiple myeloma solely using mass spectrometry data obtained from peripheral blood [152,153]. Personalized models have been shown to possess a high diagnostic capability when distinguishing between challenging conditions, such as different causes of bone marrow failure, is difficult. This is achieved by integrating patient demographics, laboratory data, and fundamental genetic information. Previous studies have also utilized similar methods to differentiate between peripheral leukemia and lymphoma [154].

The task of prognosis is widely recognized as challenging, and even commonly used clinical prognostication tools exhibit notable variability within different risk categories [155]. AI, possessing advanced capabilities in processing nonlinear and intricate data, promises to deliver more sophisticated and individualized prognostications. The aforementioned methodologies have been used within the field of benign hematology to enhance the accuracy of risk assessment for central catheter thrombosis. These methodologies have successfully identified individuals with a low risk of developing thrombosis [155]. AI has been utilized to categorize patients undergoing hematopoietic stem cell transplants into low and high-risk groups for acute graft-versus-host disease. This classification has important implications for making informed decisions regarding the administration of immunosuppressive treatments to these individuals [156] Previous studies have also been conducted in the field of autologous transplants for multiple myeloma. AI has been utilized in the field of malignant hematology to enhance the initial assessment of risk stratification for acute myeloid leukemia and myelodysplastic syndromes [157,158]. In post-treatment scenarios, where minimal residual disease is considered a negative prognostic factor, AI has exhibited the capability to attain performance comparable to that of humans. This achievement has the potential to simplify and establish a consistent approach to handling and analyzing this kind of data [159].

3.16. Intensive Care Unit

ML models have been applied within the intensive care unit (ICU) setting to anticipate pathologies such as acute kidney injury, identify symptoms such as delirium, and suggest appropriate therapeutic interventions such as vasopressors and fluid administration in cases of sepsis. The timely identification and management of sepsis is of paramount importance due to its potential to significantly decrease mortality rates. Although the management of early sepsis involves source control and the administration of broad-spectrum antibiotics, the detection of sepsis during this phase of the illness poses considerable challenges [159]. Identifying sepsis becomes increasingly feasible as the condition advances, while the treatment poses considerable challenges. Due to the diverse nature of sepsis, the existing diagnostic and prognostic methods pose a significant challenge in early sepsis detection and accurate prognosis estimation. This difficulty further complicates determining an appropriate treatment strategy for individual patients [160]. AI prediction models have the potential to provide significant value for patients diagnosed with sepsis. AI models possess the capacity to enhance the timely identification of individuals requiring antibiotic treatment. Certain AI prediction models appear to outperform existing diagnostic methods; however, these models exhibit notable limitations, such as including predictor variables like blood pressure in the present sepsis definition. The assessment of AI models' performance is exaggerated in this context. These models exhibit limited generalizability. The existence of unresolved concerns has resulted in a significant disparity between the advancement of algorithms and their practical implementation in clinical settings.

The growing utilization of Electronic Health Records within the ICU is driving the dissemination of data science and ML techniques in the critical care setting. Hemodynamic data derived from monitors, infusion data obtained from infusion pumps, and respiratory data collected from ventilators generate substantial data volumes. These datasets can be compared to other sources of big data, such as omics data encompassing genomics or proteomics. A study devised a computational model utilizing reinforcement learning techniques to propose optimal treatment strategies dynamically for adult ICU patients [161]. A study was conducted wherein pervasive monitoring and ML techniques were applied to continuously evaluate delirium and agitation in a cohort of 22 patients admitted to an ICU [162]. The patients were categorized based on the Confusion Assessment Method for the ICU scale. The researchers utilized cameras and accelerometers to capture and document facial expressions and movements. Three accelerometers were strategically positioned on the patient's wrist, ankle, and arm to discern and classify their posture. The researchers applied a pre-existing neural network model to perform facial recognition and detect facial expressions using individual facial features.

3.17. Diagnostic Methods

AI has the capability to fundamentally transform the methodologies applied in disease diagnosis and treatment. Through the examination of patient data, including medical history, symptoms, and test results, AI algorithms can provide clinicians with more precise and tailored diagnoses for individual patients. This has the potential to facilitate the identification of diseases at an earlier stage and enhance the efficacy of treatment strategies. Additionally, AI can assist in identifying potential drug interactions and adverse reactions, ensuring that patients receive the safest and most effective treatments.

Ultrasound (US) has gained widespread global adoption as a primary imaging modality in various clinical domains, owing to the continuous advancements in ultrasonic technologies and the established digital health infrastructure. Breast cancer is widely recognized as a prevalent form of cancer among women globally and continues to be the second most significant contributor to cancer-related mortality [163]. The predominant utilization of DL in breast US, as observed in the surveyed literature, pertains to diagnosing and categorizing breast masses [164]. The utilization of DL techniques for the analysis of abdominopelvic imaging has various applications within the United States. A significant portion of these applications have been specifically directed towards the examination of the liver [165]. Their research revealed that this approach exhibited superior accuracy compared to two-dimensional shear wave elastography and certain biomarkers in the evaluation of advanced fibrosis and cirrhosis in patients infected with the hepatitis B virus. In a study, the authors devised a CNN approach to predict the METAVIR score, a semi-quantitative measure of liver fibrosis [166]. The training dataset consisted of several thousands of US images obtained from two tertiary academic referral centers. This approach demonstrated a notable level of precision in forecasting the METAVIR score, surpassing radiologists' diagnostic capabilities in identifying liver fibrosis. In their study, Ta et al. developed a computer-aided diagnosis system for the classification of malignant and benign focal liver lesions using contrast-enhanced ultrasound cine recordings [167]. The researchers found that the accuracy of this method was comparable to that of an expert reader.

Deep learning algorithms have been utilized in various applications, including identifying muscle diseases [168], determining cone positioning, and segmentation of muscle imaging [169]. The diagnostic accuracy for neuro-muscular diseases was improved by using a CNN-based method, which enhanced the assessment and classification of inflammatory muscle diseases [168]. The progress in the field of AI/ML tools for the interpretation of imaging is experiencing rapid acceleration. This can be attributed to several factors, including the availability of extensive digitized image datasets, the accessibility of open-source algorithms, advancements in computing power, the emergence of cloud services, and the continuous development of DL techniques [170].

The utilization of automation in tasks frequently performed by radiologists, such as identifying rib fractures and lung nodules using CT scans, reassessing pleural effusion size via sequential chest radiographs, or conducting mammographic screening, shows potential as a favorable strategy. This has the potential to enable radiologists to dedicate additional time to more advanced interpretive tasks that may not be amenable to automation, as well as participate in endeavors such as multidisciplinary team meetings. The utilization of AI in the triage procedure demonstrates a notable ability to efficiently assign priority to critical cases that require immediate reporting. These cases may include CT scans that unveil the presence of pulmonary embolism, chest radiographs that indicate pneumothorax, or head CTs that reveal hemorrhage. The aforementioned methodology possesses the capacity to reduce patient morbidity and expedite the duration of hospitalization in emergency departments. The utilization of DL systems in synthetic MRI enables the post-processing and reconstruction of MR image data, reducing image acquisition time without significant deterioration in image quality. This advancement can potentially enhance efficiency, decrease expenses, and enhance accessibility [171].

CNNs are a fundamental element within networks utilized in orthopedic and musculoskeletal radiology [172]. In recent times, the concept of generative adversarial networks (GANs) has been introduced. GANs are models designed to generate novel data closely resembling the original dataset. These models comprise two distinct networks that engage in a competitive game-like training process. A total of 12 enhanced GAN models incorporating CNNs have been successfully utilized in the domain of radiographic images [173].

Extensive research has been conducted in the field of oral and maxillofacial radiology to investigate the potential of AI in the diagnosis of various conditions. The aforementioned conditions encompass dental caries, periodontal disease, osteosclerosis, odontogenic cysts, and tumors, as well as ailments that impact the maxillary sinus or temporomandibular joints.

AI has been used in the field of dentistry for image analysis, encompassing a range of tasks, including tooth segmentation and localization, assessment of bone quality for osteoporosis, determination of bone age through hand-wrist radiographs, and localization of cephalometric landmarks [172]. DL systems utilizing CNN architectures have been successfully applied in the domain of dentistry. Notably, a novel system has been developed that incorporates both three-dimensional cone beam computed tomography images and two-dimensional images [172].

The application of DL techniques has been used to identify and categorize teeth within both cone beam computed tomography images and panoramic images. The utilization of classification systems for teeth enables dentists to make clinical decisions and streamline their charting process by using automated computer-aided design outputs. These outputs facilitate the automatic completion of digital patient records [174].

4. Discussion

4.1. General

The aim of this literature review is to provide a comprehensive overview of existing research on AI in clinical practice. In addition to its extensive use in diagnostic techniques, AI integration has been explored in various medical disciplines, including cardiology, anesthesiology, surgery, pneumology, neurology, urology, gynecology, hematology, and pediatrics. Given the continuous influx of new articles and the exponential increase in published papers, this review focuses on indicative articles to illustrate AI's robust penetration and wide-ranging applications in clinical practice.

One of the major advantages of AI in clinical practice is its ability to improve diagnostic accuracy and treatment outcomes [10–22]. With AI-powered algorithms, healthcare providers can analyze large amounts of patient data and identify patterns that may not be immediately apparent to human clinicians. This can help to identify diseases earlier, resulting in expedited medical intervention and improved prognoses for individuals. Additionally, AI can predict which treatments are likely to be most effective for a given patient, allowing for personalized medicine that considers individual patient characteristics. AI can help reduce healthcare costs by streamlining administrative processes and reducing unnecessary tests [23].

In a study, ML techniques were applied in the analysis of high-throughput genome sequencing data, aiming to enhance comprehension of disease processes and the development of therapeutic modalities [175]. In this study, cutting-edge ML algorithms, including random forest, support vector machine radial kernel, adaptive boost, averaged neural network, and gradient boosting machine, were applied. The goal was to stratify patients with head and neck squamous cell carcinoma into early and late clinical stages and to predict the risk based on the expression profiles of miRNAs [175]. Also, quite recently, variational autoencoders (i.e., a type of neural network) were introduced as a method to effectively augment the sample size of clinical studies, thereby mitigating costs, time constraints, dropouts, and ethical considerations [176]. In a study, the efficacy of variational autoencoders in the context of data augmentation was demonstrated through the utilization of simulations encompassing multiple scenarios [176]. Also, in the field of bioequivalence studies, several ML methods were utilized to solve the old problem of defining the appropriate absorption rate metric [20–22]. Through the joint utilization of ML algorithms, non-linear mixed effect modeling, and Monte Carlo simulations, a new metric termed "average slope" was defined and introduced. It was proven that the currently used Cmax (i.e., the maximum observed plasma concentration) is unsuitable for expressing the absorption rate. On the contrary, the newly defined measure (average slope) comprises the desired properties of absorption rate, has the appropriate units of measurement (i.e., concentration units per time), exhibits ease of estimation directly from the concentration-time data of the drug, and all ML algorithms showed its supremacy over all other metrics used or proposed in bioequivalence [20–22].

Although the advantages of AI in clinical practice are evident, a number of obstacles exist that necessitate attention and resolution [177]. One of the foremost considerations revolves around the possibility for AI to sustain and propagate bias within the healthcare sector. If the algorithms used in AI systems are trained on biased data, they may produce biased results, leading to disparities in healthcare outcomes for certain patient populations. Also, there is a lack of transparency, interpretability, and explainability of the AI algorithms since the latter can be considered as a black box. This issue will be discussed later. Additionally, there is a risk that AI could dehumanize healthcare, with patients feeling disconnected from their care providers and reduced to a set of data points. There are also fears around the ethical implications of using AI in healthcare, particularly. The impact of integrating AI technologies into the field of medicine will be most pronounced among present and prospective medical trainees. Consequently, it is imperative for medical schools and graduate medical education programs to modify their curriculum in order to instruct current and future generations of physicians on the conscientious utilization of these potent and transformative technologies [178]. Integrating AI in medicine introduces several legal and ethical considerations, including medical liability issues such as training and competence, transparency/traceability, explainability, personal health data, regulatory compliance, product liability, and malpractice insurance. The fundamental aspects of these issues are delineated below.

4.2. Training of Healthcare Professionals

Healthcare professionals using AI systems need appropriate training to ensure competence in their use. Failure to properly understand and operate AI systems may lead to medical errors and subsequent liability. Furthermore, prospective medical students will be required to develop new skills, encompassing the concept of "knowledge capture, not knowledge retention". This implies a shift from a curriculum focused on rote memorization to one that prioritizes critical thinking. The field of AI places significant emphasis on two primary domains: collaboration and management of AI applications, as well as a deeper comprehension of probabilities and their consequential application in clinical decision-making involving patients and families. The aforementioned domains aim to acquire knowledge pertaining to the efficient and ethical utilization of AI, while also promoting the practical application of AI technologies in the healthcare field [179].

4.3. Transparency, Traceability, and Explainability

Drawing inspiration from various disciplines, the domains of transparency and traceability in the context of healthcare and individual patients necessitate adherence to more rigorous criteria [180]. From a legal standpoint, it is necessary for data to adhere to all applicable laws, regulations, and additional legal obligations throughout its entire lifecycle, including acquisition, storage, transfer, processing, and analysis. Furthermore, it is imperative for the law, its interpretation, and its implementation to continually adjust in response to the ever-changing advancements in technology [181]. Numerous AI algorithms, particularly those based on deep learning models, function as enigmatic "black boxes", rendering it difficult to clarify the rationale behind their decisions. In situations where the validity of a medical decision generated by an AI system is scrutinized, the absence of transparency can give rise to legal complications. Even when meeting all of these apparent prerequisites, the question persists as to whether the utilization of AI-driven solutions and tools necessitates the need for explainability. In essence, doctors and patients must possess knowledge regarding the outcomes presented and an understanding of the qualities and attributes upon which these outcomes are founded, as well as the corresponding underlying assumptions. Furthermore, the inclusion of additional stakeholders may necessitate a comprehensive comprehension and explication of algorithms and models. From a Western legal standpoint, three fundamental areas have been identified for the purpose of elucidating the concept of explainability [182]. These areas include (a) informed consent, (b) certification and approval in accordance with the regulations set forth by the FDA and the Medical Device Regulation (MDR), and (c) Liability.

4.4. Liability and Regulatory Framework

The certification and approval bodies responsible for medical devices have been relatively sluggish in implementing regulations pertaining to explainable AI and its impact on developing and marketing such products. The FDA significantly promotes the continuous development and enhancement of AI-based medical products through its comprehensive total product lifecycle approach. The concept of explainability is not explicitly referenced; however, there is a requirement for an appropriate level of transparency and clarity in the output and algorithm intended for users [183,184]. The primary focus of this inquiry pertains to the functionalities of the software and the alterations it has undergone throughout its evolution. The MDR does not explicitly address the requirement for explainability in relation to medical devices utilizing AI and ML, specifically. In this context, it is important to emphasize the significance of accountability and transparency [185]. Specifically, these requirements pertain to the provision of information that enables the tracing, transparency, and explication of the development process of ML and DL models that contribute to medical treatment. There is a strong probability that in forthcoming times, a more refined delineation of these prerequisites will emerge, necessitating manufacturers of AI-driven medical devices/software to furnish exhaustive details pertaining to the training and evaluation of the models, the data utilized, and the overarching methodologies used in their creation.

The integration of AI in healthcare frequently entails the utilization of highly sensitive patient data. Suppose a data security or privacy breach results in unauthorized access or use of patient information; healthcare providers and AI developers may face legal repercussions. This underscores the critical importance of robust data protection measures and adherence to privacy regulations in developing and deploying AI applications within the healthcare sector [185]. The processing of personal health data is permissible under the law only when the individual has provided explicit consent for its utilization. The current standard for utilizing patient data in AI applications is informed consent, as there is a lack of overarching legislation governing the use of personal data and information [185]. Healthcare organizations and AI developers must comply with various data protection regulations, such as the Health Insurance Portability and Accountability Act (HIPAA) in the United States or the General Data Protection Regulation (GDPR) in Europe [186,187]. Failure to comply with these regulations can lead to legal consequences and fines. Clear policies should be established regarding data ownership and patient consent. Patients should be informed about how their data will be used, who will have access to them, and for what purposes. Obtaining informed consent is essential for ethical and legal reasons. Implementing robust encryption methods for both data in transit and data at rest helps protect patient information from unauthorized access. Encryption adds an extra layer of security to prevent sensitive data from being intercepted or accessed by unauthorized parties. Access to healthcare AI systems and the data they process should be restricted and monitored. Role-based access controls should be implemented to ensure that only authorized personnel can access sensitive information. Also, healthcare AI systems should use secure methods for storing and transmitting data, while periodic security audits and assessments help identify vulnerabilities in the system.

The legal framework concerning AI in healthcare is still in a state of evolution. New laws and regulations may be introduced to address issues related to liability specifically, and healthcare professionals must stay informed about these changes. As the field of AI continues to advance, legal frameworks are likely to adapt to ensure that ethical, transparent, and responsible practices are followed in developing and deploying AI technologies in healthcare. Staying abreast of these evolving regulations is crucial for healthcare professionals to navigate the complex legal terrain and uphold accountability and patient care standards. Nevertheless, even to this day, disparities persist in international guidance between Europe and the United States regarding the legal challenges that may arise from the use of AI in healthcare. These regions adopt distinct approaches to addressing these challenges. The European Union (EU) has emerged as a leading force in the field of medical AI innovation and has acknowledged the specific difficulties that AI poses to current liability frameworks. In order to establish coherence in liability principles and ensure legal clarity, the European Commission has introduced the Artificial Intelligence Act, which represents one of the initial legal frameworks dedicated specifically to AI [186]. The European Commission endeavors to advance the secure utilization of AI in sectors with significant consequences, such as healthcare, while concurrently enhancing technological innovation. The United States lacks a comprehensive legal framework that specifically regulates AI, resulting in limited legal precedent concerning liability and medical AI. The regulatory aspect of AI in healthcare has been acknowledged by the FDA, which aims to facilitate the secure implementation of AI by developing a strategic plan to ensure ongoing supervision of AI as a medical device [187]. In order to foster a patient-centric approach, the FDA endeavors to enhance transparency by requesting manufacturers to provide detailed descriptions of the operational mechanisms of their AI devices. This initiative is aimed at facilitating a comprehensive comprehension of the advantages and potential hazards associated with such devices. The FDA also endeavors to address potential bias that may arise from training AI algorithms on specific populations or historical datasets [187]. The FDA has recently published a discussion paper entitled "Proposed Regulatory Framework for Modifications to Artificial Intelligence/Machine Learning-Based Software as a Medical Device" with the aim of guaranteeing the safety of medical software that utilizes AI and ML technologies.

In addition, as the use of AI in healthcare becomes more prevalent, malpractice insurance considerations evolve to address potential risks associated with these technologies [188]. Given the unique risks associated with AI in healthcare, malpractice insurance policies may need to be tailored to address liabilities arising from the use of AI technologies specifically. This could include coverage for errors or malfunctions in AI algorithms that lead to adverse patient outcomes. Insurers may face challenges in underwriting policies related to AI, as the field is rapidly evolving, and assessing risks associated with emerging technologies can be complex. Insurers may need to adapt their underwriting processes to account for the specific risks posed by AI in healthcare [189]. Malpractice insurance policies related to AI must align with evolving legal and regulatory frameworks governing the use of AI in healthcare. Insurers may need to stay informed about changes in laws and regulations to ensure that their policies remain relevant and compliant. Also, insurers may require healthcare organizations to implement monitoring and reporting mechanisms for AI-related incidents. Timely reporting adverse events can facilitate a proactive response and help mitigate potential liabilities.

In the same vein, product liability in the context of AI refers to the legal responsibility of those involved in the design, development, manufacturing, distribution, and deployment of AI systems for any harm or damages caused by the AI product [190,191]. If the design of an AI system is inherently flawed and leads to harmful consequences, the designers and developers may be held liable. AI design defects could include biased algorithms, inadequate testing, or a lack of robust safety features. Manufacturing defects pertain to issues arising during an AI system's production or deployment. These defects may lead to malfunctions, security vulnerabilities, or other problems that could result in harm. Manufacturers may be held responsible for these defects. If AI products come with inadequate warnings or instructions regarding their proper use, healthcare professionals or end-users may be unaware of potential risks. Failure to provide clear guidance on the limitations and risks of an AI system could result in liability. In cases where third-party vendors provide AI components or services, liability may extend to these vendors if their products or services contribute to harm. Determining the chain of responsibility and liability in complex AI ecosystems can be challenging. Rigorous testing and validation of AI systems are essential to identify and address potential issues before deployment. If inadequate testing contributes to harm, the parties responsible for the testing and validation process may be held liable. Maintaining detailed records of the design, development, testing, and deployment processes is important for demonstrating due diligence in the event of a product liability claim. Clear documentation can provide evidence of compliance with industry standards and best practices.

4.5. Overall

This literature review discusses the current applications of AI in various medical specialties in clinical practice. AI possesses the capacity to fundamentally transform clinical practice and enhance patient outcomes. The potential for further advancements in AI technology is vast, and the impact on patient outcomes could be significant.

Certainly, not all aspects and bibliographic references could have been included, as this literature review aims to offer a comprehensive overview of the applications of AI in as many medical specialties as possible. Consequently, it was unavoidable that some important contributions in each field could not be discussed. As an example, there has been a substantial rise in the utilization of biomarkers as early warning systems for assessing disease risk over the past decade, with extensive application evident during the recent COVID-19 pandemic [192]. In the same context, in dermatology, which is one of the fields with the widest use of AI, quite recently, a research paper has underscored the capability of machine learning to serve as a biomarker for differentiating among individuals with psoriasis, psoriatic arthritis, and those in good health [193]. Also, several investigations have been carried out, uncovering notable molecular biomarkers through miRNA expression that can differentiate between early and late stages of carcinomas [175].

It should also be emphasized that despite the concerns discussed in this review, the potential benefits of AI in healthcare cannot be ignored. AI has the ability to improve diagnosis accuracy, personalize treatment plans, and reduce healthcare costs. In the future, AI will likely become a standard tool in clinical practice, with healthcare providers working alongside AI systems to provide the best possible care for patients [185,186]. Nevertheless, to guarantee the ethical and efficient utilization of AI in the healthcare sector, we must confront these apprehensions and establish explicit protocols for advancing and integrating AI systems. The complete realization of AI's potential to enhance healthcare outcomes can only be achieved under such circumstances. Overall, it is considered that AI has the potential to greatly improve healthcare outcomes, but it is important to address ethical concerns and establish clear guidelines for its development and implementation.

5. Conclusions

Integrating AI into clinical practice brings forth many benefits and challenges with significant implications for ethical and legal considerations. AI holds the promise of improving the precision of diagnostic accuracy, streamlining administrative tasks, and personalizing treatment plans. By analyzing vast amounts of medical data, AI systems can identify patterns and correlations that might escape human observation, leading to more precise and timely interventions. Additionally, AI can contribute to cost-effective healthcare solutions and improve overall patient outcomes. This integration of AI technology facilitates informed clinical decision-making processes. Thus, AI has the potential to enhance patient outcomes, offering faster and more accurate diagnoses, personalized treatment plans, and reduced healthcare costs. While there is still much to be explored and developed in the field of AI in clinical practice, it is crucial that we continue to invest in research and development to unlock its full potential. This approach can improve our understanding of how AI can enhance healthcare and lead to the developing of new tools and technologies that benefit patients and healthcare professionals alike. Nevertheless, it is important to approach its development and implementation cautiously and collaborate with healthcare professionals to ensure ethical considerations are met.

However, these advancements come with ethical dilemmas and legal complexities. The use of AI in clinical decision-making raises concerns about transparency, accountability, and the potential for bias in algorithms. Ensuring patient privacy and data security becomes paramount, demanding robust ethical guidelines and legal frameworks. Striking the right balance between innovation and safeguarding patient rights requires careful consideration of the ethical implications of AI in clinical practice, alongside the development of legal frameworks that can adapt to the rapid pace of technological evolution in the healthcare sector. To overcome the ethical and legal challenges associated with the integration of ai in clinical practice, a multi-faceted approach is essential that includes legal frameworks and regulations, transparent and explainable AI, ethical guidelines and standards, regular audits and assessments, incentives for ethical practices, and international collaboration.

Funding: This research received no external funding.

Conflicts of Interest: The author declares no conflict of interest.

References

1. Henderson, H. *Artificial Intelligence: Mirrors for the Mind (Milestones in Discovery and Invention)*, 1st ed.; Chelsea House Publisher: New York, NY, USA, 2007; 176p.
2. Russell, S.; Norvig, P. *Artificial Intelligence: A Modern Approach*, 4th ed.; Pearson: Bloomington, MN, USA, 2021; 1136p.
3. Philippidis, A. Charles River Licenses ERS Genomics' CRISPR/Cas9 Technology. *Clin. OMICs* **2018**, *5*, 7. [CrossRef]
4. Turing, A.M. On Computable Numbers, with an Application to the Entscheidungsproblem. *Proc. Lond. Math. Soc.* **1937**, *s2-42*, 230–265. [CrossRef]
5. McCulloch, W.S.; Pitts, W. A Logical Calculus of the Ideas Immanent in Nervous Activity. *Bull. Math. Biophys.* **1943**, *5*, 115–133. [CrossRef]
6. Widrow, B.; Lehr, M.A. 30 Years of Adaptive Neural Networks: Perceptron, Madaline, and Backpropagation. *Proc. IEEE* **1990**, *78*, 1415–1442. [CrossRef]
7. Lodwick, G.S.; Keats, T.E.; Dorst, J.P. The Coding of Roentgen Images for Computer Analysis as Applied to Lung Cancer. *Radiology* **1963**, *81*, 185–200. [CrossRef] [PubMed]
8. Feigenbaum, E.A.; Buchanan, B.G. DENDRAL and Meta-DENDRAL: Roots of Knowledge Systems and Expert System Applications. *Artif. Intell.* **1993**, *59*, 233–240. [CrossRef]
9. Goldberg, J.E.; Reig, B.; Lewin, A.A.; Gao, Y.; Heacock, L.; Heller, S.L.; Moy, L. New Horizons: Artificial Intelligence for Digital Breast Tomosynthesis. *RadioGraphics* **2023**, *43*, e220060. [CrossRef]
10. Zhong, F.; Xing, J.; Li, X.; Liu, X.; Fu, Z.; Xiong, Z.; Lu, D.; Wu, X.; Zhao, J.; Tan, X.; et al. Artificial Intelligence in Drug Design. *Sci. China Life Sci.* **2018**, *61*, 1191–1204. [CrossRef]
11. Bakkar, N.; Kovalik, T.; Lorenzini, I.; Spangler, S.; Lacoste, A.; Sponaugle, K.; Ferrante, P.; Argentinis, E.; Sattler, R.; Bowser, R. Artificial Intelligence in Neurodegenerative Disease Research: Use of IBM Watson to Identify Additional RNA-Binding Proteins Altered in Amyotrophic Lateral Sclerosis. *Acta Neuropathol.* **2017**, *135*, 227–247. [CrossRef]
12. Segler, M.H.S.; Preuss, M.; Waller, M.P. Planning Chemical Syntheses with Deep Neural Networks and Symbolic AI. *Nature* **2018**, *555*, 604–610. [CrossRef]
13. Simões, M.F.; Silva, G.; Pinto, A.C.; Fonseca, M.; Silva, N.E.; Pinto, R.M.A.; Simões, S. Artificial Neural Networks Applied to Quality-by-Design: From Formulation Development to Clinical Outcome. *Eur. J. Pharm. Biopharm.* **2020**, *152*, 282–295. [CrossRef] [PubMed]
14. Gaisford, S.; Saunders, M. *Essentials of Pharmaceutical Preformulation*, 1st ed.; Wiley-Blackwell: Hoboken, NJ, USA, 2013; 268p.
15. Babu, N.J.; Nangia, A. Solubility Advantage of Amorphous Drugs and Pharmaceutical Cocrystals. *Cryst. Growth Des.* **2011**, *11*, 2662–2679. [CrossRef]
16. Damiati, S.A.; Martini, L.G.; Smith, N.W.; Lawrence, M.J.; Barlow, D.J. Application of Machine Learning in Prediction of Hydrotrope-Enhanced Solubilisation of Indomethacin. *Int. J. Pharm.* **2017**, *530*, 99–106. [CrossRef] [PubMed]

17. Hossain, S.; Kabedev, A.; Parrow, A.; Bergström, C.A.S.; Larsson, P. Molecular Simulation as a Computational Pharmaceutics Tool to Predict Drug Solubility, Solubilization Processes and Partitioning. *Eur. J. Pharm. Biopharm.* **2019**, *137*, 46–55. [CrossRef] [PubMed]
18. Ye, Z.; Yang, Y.; Li, X.; Cao, D.; Ouyang, D. An Integrated Transfer Learning and Multitask Learning Approach for Pharmacokinetic Parameter Prediction. *Mol. Pharm.* **2018**, *16*, 533–541. [CrossRef]
19. Ota, R.; Yamashita, F. Application of Machine Learning Techniques to the Analysis and Prediction of Drug Pharmacokinetics. *J. Control. Release* **2022**, *352*, 961–969. [CrossRef]
20. Karalis, V.D. Machine Learning in Bioequivalence: Towards Identifying an Appropriate Measure of Absorption Rate. *Appl. Sci.* **2022**, *13*, 418. [CrossRef]
21. Karalis, V.D. On the Interplay between Machine Learning, Population Pharmacokinetics, and Bioequivalence to Introduce Average Slope as a New Measure for Absorption Rate. *Appl. Sci.* **2023**, *13*, 2257. [CrossRef]
22. Karalis, V.D. An In Silico Approach toward the Appropriate Absorption Rate Metric in Bioequivalence. *Pharmaceuticals* **2023**, *16*, 725. [CrossRef]
23. Ferrara, P.; Battiato, S.; Polosa, R. Progress and Prospects for Artificial Intelligence in Clinical Practice: Learning from COVID-19. *Intern. Emerg. Med.* **2022**, *17*, 1855–1857. [CrossRef]
24. Beneke, F.; Mackenrodt, M.-O. Artificial Intelligence and Collusion. *IIC-Int. Rev. Intellect. Prop. Compet. Law* **2018**, *50*, 109–134.
25. Steels, L.; Brooks, R. *The Artificial Life Route to Artificial Intelligence: Building Embodied, Situated Agents*, 1st ed.; Routledge: London, UK, 2018; 300p.
26. Bielecki, A. *Models of Neurons and Perceptrons: Selected Problems and Challenges*, 1st ed.; Springer International Publishing: Berlin/Heidelberg, Germany, 2019; 156p.
27. van der Maaten, L.; Hinton, G. Visualizing Non-Metric Similarities in Multiple Maps. *Mach. Learn.* **2011**, *87*, 33–55. [CrossRef]
28. Gadd, C.; Wade, S.; Shah, A.A. Pseudo-Marginal Bayesian Inference for Gaussian Process Latent Variable Models. *Mach. Learn.* **2021**, *110*, 1105–1143. [CrossRef]
29. Gallego, V.; Naveiro, R.; Roca, C.; Ríos Insua, D.; Campillo, N.E. AI in Drug Development: A Multidisciplinary Perspective. *Mol. Divers.* **2021**, *25*, 1461–1479. [CrossRef] [PubMed]
30. Kaul, V.; Enslin, S.; Gross, S.A. History of Artificial Intelligence in Medicine. *Gastrointest. Endosc.* **2020**, *92*, 807–812. [CrossRef] [PubMed]
31. Kusunose, K. Steps to Use Artificial Intelligence in Echocardiography. *J. Echocardiogr.* **2020**, *19*, 21–27. [CrossRef] [PubMed]
32. Kusunose, K.; Haga, A.; Abe, T.; Sata, M. Utilization of Artificial Intelligence in Echocardiography. *Circ. J.* **2019**, *83*, 1623–1629. [CrossRef]
33. Dey, D.; Slomka, P.J.; Leeson, P.; Comaniciu, D.; Shrestha, S.; Sengupta, P.P.; Marwick, T.H. Artificial Intelligence in Cardiovascular Imaging. *J. Am. Coll. Cardiol.* **2019**, *73*, 1317–1335. [CrossRef]
34. van Hamersvelt, R.W.; Zreik, M.; Voskuil, M.; Viergever, M.A.; Išgum, I.; Leiner, T. Deep Learning Analysis of Left Ventricular Myocardium in CT Angiographic Intermediate-Degree Coronary Stenosis Improves the Diagnostic Accuracy for Identification of Functionally Significant Stenosis. *Eur. Radiol.* **2018**, *29*, 2350–2359. [CrossRef]
35. Zhang, N.; Yang, G.; Gao, Z.; Xu, C.; Zhang, Y.; Shi, R.; Keegan, J.; Xu, L.; Zhang, H.; Fan, Z.; et al. Deep Learning for Diagnosis of Chronic Myocardial Infarction on Nonenhanced Cardiac Cine MRI. *Radiology* **2019**, *291*, 606–617. [CrossRef]
36. Attia, Z.I.; Kapa, S.; Lopez-Jimenez, F.; McKie, P.M.; Ladewig, D.J.; Satam, G.; Pellikka, P.A.; Enriquez-Sarano, M.; Noseworthy, P.A.; Munger, T.M.; et al. Screening for Cardiac Contractile Dysfunction Using an Artificial Intelligence–Enabled Electrocardiogram. *Nat. Med.* **2019**, *25*, 70–74. [CrossRef] [PubMed]
37. Bachtiger, P.; Petri, C.F.; Scott, F.E.; Ri Park, S.; Kelshiker, M.A.; Sahemey, H.K.; Dumea, B.; Alquero, R.; Padam, P.S.; Hatrick, I.R.; et al. Point-of-Care Screening for Heart Failure with Reduced Ejection Fraction Using Artificial Intelligence during ECG-Enabled Stethoscope Examination in London, UK: A Prospective, Observational, Multicentre Study. *Lancet Digit. Health* **2022**, *4*, e117–e125. [CrossRef] [PubMed]
38. Yao, X.; Rushlow, D.R.; Inselman, J.W.; McCoy, R.G.; Thacher, T.D.; Behnken, E.M.; Bernard, M.E.; Rosas, S.L.; Akfaly, A.; Misra, A.; et al. Artificial Intelligence–Enabled Electrocardiograms for Identification of Patients with Low Ejection Fraction: A Pragmatic, Randomized Clinical Trial. *Nat. Med.* **2021**, *27*, 815–819. [CrossRef] [PubMed]
39. Duffy, G.; Cheng, P.P.; Yuan, N.; He, B.; Kwan, A.C.; Shun-Shin, M.J.; Alexander, K.M.; Ebinger, J.; Lungren, M.P.; Rader, F.; et al. High-Throughput Precision Phenotyping of Left Ventricular Hypertrophy With Cardiovascular Deep Learning. *JAMA Cardiol.* **2022**, *7*, 386–395. [CrossRef] [PubMed]
40. Narula, S.; Shameer, K.; Salem Omar, A.M.; Dudley, J.T.; Sengupta, P.P. Machine-Learning Algorithms to Automate Morphological and Functional Assessments in 2D Echocardiography. *J. Am. Coll. Cardiol.* **2016**, *68*, 2287–2295. [CrossRef] [PubMed]
41. Frazer, J.; Notin, P.; Dias, M.; Gomez, A.; Min, J.K.; Brock, K.; Gal, Y.; Marks, D.S. Disease Variant Prediction with Deep Generative Models of Evolutionary Data. *Nature* **2021**, *599*, 91–95. [CrossRef] [PubMed]
42. Zhou, H.; Li, L.; Liu, Z.; Zhao, K.; Chen, X.; Lu, M.; Yin, G.; Song, L.; Zhao, S.; Zheng, H.; et al. Deep Learning Algorithm to Improve Hypertrophic Cardiomyopathy Mutation Prediction Using Cardiac Cine Images. *Eur. Radiol.* **2020**, *31*, 3931–3940. [CrossRef] [PubMed]

43. Hathaway, J.; Heliö, K.; Saarinen, I.; Tallila, J.; Seppälä, E.H.; Tuupanen, S.; Turpeinen, H.; Kangas-Kontio, T.; Schleit, J.; Tommiska, J.; et al. Diagnostic Yield of Genetic Testing in a Heterogeneous Cohort of 1376 HCM Patients. *BMC Cardiovasc. Disord.* **2021**, *21*, 126. [CrossRef]
44. Li, X.; Yang, L.; Yuan, Z.; Lou, J.; Fan, Y.; Shi, A.; Huang, J.; Zhao, M.; Wu, Y. Multi-Institutional Development and External Validation of Machine Learning-Based Models to Predict Relapse Risk of Pancreatic Ductal Adenocarcinoma after Radical Resection. *J. Transl. Med.* **2021**, *19*, 281. [CrossRef]
45. Kaissis, G.A.; Jungmann, F.; Ziegelmayer, S.; Lohöfer, F.K.; Harder, F.N.; Schlitter, A.M.; Muckenhuber, A.; Steiger, K.; Schirren, R.; Friess, H.; et al. Multiparametric Modelling of Survival in Pancreatic Ductal Adenocarcinoma Using Clinical, Histomorphological, Genetic and Image-Derived Parameters. *J. Clin. Med.* **2020**, *9*, 1250. [CrossRef]
46. Zhang, Y.; Zhu, S.; Yuan, Z.; Li, Q.; Ding, R.; Bao, X.; Zhen, T.; Fu, Z.; Fu, H.; Xing, K.; et al. Risk Factors and Socio-Economic Burden in Pancreatic Ductal Adenocarcinoma Operation: A Machine Learning Based Analysis. *BMC Cancer* **2020**, *20*, 1161. [CrossRef] [PubMed]
47. Pépin, J.-L.; Letesson, C.; Le-Dong, N.N.; Dedave, A.; Denison, S.; Cuthbert, V.; Martinot, J.-B.; Gozal, D. Assessment of Mandibular Movement Monitoring With Machine Learning Analysis for the Diagnosis of Obstructive Sleep Apnea. *JAMA Netw. Open* **2020**, *3*, e1919657. [CrossRef] [PubMed]
48. Zhou, C.-M.; Xue, Q.; Ye, H.-T.; Wang, Y.; Tong, J.; Ji, M.-H.; Yang, J.-J. Constructing a Prediction Model for Difficult Intubation of Obese Patients Based on Machine Learning. *J. Clin. Anesth.* **2021**, *72*, 110278. [CrossRef] [PubMed]
49. Carron, M.; Safaee Fakhr, B.; Ieppariello, G.; Foletto, M. Perioperative Care of the Obese Patient. *Br. J. Surg.* **2020**, *107*, e39–e55. [CrossRef] [PubMed]
50. Ermer, S.C.; Farney, R.J.; Johnson, K.B.; Orr, J.A.; Egan, T.D.; Brewer, L.M. An Automated Algorithm Incorporating Poincaré Analysis Can Quantify the Severity of Opioid-Induced Ataxic Breathing. *Anesth. Analg.* **2020**, *130*, 1147–1156. [CrossRef] [PubMed]
51. Ingrande, J.; Gabriel, R.A.; McAuley, J.; Krasinska, K.; Chien, A.; Lemmens, H.J.M. The Performance of an Artificial Neural Network Model in Predicting the Early Distribution Kinetics of Propofol in Morbidly Obese and Lean Subjects. *Anesth. Analg.* **2020**, *131*, 1500–1509. [CrossRef] [PubMed]
52. Yang, Z.; Zhang, M.; Li, C.; Meng, Z.; Li, Y.; Chen, Y.; Liu, L. Image Classification for Automobile Pipe Joints Surface Defect Detection Using Wavelet Decomposition and Convolutional Neural Network. *IEEE Access* **2022**, *10*, 77191–77204. [CrossRef]
53. Bellini, V.; Guzzon, M.; Bigliardi, B.; Mordonini, M.; Filippelli, S.; Bignami, E. Artificial Intelligence: A New Tool in Operating Room Management. Role of Machine Learning Models in Operating Room Optimization. *J. Med. Syst.* **2020**, *44*, 20. [CrossRef]
54. Rozario, N.; Rozario, D. Can Machine Learning Optimize the Efficiency of the Operating Room in the Era of COVID-19? *Can. J. Surg.* **2020**, *63*, E527–E529. [CrossRef]
55. Brennan, M.; Puri, S.; Ozrazgat-Baslanti, T.; Feng, Z.; Ruppert, M.; Hashemighouchani, H.; Momcilovic, P.; Li, X.; Wang, D.Z.; Bihorac, A. Comparing Clinical Judgment with the MySurgeryRisk Algorithm for Preoperative Risk Assessment: A Pilot Usability Study. *Surgery* **2019**, *165*, 1035–1045. [CrossRef]
56. Xue, B.; Li, D.; Lu, C.; King, C.R.; Wildes, T.; Avidan, M.S.; Kannampallil, T.; Abraham, J. Use of Machine Learning to Develop and Evaluate Models Using Preoperative and Intraoperative Data to Identify Risks of Postoperative Complications. *JAMA Netw. Open* **2021**, *4*, e212240. [CrossRef] [PubMed]
57. Tavolara, T.E.; Gurcan, M.N.; Segal, S.; Niazi, M.K.K. Identification of Difficult to Intubate Patients from Frontal Face Images Using an Ensemble of Deep Learning Models. *Comput. Biol. Med.* **2021**, *136*, 104737. [CrossRef] [PubMed]
58. Cheney, F.W. The American Society of Anesthesiologists Closed Claims Project. *Anesthesiology* **1999**, *91*, 552–556. [CrossRef] [PubMed]
59. León, M.A.; Räsänen, J. Neural Network-Based Detection of Esophageal Intubation in Anesthetized Patients. *J. Clin. Monit.* **1996**, *12*, 165–169. [CrossRef] [PubMed]
60. Ahn, J.C.; Connell, A.; Simonetto, D.A.; Hughes, C.; Shah, V.H. Application of Artificial Intelligence for the Diagnosis and Treatment of Liver Diseases. *Hepatology* **2021**, *73*, 2546–2563. [CrossRef] [PubMed]
61. Hashimoto, R.; Requa, J.; Dao, T.; Ninh, A.; Tran, E.; Mai, D.; Lugo, M.; El-Hage Chehade, N.; Chang, K.J.; Karnes, W.E.; et al. Artificial Intelligence Using Convolutional Neural Networks for Real-Time Detection of Early Esophageal Neoplasia in Barrett's Esophagus (with Video). *Gastrointest. Endosc.* **2020**, *91*, 1264–1271.e1. [CrossRef] [PubMed]
62. Ebigbo, A.; Mendel, R.; Probst, A.; Manzeneder, J.; Prinz, F.; de Souza, L.A., Jr.; Papa, J.; Palm, C.; Messmann, H. Real-Time Use of Artificial Intelligence in the Evaluation of Cancer in Barrett's Oesophagus. *Gut* **2019**, *69*, 615–616. [CrossRef]
63. Kohli, A.; Holzwanger, E.A.; Levy, A.N. Emerging Use of Artificial Intelligence in Inflammatory Bowel Disease. *World J. Gastroenterol.* **2020**, *26*, 6923–6928. [CrossRef]
64. Waljee, A.K.; Liu, B.; Sauder, K.; Zhu, J.; Govani, S.M.; Stidham, R.W.; Higgins, P.D.R. Predicting Corticosteroid-Free Biologic Remission with Vedolizumab in Crohn's Disease. *Inflamm. Bowel Dis.* **2018**, *24*, 1185–1192. [CrossRef]
65. Le Berre, C.; Sandborn, W.J.; Aridhi, S.; Devignes, M.-D.; Fournier, L.; Smaïl-Tabbone, M.; Danese, S.; Peyrin-Biroulet, L. Application of Artificial Intelligence to Gastroenterology and Hepatology. *Gastroenterology* **2020**, *158*, 76–94.e2. [CrossRef]
66. Spann, A.; Yasodhara, A.; Kang, J.; Watt, K.; Wang, B.; Goldenberg, A.; Bhat, M. Applying Machine Learning in Liver Disease and Transplantation: A Comprehensive Review. *Hepatology* **2020**, *71*, 1093–1105. [CrossRef] [PubMed]

67. Repici, A.; Badalamenti, M.; Maselli, R.; Correale, L.; Radaelli, F.; Rondonotti, E.; Ferrara, E.; Spadaccini, M.; Alkandari, A.; Fugazza, A.; et al. Efficacy of Real-Time Computer-Aided Detection of Colorectal Neoplasia in a Randomized Trial. *Gastroenterology* **2020**, *159*, 512–520.e7. [CrossRef]
68. Hassan, C.; Spadaccini, M.; Iannone, A.; Maselli, R.; Jovani, M.; Chandrasekar, V.T.; Antonelli, G.; Yu, H.; Areia, M.; Dinis-Ribeiro, M.; et al. Performance of Artificial Intelligence in Colonoscopy for Adenoma and Polyp Detection: A Systematic Review and Meta-Analysis. *Gastrointest. Endosc.* **2021**, *93*, 77–85.e6.
69. Lahner, E. Possible Contribution of Advanced Statistical Methods (Artificial Neural Networks and Linear Discriminant Analysis) in Recognition of Patients with Suspected Atrophic Body Gastritis. *World J. Gastroenterol.* **2005**, *11*, 5867. [CrossRef] [PubMed]
70. Rotondano, G.; Cipolletta, L.; Grossi, E.; Koch, M.; Intraligi, M.; Buscema, M.; Marmo, R. Artificial Neural Networks Accurately Predict Mortality in Patients with Nonvariceal Upper GI Bleeding. *Gastrointest. Endosc.* **2011**, *73*, 2018–2026.e2. [CrossRef]
71. Sung, H.; Ferlay, J.; Siegel, R.L.; Laversanne, M.; Soerjomataram, I.; Jemal, A.; Bray, F. Global Cancer Statistics 2020: GLOBOCAN Estimates of Incidence and Mortality Worldwide for 36 Cancers in 185 Countries. *CA Cancer J. Clin.* **2021**, *71*, 209–249. [CrossRef] [PubMed]
72. Zhao, W.; Yang, J.; Sun, Y.; Li, C.; Wu, W.; Jin, L.; Yang, Z.; Ni, B.; Gao, P.; Wang, P.; et al. 3D Deep Learning from CT Scans Predicts Tumor Invasiveness of Subcentimeter Pulmonary Adenocarcinomas. *Cancer Res.* **2018**, *78*, 6881–6889. [CrossRef] [PubMed]
73. Ardila, D.; Kiraly, A.P.; Bharadwaj, S.; Choi, B.; Reicher, J.J.; Peng, L.; Tse, D.; Etemadi, M.; Ye, W.; Corrado, G.; et al. End-to-End Lung Cancer Screening with Three-Dimensional Deep Learning on Low-Dose Chest Computed Tomography. *Nat. Med.* **2019**, *25*, 954–961. [CrossRef] [PubMed]
74. Zhang, Y.; Jiang, B.; Zhang, L.; Greuter, M.J.W.; de Bock, G.H.; Zhang, H.; Xie, X. Lung Nodule Detectability of Artificial Intelligence-Assisted CT Image Reading in Lung Cancer Screening. *Curr. Med. Imaging* **2022**, *18*, 327–334. [CrossRef]
75. Esteva, H.; Marchevsky, A.; Núñez, T.; Luna, C.; Esteva, M. Neural Networks as a Prognostic Tool of Surgical Risk in Lung Resections. *Ann. Thorac. Surg.* **2002**, *73*, 1576–1581. [CrossRef]
76. Bendixen, M.; Jørgensen, O.D.; Kronborg, C.; Andersen, C.; Licht, P.B. Postoperative Pain and Quality of Life after Lobectomy via Video-Assisted Thoracoscopic Surgery or Anterolateral Thoracotomy for Early Stage Lung Cancer: A Randomised Controlled Trial. *Lancet Oncol.* **2016**, *17*, 836–844. [CrossRef]
77. Somashekhar, S.P.; Sepúlveda, M.-J.; Puglielli, S.; Norden, A.D.; Shortliffe, E.H.; Rohit Kumar, C.; Rauthan, A.; Arun Kumar, N.; Patil, P.; Rhee, K.; et al. Watson for Oncology and Breast Cancer Treatment Recommendations: Agreement with an Expert Multidisciplinary Tumor Board. *Ann. Oncol.* **2018**, *29*, 418–423. [CrossRef] [PubMed]
78. Gonem, S.; Janssens, W.; Das, N.; Topalovic, M. Applications of Artificial Intelligence and Machine Learning in Respiratory Medicine. *Thorax* **2020**, *75*, 695–701. [CrossRef] [PubMed]
79. Geddes, C.C.; Fox, J.G.; Allison, M.E.M.; Boulton-Jones, J.M.; Simpson, K. An Artificial Neural Network Can Select Patients at High Risk of Developing Progressive IgA Nephropathy More Accurately than Experienced Nephrologists. *Nephrol. Dial. Transplant.* **1998**, *13*, 67–71. [CrossRef] [PubMed]
80. Niel, O.; Boussard, C.; Bastard, P. Artificial Intelligence Can Predict GFR Decline During the Course of ADPKD. *Am. J. Kidney Dis.* **2018**, *71*, 911–912. [CrossRef] [PubMed]
81. Barbieri, C.; Molina, M.; Ponce, P.; Tothova, M.; Cattinelli, I.; Ion Titapiccolo, J.; Mari, F.; Amato, C.; Leipold, F.; Wehmeyer, W.; et al. An International Observational Study Suggests That Artificial Intelligence for Clinical Decision Support Optimizes Anemia Management in Hemodialysis Patients. *Kidney Int.* **2016**, *90*, 422–429. [CrossRef] [PubMed]
82. Zhang, J.; Friberg, I.M.; Kift-Morgan, A.; Parekh, G.; Morgan, M.P.; Liuzzi, A.R.; Lin, C.-Y.; Donovan, K.L.; Colmont, C.S.; Morgan, P.H.; et al. Machine-Learning Algorithms Define Pathogen-Specific Local Immune Fingerprints in Peritoneal Dialysis Patients with Bacterial Infections. *Kidney Int.* **2017**, *92*, 179–191. [CrossRef]
83. Niel, O.; Bastard, P.; Boussard, C.; Hogan, J.; Kwon, T.; Deschênes, G. Artificial Intelligence Outperforms Experienced Nephrologists to Assess Dry Weight in Pediatric Patients on Chronic Hemodialysis. *Pediatr. Nephrol.* **2018**, *33*, 1799–1803. [CrossRef]
84. Checcucci, E.; Amparore, D.; De Luca, S.; Autorino, R.; Fiori, C.; Porpiglia, F. Precision Prostate Cancer Surgery: An Overview of New Technologies and Techniques. *Minerva Urol. Nefrol.* **2019**, *71*, 487–501. [CrossRef]
85. Cicione, A.; De Nunzio, C.; Manno, S.; Damiano, R.; Posti, A.; Lima, E.; Tubaro, A.; Balloni, F. An Update on Prostate Biopsy in the Era of Magnetic Resonance Imaging. *Minerva Urol. Nephrol.* **2018**, *70*, 264–274. [CrossRef]
86. Snow, P.B.; Smith, D.S.; Catalona, W.J. Artificial Neural Networks in the Diagnosis and Prognosis of Prostate Cancer: A Pilot Study. *J. Urol.* **1994**, *152 Pt 2*, 1923–1926. [CrossRef]
87. Djavan, B.; Remzi, M.; Zlotta, A.; Seitz, C.; Snow, P.; Marberger, M. Novel Artificial Neural Network for Early Detection of Prostate Cancer. *J. Clin. Oncol.* **2002**, *20*, 921–929. [CrossRef] [PubMed]
88. Hung, A.J.; Chen, J.; Che, Z.; Nilanon, T.; Jarc, A.; Titus, M.; Oh, P.J.; Gill, I.S.; Liu, Y. Utilizing Machine Learning and Automated Performance Metrics to Evaluate Robot-Assisted Radical Prostatectomy Performance and Predict Outcomes. *J. Endourol.* **2018**, *32*, 438–444. [CrossRef] [PubMed]
89. Kattan, M.W. Comparison of Cox Regression with Other Methods for Determining Prediction Models and Nomograms. *J. Urol.* **2003**, *170*, S6–S10. [CrossRef] [PubMed]
90. Buchner, A.; Kendlbacher, M.; Nuhn, P.; Tüllmann, C.; Haseke, N.; Stief, C.G.; Staehler, M. Outcome Assessment of Patients with Metastatic Renal Cell Carcinoma Under Systemic Therapy Using Artificial Neural Networks. *Clin. Genitourin. Cancer* **2012**, *10*, 37–42. [CrossRef] [PubMed]

91. Ozkan, I.A.; Koklu, M.; Sert, I.U. Diagnosis of Urinary Tract Infection Based on Artificial Intelligence Methods. *Comput. Methods Programs Biomed.* **2018**, *166*, 51–59. [CrossRef] [PubMed]
92. Serati, M.; Salvatore, S.; Siesto, G.; Cattoni, E.; Braga, A.; Sorice, P.; Cromi, A.; Ghezzi, F.; Bolis, P. Urinary Symptoms and Urodynamic Findings in Women with Pelvic Organ Prolapse: Is There a Correlation? Results of an Artificial Neural Network Analysis. *Eur. Urol.* **2011**, *60*, 253–260. [CrossRef]
93. Tapak, L.; Hamidi, O.; Amini, P.; Poorolajal, J. Prediction of Kidney Graft Rejection Using Artificial Neural Network. *Healthc. Inform. Res.* **2017**, *23*, 277. [CrossRef]
94. Berk-Krauss, J.; Polsky, D.; Stein, J.A. Mole Mapping for Management of Pigmented Skin Lesions. *Dermatol. Clin.* **2017**, *35*, 439–445. [CrossRef]
95. Demers, A.A.; Nugent, Z.; Mihalcioiu, C.; Wiseman, M.C.; Kliewer, E.V. Trends of Nonmelanoma Skin Cancer from 1960 through 2000 in a Canadian Population. *J. Am. Acad. Dermatol.* **2005**, *53*, 320–328. [CrossRef]
96. Kaddu, S.; Soyer, H.P.; Gabler, G.; Kovarik, C. The Africa Teledermatology Project: Preliminary Experience with a Sub-Saharan Teledermatology and e-Learning Program. *J. Am. Acad. Dermatol.* **2009**, *61*, 155–157. [CrossRef]
97. Gaffney, R.; Rao, B. Global Teledermatology. *Glob. Dermatol.* **2016**, *2*, 209–214. [CrossRef]
98. Kaliyadan, F.; Ashique, K. Use of Mobile Applications in Dermatology. *Indian J. Dermatol.* **2020**, *65*, 371. [CrossRef]
99. Freeman, K.; Dinnes, J.; Chuchu, N.; Takwoingi, Y.; Bayliss, S.E.; Matin, R.N.; Jain, A.; Walter, F.M.; Williams, H.C.; Deeks, J.J. Algorithm Based Smartphone Apps to Assess Risk of Skin Cancer in Adults: Systematic Review of Diagnostic Accuracy Studies. *BMJ* **2020**, *368*, m127. [CrossRef] [PubMed]
100. Veronese, F.; Branciforti, F.; Zavattaro, E.; Tarantino, V.; Romano, V.; Meiburger, K.M.; Salvi, M.; Seoni, S.; Savoia, P. The Role in Teledermoscopy of an Inexpensive and Easy-to-Use Smartphone Device for the Classification of Three Types of Skin Lesions Using Convolutional Neural Networks. *Diagnostics* **2021**, *11*, 451. [CrossRef] [PubMed]
101. Kagian, A.; Dror, G.; Leyvand, T.; Meilijson, I.; Cohen-Or, D.; Ruppin, E. A Machine Learning Predictor of Facial Attractiveness Revealing Human-like Psychophysical Biases. *Vis. Res.* **2008**, *48*, 235–243. [CrossRef]
102. Potter, B.; Ronan, S.G. Computerized Dermatopathologic Diagnosis. *J. Am. Acad. Dermatol.* **1987**, *17*, 119–131. [CrossRef]
103. Wells, A.; Patel, S.; Lee, J.B.; Motaparthi, K. Artificial Intelligence in Dermatopathology: Diagnosis, Education, and Research. *J. Cutan. Pathol.* **2021**, *48*, 1061–1068. [CrossRef]
104. Ianni, J.D.; Soans, R.E.; Sankarapandian, S.; Chamarthi, R.V.; Ayyagari, D.; Olsen, T.G.; Bonham, M.J.; Stavish, C.C.; Motaparthi, K.; Cockerell, C.J.; et al. Tailored for Real-World: A Whole Slide Image Classification System Validated on Uncurated Multi-Site Data Emulating the Prospective Pathology Workload. *Sci. Rep.* **2020**, *10*, 3217. [CrossRef]
105. Hekler, A.; Utikal, J.S.; Enk, A.H.; Berking, C.; Klode, J.; Schadendorf, D.; Jansen, P.; Franklin, C.; Holland-Letz, T.; Krahl, D.; et al. Pathologist-Level Classification of Histopathological Melanoma Images with Deep Neural Networks. *Eur. J. Cancer* **2019**, *115*, 79–83. [CrossRef]
106. Olsen, T.G.; Jackson, B.H.; Feeser, T.A.; Kent, M.N.; Moad, J.C.; Krishnamurthy, S.; Lunsford, D.D.; Soans, R.E. Diagnostic Performance of Deep Learning Algorithms Applied to Three Common Diagnoses in Dermatopathology. *J. Pathol. Inform.* **2018**, *9*, 32. [CrossRef]
107. Tschandl, P.; Rosendahl, C.; Akay, B.N.; Argenziano, G.; Blum, A.; Braun, R.P.; Cabo, H.; Gourhant, J.-Y.; Kreusch, J.; Lallas, A.; et al. Expert-Level Diagnosis of Nonpigmented Skin Cancer by Combined Convolutional Neural Networks. *JAMA Dermatol.* **2019**, *155*, 58–65. [CrossRef] [PubMed]
108. Fayaz, A.; Croft, P.; Langford, R.M.; Donaldson, L.J.; Jones, G.T. Prevalence of Chronic Pain in the UK: A Systematic Review and Meta-Analysis of Population Studies. *BMJ Open* **2016**, *6*, e010364. [CrossRef] [PubMed]
109. Lötsch, J.; Ultsch, A. Machine Learning in Pain Research. *Pain* **2017**, *159*, 623–630. [CrossRef] [PubMed]
110. Culvenor, A.G.; Schache, A.G.; Vicenzino, B.; Pandy, M.G.; Collins, N.J.; Cook, J.L.; Crossley, K.M. Are Knee Biomechanics Different in Those With and Without Patellofemoral Osteoarthritis After Anterior Cruciate Ligament Reconstruction? *Arthritis Care Res.* **2014**, *66*, 1566–1570. [CrossRef] [PubMed]
111. Kianifar, R.; Lee, A.; Raina, S.; Kulic, D. Automated Assessment of Dynamic Knee Valgus and Risk of Knee Injury During the Single Leg Squat. *IEEE J. Transl. Eng. Health Med.* **2017**, *5*, 2100213. [CrossRef] [PubMed]
112. Navani, A.; Li, G. Chronic Pain Challenge: A Statistical Machine-Learning Method for Chronic Pain Assessment. *J. Recent Adv. Pain* **2016**, *2*, 82–86. [CrossRef]
113. Louis, D.N.; Perry, A.; Reifenberger, G.; von Deimling, A.; Figarella-Branger, D.; Cavenee, W.K.; Ohgaki, H.; Wiestler, O.D.; Kleihues, P.; Ellison, D.W. The 2016 World Health Organization Classification of Tumors of the Central Nervous System: A Summary. *Acta Neuropathol.* **2016**, *131*, 803–820. [CrossRef]
114. Chang, K.; Bai, H.X.; Zhou, H.; Su, C.; Bi, W.L.; Agbodza, E.; Kavouridis, V.K.; Senders, J.T.; Boaro, A.; Beers, A.; et al. Residual Convolutional Neural Network for the Determination of IDH Status in Low- and High-Grade Gliomas from MR Imaging. *Clin. Cancer Res.* **2018**, *24*, 1073–1081. [CrossRef]
115. Chang, P.; Grinband, J.; Weinberg, B.D.; Bardis, M.; Khy, M.; Cadena, G.; Su, M.-Y.; Cha, S.; Filippi, C.G.; Bota, D.; et al. Deep-Learning Convolutional Neural Networks Accurately Classify Genetic Mutations in Gliomas. *Am. J. Neuroradiol.* **2018**, *39*, 1201–1207. [CrossRef]

116. Akbari, H.; Bakas, S.; Pisapia, J.M.; Nasrallah, M.P.; Rozycki, M.; Martinez-Lage, M.; Morrissette, J.J.D.; Dahmane, N.; O'Rourke, D.M.; Davatzikos, C. In Vivoevaluation of EGFRvIII Mutation in Primary Glioblastoma Patients via Complex Multiparametric MRI Signature. *Neuro-Oncology* **2018**, *20*, 1068–1079. [CrossRef]
117. Laukamp, K.R.; Thiele, F.; Shakirin, G.; Zopfs, D.; Faymonville, A.; Timmer, M.; Maintz, D.; Perkuhn, M.; Borggrefe, J. Fully Automated Detection and Segmentation of Meningiomas Using Deep Learning on Routine Multiparametric MRI. *Eur. Radiol.* **2018**, *29*, 124–132. [CrossRef] [PubMed]
118. Kickingereder, P.; Isensee, F.; Tursunova, I.; Petersen, J.; Neuberger, U.; Bonekamp, D.; Brugnara, G.; Schell, M.; Kessler, T.; Foltyn, M.; et al. Automated Quantitative Tumour Response Assessment of MRI in Neuro-Oncology with Artificial Neural Networks: A Multicentre, Retrospective Study. *Lancet Oncol.* **2019**, *20*, 728–740. [CrossRef] [PubMed]
119. Kang, H.; Witanto, J.N.; Pratama, K.; Lee, D.; Choi, K.; Choi, S.; Kim, K.; Kim, M.; Kim, J.; Kim, Y.; et al. P13.02.B Fully Automated Segmentation and Volumetric Measurement of Intracranial Meningioma Using Deep Learning. *Neuro-Oncology* **2022**, *24* (Suppl. S2), ii80–ii81. [CrossRef]
120. Gates, E.D.H.; Lin, J.S.; Weinberg, J.S.; Hamilton, J.; Prabhu, S.S.; Hazle, J.D.; Fuller, G.N.; Baladandayuthapani, V.; Fuentes, D.; Schellingerhout, D. Guiding the First Biopsy in Glioma Patients Using Estimated Ki-67 Maps Derived from MRI: Conventional versus Advanced Imaging. *Neuro-Oncology* **2019**, *21*, 527–536. [CrossRef] [PubMed]
121. Artzi, M.; Bressler, I.; Ben Bashat, D. Differentiation between Glioblastoma, Brain Metastasis and Subtypes Using Radiomics Analysis. *J. Magn. Reson. Imaging* **2019**, *50*, 519–528. [CrossRef]
122. Emin, E.I.; Emin, E.; Papalois, A.; Willmott, F.; Clarke, S.; Sideris, M. Artificial Intelligence in Obstetrics and Gynaecology: Is This the Way Forward? *In Vivo* **2019**, *33*, 1547–1551. [CrossRef]
123. Dong, H.-C.; Dong, H.-K.; Yu, M.-H.; Lin, Y.-H.; Chang, C.-C. Using Deep Learning with Convolutional Neural Network Approach to Identify the Invasion Depth of Endometrial Cancer in Myometrium Using MR Images: A Pilot Study. *Int. J. Environ. Res. Public Health* **2020**, *17*, 5993. [CrossRef]
124. Arsalan, M.; Haider, A.; Choi, J.; Park, K.R. Detecting Blastocyst Components by Artificial Intelligence for Human Embryological Analysis to Improve Success Rate of In Vitro Fertilization. *J. Pers. Med.* **2022**, *12*, 124. [CrossRef]
125. Wu, L.; Wei, D.; Yang, N.; Lei, H.; Wang, Y. Artificial Intelligence Algorithm-Based Analysis of Ultrasonic Imaging Features for Diagnosis of Pregnancy Complicated with Brain Tumor. *J. Healthc. Eng.* **2021**, *2021*, 4022312. [CrossRef]
126. Wu, Y.; Shen, Y.; Sun, H. Intelligent Algorithm-Based Analysis on Ultrasound Image Characteristics of Patients with Lower Extremity Arteriosclerosis Occlusion and Its Correlation with Diabetic Mellitus Foot. *J. Healthc. Eng.* **2021**, *2021*, 7758206. [CrossRef]
127. Burgos-Artizzu, X.P.; Coronado-Gutiérrez, D.; Valenzuela-Alcaraz, B.; Vellvé, K.; Eixarch, E.; Crispi, F.; Bonet-Carne, E.; Bennasar, M.; Gratacos, E. Analysis of Maturation Features in Fetal Brain Ultrasound via Artificial Intelligence for the Estimation of Gestational Age. *Am. J. Obstet. Gynecol. MFM* **2021**, *3*, 100462. [CrossRef] [PubMed]
128. Sakai, A.; Komatsu, M.; Komatsu, R.; Matsuoka, R.; Yasutomi, S.; Dozen, A.; Shozu, K.; Arakaki, T.; Machino, H.; Asada, K.; et al. Medical Professional Enhancement Using Explainable Artificial Intelligence in Fetal Cardiac Ultrasound Screening. *Biomedicines* **2022**, *10*, 551. [CrossRef] [PubMed]
129. Lin, M.; He, X.; Guo, H.; He, M.; Zhang, L.; Xian, J.; Lei, T.; Xu, Q.; Zheng, J.; Feng, J.; et al. Use of Real-time Artificial Intelligence in Detection of Abnormal Image Patterns in Standard Sonographic Reference Planes in Screening for Fetal Intracranial Malformations. *Ultrasound Obstet. Gynecol.* **2022**, *59*, 304–316. [CrossRef]
130. Scanzera, A.C.; Shorter, E.; Kinnaird, C.; Valikodath, N.; Al-Khaled, T.; Cole, E.; Kravets, S.; Hallak, J.A.; McMahon, T.; Chan, R.V.P. Optometrist's Perspectives of Artificial Intelligence in Eye Care. *J. Optom.* **2022**, *15*, S91–S97. [CrossRef] [PubMed]
131. Channa, R.; Wolf, R.; Abramoff, M.D. Autonomous Artificial Intelligence in Diabetic Retinopathy: From Algorithm to Clinical Application. *J. Diabetes Sci. Technol.* **2020**, *15*, 695–698. [CrossRef] [PubMed]
132. Sabanayagam, C.; Banu, R.; Chee, M.L.; Lee, R.; Wang, Y.X.; Tan, G.; Jonas, J.B.; Lamoureux, E.L.; Cheng, C.-Y.; Klein, B.E.K.; et al. Incidence and Progression of Diabetic Retinopathy: A Systematic Review. *Lancet Diabetes Endocrinol.* **2019**, *7*, 140–149. [CrossRef] [PubMed]
133. Li, Z.; Keel, S.; Liu, C.; He, Y.; Meng, W.; Scheetz, J.; Lee, P.Y.; Shaw, J.; Ting, D.; Wong, T.Y.; et al. An Automated Grading System for Detection of Vision-Threatening Referable Diabetic Retinopathy on the Basis of Color Fundus Photographs. *Diabetes Care* **2018**, *41*, 2509–2516. [CrossRef] [PubMed]
134. Ferris, F.L.; Wilkinson, C.P.; Bird, A.; Chakravarthy, U.; Chew, E.; Csaky, K.; Sadda, S.R. Clinical Classification of Age-Related Macular Degeneration. *Ophthalmology* **2013**, *120*, 844–851. [CrossRef]
135. Chen, Q.; Leng, T.; Zheng, L.; Kutzscher, L.; Ma, J.; de Sisternes, L.; Rubin, D.L. Automated Drusen Segmentation and Quantification in SD-OCT Images. *Med. Image Anal.* **2013**, *17*, 1058–1072. [CrossRef]
136. Schlanitz, F.G.; Baumann, B.; Kundi, M.; Sacu, S.; Baratsits, M.; Scheschy, U.; Shahlaee, A.; Mittermüller, T.J.; Montuoro, A.; Roberts, P.; et al. Drusen Volume Development over Time and Its Relevance to the Course of Age-Related Macular Degeneration. *Br. J. Ophthalmol.* **2016**, *101*, 198–203. [CrossRef]
137. Peng, Y.; Dharssi, S.; Chen, Q.; Keenan, T.D.; Agrón, E.; Wong, W.T.; Chew, E.Y.; Lu, Z. DeepSeeNet: A Deep Learning Model for Automated Classification of Patient-Based Age-Related Macular Degeneration Severity from Color Fundus Photographs. *Ophthalmology* **2019**, *126*, 565–575. [CrossRef]

138. Asaoka, R.; Murata, H.; Hirasawa, K.; Fujino, Y.; Matsuura, M.; Miki, A.; Kanamoto, T.; Ikeda, Y.; Mori, K.; Iwase, A.; et al. Using Deep Learning and Transfer Learning to Accurately Diagnose Early-Onset Glaucoma From Macular Optical Coherence Tomography Images. *Am. J. Ophthalmol.* **2019**, *198*, 136–145. [CrossRef] [PubMed]
139. Pringle, C.; Kilday, J.-P.; Kamaly-Asl, I.; Stivaros, S.M. The Role of Artificial Intelligence in Paediatric Neuroradiology. *Pediatr. Radiol.* **2022**, *52*, 2159–2172. [CrossRef] [PubMed]
140. Quon, J.L.; Han, M.; Kim, L.H.; Koran, M.E.; Chen, L.C.; Lee, E.H.; Wright, J.; Ramaswamy, V.; Lober, R.M.; Taylor, M.D.; et al. Artificial Intelligence for Automatic Cerebral Ventricle Segmentation and Volume Calculation: A Clinical Tool for the Evaluation of Pediatric Hydrocephalus. *J. Neurosurg. Pediatr.* **2021**, *27*, 131–138. [CrossRef] [PubMed]
141. Grimm, F.; Edl, F.; Kerscher, S.R.; Nieselt, K.; Gugel, I.; Schuhmann, M.U. Semantic Segmentation of Cerebrospinal Fluid and Brain Volume with a Convolutional Neural Network in Pediatric Hydrocephalus—Transfer Learning from Existing Algorithms. *Acta Neurochir.* **2020**, *162*, 2463–2474. [CrossRef]
142. Stivaros, S.M.; Radon, M.R.; Mileva, R.; Connolly, D.J.A.; Cowell, P.E.; Hoggard, N.; Wright, N.B.; Tang, V.; Gledson, A.; Batty, R.; et al. Quantification of Structural Changes in the Corpus Callosumin Children with Profound Hypoxic–Ischaemic Brain Injury. *Pediatr. Radiol.* **2015**, *46*, 73–81. [CrossRef]
143. Raji, C.A.; Wang, M.B.; Nguyen, N.; Owen, J.P.; Palacios, E.M.; Yuh, E.L.; Mukherjee, P. Connectome Mapping with Edge Density Imaging Differentiates Pediatric Mild Traumatic Brain Injury from Typically Developing Controls: Proof of Concept. *Pediatr. Radiol.* **2020**, *50*, 1594–1601. [CrossRef]
144. Attallah, O.; Sharkas, M.A.; Gadelkarim, H. Fetal Brain Abnormality Classification from MRI Images of Different Gestational Age. *Brain Sci.* **2019**, *9*, 231. [CrossRef]
145. Bull, J.G.; Saunders, D.E.; Clark, C.A. Discrimination of Paediatric Brain Tumours Using Apparent Diffusion Coefficient Histograms. *Eur. Radiol.* **2011**, *22*, 447–457. [CrossRef]
146. Orphanidou-Vlachou, E.; Vlachos, N.; Davies, N.P.; Arvanitis, T.N.; Grundy, R.G.; Peet, A.C. Texture Analysis ofT1- andT2-Weighted MR Images and Use of Probabilistic Neural Network to Discriminate Posterior Fossa Tumours in Children. *NMR Biomed.* **2014**, *27*, 632–639. [CrossRef]
147. Stivaros, S.; Garg, S.; Tziraki, M.; Cai, Y.; Thomas, O.; Mellor, J.; Morris, A.A.; Jim, C.; Szumanska-Ryt, K.; Parkes, L.M.; et al. Randomised Controlled Trial of Simvastatin Treatment for Autism in Young Children with Neurofibromatosis Type 1 (SANTA). *Mol. Autism* **2018**, *9*, 12. [CrossRef] [PubMed]
148. Wang, Q.; Bi, S.; Sun, M.; Wang, Y.; Wang, D.; Yang, S. Deep Learning Approach to Peripheral Leukocyte Recognition. *PLoS ONE* **2019**, *14*, e0218808. [CrossRef] [PubMed]
149. Chandradevan, R.; Aljudi, A.A.; Drumheller, B.R.; Kunananthaseelan, N.; Amgad, M.; Gutman, D.A.; Cooper, L.A.D.; Jaye, D.L. Machine-Based Detection and Classification for Bone Marrow Aspirate Differential Counts: Initial Development Focusing on Nonneoplastic Cells. *Lab. Investig.* **2020**, *100*, 98–109. [CrossRef] [PubMed]
150. Xu, M.; Papageorgiou, D.P.; Abidi, S.Z.; Dao, M.; Zhao, H.; Karniadakis, G.E. A Deep Convolutional Neural Network for Classification of Red Blood Cells in Sickle Cell Anemia. *PLoS Comput. Biol.* **2017**, *13*, e1005746. [CrossRef]
151. Alsalem, M.A.; Zaidan, A.A.; Zaidan, B.B.; Hashim, M.; Madhloom, H.T.; Azeez, N.D.; Alsyisuf, S. A Review of the Automated Detection and Classification of Acute Leukaemia: Coherent Taxonomy, Datasets, Validation and Performance Measurements, Motivation, Open Challenges and Recommendations. *Comput. Methods Programs Biomed.* **2018**, *158*, 93–112. [CrossRef] [PubMed]
152. Deulofeu, M.; Kolářová, L.; Salvadó, V.; María Peña-Méndez, E.; Almáši, M.; Štork, M.; Pour, L.; Boadas-Vaello, P.; Ševčíková, S.; Havel, J.; et al. Rapid Discrimination of Multiple Myeloma Patients by Artificial Neural Networks Coupled with Mass Spectrometry of Peripheral Blood Plasma. *Sci. Rep.* **2019**, *9*, 7975. [CrossRef] [PubMed]
153. Moraes, L.O.; Pedreira, C.E.; Barrena, S.; Lopez, A.; Orfao, A. A Decision-Tree Approach for the Differential Diagnosis of Chronic Lymphoid Leukemias and Peripheral B-Cell Lymphomas. *Comput. Methods Programs Biomed.* **2019**, *178*, 85–90. [CrossRef]
154. Patel, S.S.; Sekeres, M.A.; Nazha, A. Prognostic Models in Predicting Outcomes in Myelodysplastic Syndromes after Hypomethylating Agent Failure. *Leuk. Lymphoma* **2017**, *58*, 2532–2539. [CrossRef]
155. Liu, S.; Zhang, F.; Xie, L.; Wang, Y.; Xiang, Q.; Yue, Z.; Feng, Y.; Yang, Y.; Li, J.; Luo, L.; et al. Machine Learning Approaches for Risk Assessment of Peripherally Inserted Central Catheter-Related Vein Thrombosis in Hospitalized Patients with Cancer. *Int. J. Med. Inform.* **2019**, *129*, 175–183. [CrossRef]
156. Arai, Y.; Kondo, T.; Fuse, K.; Shibasaki, Y.; Masuko, M.; Sugita, J.; Teshima, T.; Uchida, N.; Fukuda, T.; Kakihana, K.; et al. Using a Machine Learning Algorithm to Predict Acute Graft-versus-Host Disease Following Allogeneic Transplantation. *Blood Adv.* **2019**, *3*, 3626–3634. [CrossRef]
157. Nazha, A.; Komrokji, R.S.; Meggendorfer, M.; Mukherjee, S.; Al Ali, N.; Walter, W.; Hutter, S.; Padron, E.; Madanat, Y.F.; Sallman, D.A.; et al. A Personalized Prediction Model to Risk Stratify Patients with Myelodysplastic Syndromes. *Blood* **2018**, *132* (Suppl. S1), 793. [CrossRef]
158. Ko, B.-S.; Wang, Y.-F.; Li, J.-L.; Li, C.-C.; Weng, P.-F.; Hsu, S.-C.; Hou, H.-A.; Huang, H.-H.; Yao, M.; Lin, C.-T.; et al. Clinically Validated Machine Learning Algorithm for Detecting Residual Diseases with Multicolor Flow Cytometry Analysis in Acute Myeloid Leukemia and Myelodysplastic Syndrome. *EBioMedicine* **2018**, *37*, 91–100. [CrossRef] [PubMed]
159. Dellinger, R.P.; Levy, M.M.; Rhodes, A.; Annane, D.; Gerlach, H.; Opal, S.M.; Sevransky, J.E.; Sprung, C.L.; Douglas, I.S.; Jaeschke, R.; et al. Surviving Sepsis Campaign. *Crit. Care Med.* **2013**, *41*, 580–637. [CrossRef] [PubMed]

160. Quinten, V.M.; van Meurs, M.; Wolffensperger, A.E.; ter Maaten, J.C.; Ligtenberg, J.J.M. Sepsis Patients in the Emergency Department. *Eur. J. Emerg. Med.* **2018**, *25*, 328–334. [CrossRef] [PubMed]
161. Komorowski, M.; Celi, L.A.; Badawi, O.; Gordon, A.C.; Faisal, A.A. The Artificial Intelligence Clinician Learns Optimal Treatment Strategies for Sepsis in Intensive Care. *Nat. Med.* **2018**, *24*, 1716–1720. [CrossRef] [PubMed]
162. Davoudi, A.; Malhotra, K.R.; Shickel, B.; Siegel, S.; Williams, S.; Ruppert, M.; Bihorac, E.; Ozrazgat-Baslanti, T.; Tighe, P.J.; Bihorac, A.; et al. Intelligent ICU for Autonomous Patient Monitoring Using Pervasive Sensing and Deep Learning. *Sci. Rep.* **2019**, *9*, 8020. [CrossRef]
163. Siegel, R.L.; Miller, K.D.; Jemal, A. Cancer Statistics, 2018. *CA Cancer J. Clin.* **2018**, *68*, 7–30. [CrossRef]
164. Cao, Z.; Duan, L.; Yang, G.; Yue, T.; Chen, Q. An Experimental Study on Breast Lesion Detection and Classification from Ultrasound Images Using Deep Learning Architectures. *BMC Med. Imaging* **2019**, *19*, 51. [CrossRef]
165. Wang, K.; Lu, X.; Zhou, H.; Gao, Y.; Zheng, J.; Tong, M.; Wu, C.; Liu, C.; Huang, L.; Jiang, T.; et al. Deep Learning Radiomics of Shear Wave Elastography Significantly Improved Diagnostic Performance for Assessing Liver Fibrosis in Chronic Hepatitis B: A Prospective Multicentre Study. *Gut* **2018**, *68*, 729–741. [CrossRef]
166. Lee, J.H.; Joo, I.; Kang, T.W.; Paik, Y.H.; Sinn, D.H.; Ha, S.Y.; Kim, K.; Choi, C.; Lee, G.; Yi, J.; et al. Deep Learning with Ultrasonography: Automated Classification of Liver Fibrosis Using a Deep Convolutional Neural Network. *Eur. Radiol.* **2019**, *30*, 1264–1273. [CrossRef]
167. Ta, C.N.; Kono, Y.; Eghtedari, M.; Oh, Y.T.; Robbin, M.L.; Barr, R.G.; Kummel, A.C.; Mattrey, R.F. Focal Liver Lesions: Computer-Aided Diagnosis by Using Contrast-Enhanced US Cine Recordings. *Radiology* **2018**, *286*, 1062–1071. [CrossRef] [PubMed]
168. Cunningham, R.J.; Loram, I.D. Estimation of Absolute States of Human Skeletal Muscle via Standard B-Mode Ultrasound Imaging and Deep Convolutional Neural Networks. *J. R. Soc. Interface* **2020**, *17*, 20190715. [CrossRef] [PubMed]
169. Noort, F.; Vaart, C.H.; Grob, A.T.M.; Waarsenburg, M.K.; Slump, C.H.; Stralen, M. Deep Learning Enables Automatic Quantitative Assessment of Puborectalis Muscle and Urogenital Hiatus in Plane of Minimal Hiatal Dimensions. *Ultrasound Obstet. Gynecol.* **2019**, *54*, 270–275. [CrossRef] [PubMed]
170. Goergen, S.K.; Frazer, H.M.; Reddy, S. Quality Use of Artificial Intelligence in Medical Imaging: What Do Radiologists Need to Know? *J. Med. Imaging Radiat. Oncol.* **2022**, *66*, 225–232. [CrossRef] [PubMed]
171. Fayad, L.M.; Parekh, V.S.; de Castro Luna, R.; Ko, C.C.; Tank, D.; Fritz, J.; Ahlawat, S.; Jacobs, M.A. A Deep Learning System for Synthetic Knee Magnetic Resonance Imaging. *Investig. Radiol.* **2020**, *56*, 357–368. [CrossRef] [PubMed]
172. Hwang, J.-J.; Jung, Y.-H.; Cho, B.-H.; Heo, M.-S. An Overview of Deep Learning in the Field of Dentistry. *Imaging Sci. Dent.* **2019**, *49*, 1. [CrossRef]
173. Costa, P.; Galdran, A.; Meyer, M.I.; Niemeijer, M.; Abramoff, M.; Mendonca, A.M.; Campilho, A. End-to-End Adversarial Retinal Image Synthesis. *IEEE Trans. Med. Imaging* **2018**, *37*, 781–791. [CrossRef]
174. Tuzoff, D.V.; Tuzova, L.N.; Bornstein, M.M.; Krasnov, A.S.; Kharchenko, M.A.; Nikolenko, S.I.; Sveshnikov, M.M.; Bednenko, G.B. Tooth Detection and Numbering in Panoramic Radiographs Using Convolutional Neural Networks. *Dentomaxillofac. Radiol.* **2019**, *48*, 20180051. [CrossRef]
175. Kumar, S.; Patnaik, S.; Dixit, A. Predictive Models for Stage and Risk Classification in Head and Neck Squamous Cell Carcinoma (HNSCC). *PeerJ* **2020**, *8*, e9656. [CrossRef]
176. Papadopoulos, D.N.; Karalis, V. Variational Autoencoders for Data Augmentation in Clinical Studies. *Appl. Sci.* **2023**, *13*, 8793. [CrossRef]
177. Kumar, P.; Chauhan, S.; Awasthi, L.K. Artificial Intelligence in Healthcare: Review, Ethics, Trust Challenges & Future Research Directions. *Eng. Appl. Artif. Intell.* **2023**, *120*, 105894.
178. Denecke, K.; Gabarron, E. How Artificial Intelligence for Healthcare Look Like in the Future? In *Studies in Health Technology and Informatics*; IOS Press: Amsterdam, The Netherlands, 2021; Volume 281, pp. 860–864.
179. Davenport, T.; Kalakota, R. The Potential for Artificial Intelligence in Healthcare. *Future Healthc. J.* **2019**, *6*, 94–98. [CrossRef] [PubMed]
180. Zech, J.R.; Badgeley, M.A.; Liu, M.; Costa, A.B.; Titano, J.J.; Oermann, E.K. Variable Generalization Performance of a Deep Learning Model to Detect Pneumonia in Chest Radiographs: A Cross-Sectional Study. *PLoS Med.* **2018**, *15*, e1002683. [CrossRef] [PubMed]
181. Olsen, H.P.; Slosser, J.L.; Hildebrandt, T.T.; Wiesener, C. What's in the Box? The Legal Requirement of Explainability in Computationally Aided Decision-Making in Public Administration. *SSRN Electron. J.* **2019**. [CrossRef]
182. Amann, J.; Blasimme, A.; Vayena, E.; Frey, D.; Madai, V.I. Explainability for Artificial Intelligence in Healthcare: A Multidisciplinary Perspective. *BMC Med. Inform. Decis. Mak.* **2020**, *20*, 310. [CrossRef] [PubMed]
183. US Food and Drug Administration. Proposed Regulatory Framework for Modifications to Artificial Intelligence/Machine Learning (AI/ML)-Based Software as a Medical Device (SAMD): Discussion Paper and Request for Feedback. 2019. Available online: https://www.fda.gov/files/medical%20devices/published/US-FDA-Artificial-Intelligence-and-Machine-Learning-Discussion-Paper.pdf (accessed on 24 September 2023).
184. Chaddad, A.; Peng, J.; Xu, J.; Bouridane, A. Survey of Explainable AI Techniques in Healthcare. *Sensors* **2023**, *23*, 634. [CrossRef]
185. Cohen, I.G. Informed Consent and Medical Artificial Intelligence: What to Tell the Patient? *SSRN Electron. J.* **2020**. [CrossRef]

186. Proposal for a Regulation of The European Parliament and of The Council Laying Down Harmonised Rules on Artificial Intelligence (Artificial Intelligence Act) and Amending Certain Union Legislative Acts; European Commission: Brussels, Belgium. 2021. Available online: https://eur-lex.europa.eu/legal-content/EN/TXT/?uri=celex:52021PC0206 (accessed on 24 September 2023).
187. US FDA. Artificial Intelligence and Machine Learning in Software as a Medical Device. Available online: https://www.fda.gov/medical-devices/software-medical-device-samd/artificial-intelligence-and-machine-learning-software-medical-device (accessed on 24 September 2023).
188. Takshi, S. Unexpected Inequality: Disparate-Impact from Artificial Intelligence in Healthcare Decisions. *J. Law Health* **2021**, *34*, 215–251.
189. Galvani, A.P.; Parpia, A.S.; Foster, E.M.; Singer, B.H.; Fitzpatrick, M.C. Improving the Prognosis of Health Care in the USA. *Lancet* **2020**, *395*, 524–533. [CrossRef]
190. O'Sullivan, S.; Nevejans, N.; Allen, C.; Blyth, A.; Leonard, S.; Pagallo, U.; Holzinger, K.; Holzinger, A.; Sajid, M.I.; Ashrafian, H. Legal, Regulatory, and Ethical Frameworks for Development of Standards in Artificial Intelligence (AI) and Autonomous Robotic Surgery. *Int. J. Med. Robot. Comput. Assist. Surg.* **2019**, *15*, e1968. [CrossRef]
191. Harvey, H.B.; Gowda, V. Clinical Applications of AI in MSK Imaging: A Liability Perspective. *Skelet. Radiol.* **2022**, *51*, 235–238. [CrossRef] [PubMed]
192. Bodaghi, A.; Fattahi, N.; Ramazani, A. Biomarkers: Promising and Valuable Tools towards Diagnosis, Prognosis and Treatment of COVID-19 and Other Diseases. *Heliyon* **2023**, *9*, e13323. [CrossRef] [PubMed]
193. Alber, S.; Kumar, S.; Liu, J.; Huang, Z.-M.; Paez, D.; Hong, J.; Chang, H.-W.; Bhutani, T.; Gensler, L.S.; Liao, W. Single Cell Transcriptome and Surface Epitope Analysis of Ankylosing Spondylitis Facilitates Disease Classification by Machine Learning. *Front. Immunol.* **2022**, *13*, 838636. [CrossRef] [PubMed]

MDPI AG
Grosspeteranlage 5
4052 Basel
Switzerland
Tel.: +41 61 683 77 34

Applied Biosciences Editorial Office
E-mail: applbiosci@mdpi.com
www.mdpi.com/journal/applbiosci

www.ingramcontent.com/pod-product-compliance
Lightning Source LLC
Chambersburg PA
CBHW0616 2210926
51850CB00009B/2295
9783725817511